SETON HALL UNIVERSITY
QH505 .N9 MAIN
Intermediate biophysical mechani
3 3073 00255389 7
D1777597

INTERMEDIATE BIOPHYSICAL MECHANICS

INTERMEDIATE BIOPHYSICAL MECHANICS

WESLEY L. NYBORG
University of Vermont

CUMMINGS PUBLISHING COMPANY

Menlo Park, California • Reading, Massachusetts
London • Amsterdam • Don Mills, Ontario • Sydney

SETON HALL UNIVERSITY
McLAUGHLIN LIBRARY
SO. ORANGE. N. J.

Science
QH
505
N9

Copyright © 1975 by Cummings Publishing Company, Inc.
Philippines Copyright 1975

All rights reserved. No part of this publication may be reproduced, stored in a retrieval system, or transmitted, in any form or by any means, electronic, mechanical, photocopying, recording, or otherw without the prior written permission of the publisher.
Printed in the United States of America.
Published simultaneously in Canada.
Library of Congress Catalog Card Number 75–14973

ISBN: 0–8465–4860–7
ABCDEFGHIJ–MA–798765

Cummings Publishing Company, Inc.
2727 Sand Hill Road
Menlo Park, California 94025

ISBN 0-8465-4860-7
ABCDEFGHIJ-MA-798765

PREFACE

This text has evolved from notes for courses taught in the physics department at the University of Vermont, over a period of more than ten years. It is especially designed for a one-semester course, with laboratory, taught at an intermediate college level. A year each of college level physics and chemistry, and a working knowledge of calculus (differential and integral) suffice as background. Elementary differential equations and the notation of Cartesian tensors are introduced, but with no assumption of previous knowledge. Students in the course have come from nearly all departments of physical and biological sciences, pure and applied, as well as from mathematics. For a student from an area of physical science this material offers an extension of his physical background, together with applications of physical language and ideas to biology; to him these applications may be quite novel. A student from the life sciences is offered, at an intermediate level, physical topics which are deliberately selected for their relevance to biology.

In its arrangement, the text tends to follow the style of a typical physics presentation in spite of its orientation towards biology. Accordingly, the organization is primarily in respect to the physical concept (such as "stress" or "viscosity") rather than the biological topic (such as "muscle" or "the ear"). Terms are defined, principles are stated and results are deduced, usually in mathematical form. Derivations are emphasized, although they are frequently given only in outline form, with detailed steps left for the student, in problems at the end of each chapter. Examples are sometimes chosen for their suitability in illustrating an idea, not always for their inherent significance.

However, this is obviously not a text on "pure" physics. While it is organized according to physical topics, the latter are selected for their relevance to biology. For example, continuum mechanics, which is often given little attention in physics curricula, is emphasized here because of its utility in the description of life processes and in the characterization of biomaterials. Selected topics in thermal physics are introduced because of their relevance to diffusion and other transport processes, and also to bioelasticity. Principles of acoustics are taken up (briefly) since shear waves and ultrasound are used in characterizing biomaterials, and ultrasound is increasingly employed for medical purposes.

In relating the physics to biological topics there is much reference to the literature, including research reports in professional and semi-popular journals, specialized review articles, and textbooks in various areas. This emphasis on current literature (somewhat unusual for a physics textbook) is required by the nature of the subject. Biological physics is an area of active research, in which new facts and ideas are continually being generated. (Because of this, it is likely that some of the more tentative data and concepts presented in this text will soon be superseded by others). Students should gain some appreciation of how knowledge is arrived at, and learn to evaluate scientific reports. It is hoped that as a student masters the material of this text, he will be increasingly able to read the literature, and can thus continuously up-date his own knowledge, especially on biological applications. Of course, there are difficulties in doing this: many of our journals and books are written for specialists, and hence it is necessary to be selective. Most of the publications cited here contain material which is meaningful to students who have mastered the subject matter in this book. But in general they also contain material which is unintelligible without more specialized background. In taking nourishment from the litarature the student must learn (quoting freely from Bacon) to taste some articles, devour parts of others, and thoroughly digest a selected few.

It is perhaps unnecessary to say that this volume provides only a sampling of interesting and important topics in biological physics. Other topics, just as significant, have been omitted because of a desire to limit the size of the book. A decision was made to emphasize the physical topic of mechanics (and relevant thermophysics) partly because biological applications are especially well developed in this area, and partly because mechanics is traditionally taught first in a physics sequence. At the University of Vermont our course in biophysical mechanics is followed by another course in biological physics, in which other areas of physics are emphasized.

We have found that it is not difficult to set up laboratory experiments to illustrate topics and test theories discussed in this text. For example, students in our course have done experiments on viscosity measurements in suspensions, stress-strain relations for elastomers, osmotic swelling of cells, statistics of cell counts, measurements of Brownian movement, and other topics. Such experiments, especially when done with biological materials (e.g., blood and its constituents) add considerable interest to the course. They also provide common experience to which reference can be made in lectures and discussions.

I am indebted to the authors of many books, review articles and research papers for resource material, and hope that my "borrowing" from these will be regarded as high flattery. For example, in exploitation of simple tensor notation (Chapters 2-4) much was learned from R.R. Long (Reference 2.1), while the treatment of irreversible thermodynamics (Chapter 10) owes much to A. Katchalsky and P.F. Curran (Reference 10.4). A number of authors have given me helpful information on special topics; these include E.L. Carstensen, F. Dunn, R.W. Fields, E.R. Fitzgerald, J.L. Katz, A.L. King, P.P. Lele, S.G. Mason, R.P. Rand, R. Schor, A.K. Solomon, A. Veis, and A.R. Williams. My colleagues at the University of Vermont and elsewhere have read parts of this text; by their thoughtful comments they have brought about improvements and reduced the number of errors. I am especially grateful to Drs. A. Gershoy, J.A. Rooney (Univ. of Maine), J.E. Krizan, N.R. Alpert, Wm. Halpern, H.M. Frost, R.M. Schnitzler, D.L. Storm, and C.A. Taylor (University College, Cardiff). Students in my courses have also cheerfully offered suggestions, many of which were accepted.

The editors and reviewers of W.A. Benjamin, Inc., and Cummings Publishing Company, Inc. offered candid advice on this text, as well as encouragement. Both were invaluable and much appreciated.

I am grateful to Mrs. Patricia Barber and Mrs. Hazel Pitkin for their care, skill and good humor in typing the entire text of this book. Constance Ireland is responsible for many of the better figures. Michael Wentzell and Elsa Nyborg did much of the work in preparing indices. Also those of the Cummings production department have been graciously responsive to numerous requests, and most understanding of an author's foibles.

Finally, I am thankful to colleagues, students, family and friends for tolerance and good wishes during the long gestation period of this manuscript, and for their many inquiries (How's the book coming?") about the health of the unborn.

Wesley L. Nyborg
Burlington, Vermont
July, 1975

CONTENTS

To Beth and Elsa

For constant encouragement and incredible patience.

CHAPTER 1
INTRODUCTION

1.1 NEWTON'S LAWS

Our aim in this book is to take up a selection of topics in physics which are relevant to biology. These topics are mainly from mechanics, but are supplemented with laws and concepts from thermal physics. The mechanics is primarily of the "classical" variety represented, for example, by three postulates, or "laws", enunciated by Newton in the late 1600's:

> I. Every body continues in its state of rest or of uniform motion in a straight line, unless it is compelled to change that state by forces impressed upon it.
>
> II. Change of motion is proportional to, and in the direction of, motive force impressed.
>
> III. To every action, there is always an equal and opposite reaction.

With these principles expressed in the appropriate mathematical form Newton and his contemporaries were able to predict planetary motions, as well as terrestrial events, to a degree previously unimagined. So successful were their predictive methods that deterministic philosophies arose, in which the entire course of nature was assumed subject to Newton's laws. The ultimate possibilities (for good or bad, depending on one's personal tastes) were dramatized by Laplace, with Newtonian mechanics in mind (Margenau, 1950):

> "An intelligence which knows at a given instant all forces acting in nature, as well as the momentary positions of all things of which the universe consists, would be able to comprehend the motions of the largest bodies of the world and those of the smallest atoms in one single formula; provided it were powerful enough to subject all data to analysis. To it, nothing would be uncertain, both future and past would be present before its eyes."

For those who fully accepted Laplace's statement biology was also subject to Newton's laws and was in principle fully predictable. Of course the "intelligence" required is assumed to be superhuman. Even with the use of a computer there would be "problems". A computer with the necessary storage and calculating facility would be very large; in fact it would probably constitute a significant fraction of the universe! Practical difficulties in subjecting "all data to analysis" are very real and the fact is that after several centuries we still are ignorant of many (most?) of the implications of Newton's laws, simply because of difficulties in carrying out calculations.

But apart from practical difficulties in Laplace's determinism, faith in classical mechanics as an embodiment of absolute Truth began to fail at about the beginning of the 20th century, especially with the development of quantum physics. Inherent in quantum mechanics is the famous uncertainty principle of Heisenberg, according to which there is a limit to the precision with which physical quantities can be known. Clearly such inescapable vagueness about positions and velocities makes Laplace's prediction scheme quite hopeless. (Other deterministic philosophies have arisen, however, based on statistical statements (Margenau, 1950)).

Hence classical mechanics is not omnipotent, contrary to the earlier suggestions. Nevertheless it constitutes an exceedingly powerful approach to an understanding of Nature. In particular it finds many applications in biology and medicine, and hence we give it emphasis in the present volume. For our purposes we shall find little need for the modifications inherent in relativistic mechanics, which must be used when speeds approach that of light. Also our selection of material is such that we need give little attention to the quantum effects which are so important in atomic and subatomic physics. But there is an exception here, for the shear wave resonances (in bone and intervertebral discs) discussed in Chap. 13. These resonances are explained by Fitzgerald in terms of "momentum wave modes" derived from the concept of de Broglie waves.

1.2 APPROXIMATIONS FOR WIDELY SEPARATED PARTICLES

Newton's laws, together with a postulate for gravitational force, found immediate application in deriving equations for the motion of planets about the sun. In particular, the "laws of planetary motion" which had been discovered empirically by Kepler were then readily explained. Kepler's laws are not exact, even within the framework of Newtonian mechanics. For when forces of attraction between all the bodies in the solar system (including the sun, the planets, and their moons) are taken into account the mathematical problem is very complicated. To obtain Kepler's laws the "many-body problem" was greatly simplified by introducing the following approximations:

(i) The sun is fixed;

(ii) The force on a planet is from the sun only; and

(iii) The dimensions of the sun and planets are small compared to the distance of a planet from the sun.

Approximations (i) and (ii) reduce the many-body problem to a problem involving only one body, namely, the planet in question, moving in a fixed force field (the sun's gravitational field). Approximation (iii) allows simplification in that each body becomes a "particle", localized at a "point" and lacking internal structure. It is part of approximation (ii) that whatever matter occupies the (not quite "empty") space between the bodies has no effect on the motion. When all of these approximations are used the problem of planetary motion consists of predicting the motion of a particle in vacuum under the action of a known force field. If a particle has mass m and the force in the x direction is F_x it follows from Newton's second law that the particle is accelerated in the x direction, and that the acceleration is given by

$$\frac{d^2x}{dt^2} = \frac{F_x}{m} . \qquad (1.1)$$

Similar equations apply in the y and z directions. Equations for planetary orbits are obtained as solutions of differential equations such as Eq. (1.1). The particle-in-a-vacuum approximation to Newton's laws proves quite adequate for explaining planetary motion, at least, to the extent that Kepler's laws apply.

Approximations similar to those used for planetary motion are also useful in other areas of physics, especially for the kinetic theory of gases. The "ideal gas" is a dilute gas in which the molecules are widely separated so that any given molecule travels freely in a "vacuum" much of the time. Of course, collisions occur at high frequency but nevertheless in a dilute gas a given molecule spends only a rather small fraction of its time interacting with other molecules. And when collisions do occur they involve two molecules only; a triple collision is very rare.

In the kinetic theory of gases it is assumed that the molecules are like macroscopic bodies in that their motions are governed by Newton's laws. For a dilute gas in a container it is shown in elementary physics that the pressure P exerted on the container walls is proportional to the average kinetic energy $< K.E. >$ per molecule. When an extra postulate is added, that $< K.E. >$ is proportional to the Kelvin temperature T, it is easy to derive the ideal gas law, namely,

$$PV = R_g T , \qquad (1.2)$$

where V is the volume of one mole of gas and R_g is a constant.

1.3 INTERACTING PARTICLES; CONTINUUM APPROXIMATION

In biology the motion of small bodies or particles is fully as important as in physical situations. Such fundamental life processes as synthesis of macromolecules, propagation of nervous impulses, and muscular contraction all depend on movements of ions, molecules, and other particles in and between cells. But here the particles are not isolated and do not travel in vacua! Instead they are characteristically surrounded by waterlike media; these media are usually complicated by the presence of delicate membranes and fibers to which the particles may be contiguous. As a special feature of the situation, ions and molecules suspended in water are normally hydrated, that is, surrounded by a layer of oriented H_2O molecules, the hydration layer. Clearly these suspended particles whose transport is so important in biology do not move in free space but are always interacting with a number of neighboring molecules. In biophysical mechanics we must proceed beyond the isolated-particle and dilute-gas approximations and face the more difficult "many-body problem."

A biological object such as a red cell, bacterium or sperm is usually immersed in an aqueous medium, that is, a medium consisting of water in which small percentages of ions and various molecules are dissolved or suspended. The object is then in contact with a large number of "particles", mostly H_2O molecules. For example, a sphere of 1 μm diameter (representative of some bacteria) has a surface area of 12.5 square microns; in an aqueous medium there would be more than 10^8 molecules (roughly, 10^7 per square micron) in contact with the surface. These molecules are in constant motion of a random nature and because of this, the immersed object is subjected to a large number of molecular encounters per second.

In a biological situation, then, the force on an object arises from interactions between the object and a large number of neighboring bodies. A "particle" approach to this situation would be to solve equations like Eq. (1.1) for the biological object and each of its neighbors. This is hardly feasible when there are many neighbors. Instead we follow another approach, that of continuum mechanics. Representative of this approach are the theories of hydrodynamics, aerodynamics, elasticity, acoustics, rheology and other fields of science.

In continuum mechanics the molecules are not recognized as individuals, but their effects are taken into account by imagining a continuous medium (or, simply, a continuum) which replaces them. This continuum is characterized by a density ρ which represents a related property of the "real" (molecular) medium. In defining the density $\rho(P)$ at any point P of the hypothetical continuous medium we proceed in a manner that is characteristic of continuum physics; we select a small region of space in the vicinity of P and give it our attention. We let the volume of this region be Δv, let a typical linear dimension of the region be ε, and refer to the region as a "volume element". An estimate of $\rho(P)$ is then given by $\Delta m/\Delta v$, where Δm is the sum of masses for all molecules contained in volume Δv of the real medium, in the vicinity of P. This estimate will provide an accurate value for $\rho(P)$ if Δv can be made large enough to include a sufficient number of molecules. In general, there will be a limitation on its size since $\rho(P)$, as just defined, may vary from one part of space to another. Suppose there are maxima and minima in the spatial distribution of ρ, and that the typical distance from a maximum to the nearest minimum is ℓ. Then we realize that the linear dimension ε of the volume element should be considerably less than the characteristic distance ℓ, in order for ρ to be representative of the density near P. If the typical distance of separation between nearest molecules is a we conclude that an accurate value for $\rho(P)$ can be defined by choosing ε such that

$$a \ll \varepsilon \ll \ell . \qquad (1.3)$$

This kind of consideration in respect to volume elements is characteristic of continuum mechanics. Of course, a suitable value of ε cannot be found unless $\ell >> a$; if this condition is not satisfied we should not expect the approach of continuum mechanics to yield accurate results. (But, rather surprisingly, it has sometimes proved useful even for situations where ℓ is comparable to a ; see Section 11.6.3).

An effect of the molecules (in the real medium) is to provide continual bombardment of any surface that adjoins them. In continuum mechanics this molecular bombardment is represented by stresses in the (hypothetical) continuous medium. At and near a point P on any surface (real or imaginary) stresses may exist in the adjoining continuum, both normal to the surface and tangential to it. The normal component of stress (pressure or tension) represents the average normal force per unit area exerted on the surface at and near P by the adjoining molecules. Similarly a tangential component of the stress (or shear stress) represents the average force per unit area exerted by molecules as they "slide" along the surface.

In general, as with the density ρ , the components of stress are not constant throughout a region of interest; however, if the continuum approach is valid, the spatial distribution is smooth relative to a molecular scale. Thus if the characteristic length is ℓ for the spatial distribution of a stress component, we expect that ℓ will be much greater than a typical molecular dimension.

1.4 CONCEPTS OF CONTINUUM MECHANICS

In continuum mechanics our aim is to predict properties of a continuous medium under specified conditions. Among these properties are quantities descriptive of the state of "stress" throughout the region of interest, and others describing motions set up in the medium. For example, in considering a stretched fiber we might wish to know the tension at any point along the fiber, and the displacement of each point from its starting position. As in particle mechanics, the methods for making predictions are based on the three laws of Newton listed in Section 1.1 . However, the differential equations typified by Eq. (1.1) are not suitable since the particle concept has been abandoned. Other differential equations are derived, suitable for continua, and results are obtained by solving these. In the next few chapters we discuss fundamental ideas of continuum mechanics, arriving at basic equations in Chap. 5 . Solutions of the equations are then obtained and applications taken up in subsequent chapters.

As explained in Section 1.3, the concept of "stress" is used for a continuous medium to represent force exerted by the medium on unit area of a surface; this surface may be a real one (as provided by a container wall) or may be an imaginary one passing through any part of the medium. In Chap. 2 the concept of stress is treated in some detail; it is shown that there are six independent components of stress and that, when expressed relative to Cartesian or rectangular coordinates, these form the elements of a Cartesian tensor of second order. (A displacement vector expressed relative to rectangular coordinates is a Cartesian tensor of first order.) A few elementary properties of tensors are then taken up which lead to helpful insight on the stress components. Simple tensor notation is used, and proves very convenient.

Just as forces applied to a system of particles give rise to displacements which vary with time, and from particle to particle, so stresses applied to a continuum give rise to a space-time pattern of displacement. But in continuum mechanics, we are not usually concerned with absolute values of displacements. Instead our interest is in how the displacements vary in space and/or time. Quantities for specifying these variations are defined and discussed in Chap. 3 .

Of greatest interest among these are the strain components, which serve to measure the change in shape or size of a portion of the medium as the result of displacements. It is shown that, referred to rectangular coordinates, there are six independent strain components and these (like the stress components) form the elements of a Cartesian tensor of second order.

In elementary physics we often take up Hooke's law, expressing it in a simple way by stating that for a linear solid the "strain is proportional to the stress". Since we now know that, in general, there are six components each of stress and strain, we realize that a more elaborate statement is required. When displacements in a given solid medium are small it is usually possible to write a "generalized Hooke's law" for the medium, consisting of a linear equation relating stress and strain components. (See Eq. (4.1)). Any equation, linear or not, which relates stress and strain components for a given medium is called a constitutive relation for that medium. In Chap. 4 linear constitutive relations are given for various media. Tensor notation provides an elegant means for expressing these relations. In the constitutive equations for linear media (solids, liquids and viscoelastic media) the constants which appear are the coefficients of elasticity and viscosity. Values for some of these coefficients are given in Chap. 7 for various biological media.

Using the concepts of stress and strain, and knowing constitutive relations for media of interest, we proceed in Chap. 5 with procedures for obtaining stress distributions. A basic step here is to obtain an expression for the net force on a volume element of the medium. This force arises partly from gravitation or other fields of external origin. But we give special attention to another contribution, which arises from stresses exerted on the boundary of the element by the external medium. When a net force exists (either from external fields or from stresses) its magnitude proves to be proportional to the volume Δv of the element when the latter is sufficiently small. Hence the net force per unit volume is a quantity of interest. Suitable expressions are obtained by limiting procedures, familiar from calculus. For example, f_x , the x component of net force per unit volume in a given region, is obtained by first finding the sum ΔF_x of forces along x applied to an element of finite volume Δv ; the ratio $\Delta F_x/\Delta v$ is then formed and f_x set equal to its limit as Δv approaches zero. Similar considerations apply to the corresponding quantities f_y and f_z . The contributions to f_x , etc., from stresses are written as differential expressions, namely, as derivatives of stress components with respect to the coordinates x, y or z .

In Chap. 5 we consider primarily equilibrium situations, in which the acceleration is zero. It then follows from Newton's laws (Sect. 1.1) that the net force on a volume element must be zero in every direction. Hence we can write

$$f_x = 0 ; \quad f_y = 0 ; \quad f_z = 0 . \tag{1.4}$$

When f_x , etc., are expressed explicitly in terms of the stresses and external force fields Eqs. (1.4) become differential equations. These equations apply at all parts of the medium and provide a fundamental basis for determining spatial distributions of stress in media under equilibrium conditions.

When the acceleration is not zero Eqs. (1.4) are modified. For motion along the x direction we can write

$$f_x = \rho a_x , \tag{1.5}$$

where ρ is the density and a_x the acceleration in the x direction. This equation is just an expression of Newton's second law; the force (per unit volume) on a volume element is equal to its mass (per unit volume) times its acceleration. Equation (1.5) is an "equation of motion" for continuum mechanics just as Eq. (1.1) is for particle mechanics. Explicit examples of equations typified by Eqs. (1.4) and (1.5) are taken up in Chaps. 5 and 13 .

1.5 APPLICATIONS

There are many situations important to biology in which it is desirable to know the stress components at various points, as well as the components of strain and/or velocity. In getting this information a typical procedure is to seek expressions which satisfy differential equations of the type represented by Eq. (1.4) or Eq. (1.5) . Techniques for finding such solutions have been developed to a high degree of sophistication in the centuries since Newton; many volumes have been written on the methods and results of continuum mechanics. In this book we select basic situations in biophysical mechanics for which the required mathematical language includes differential and integral calculus but not, for example, the more advanced methods of vector analysis and partial differential equations. We shall be discussing elementary differential equations, but shall not assume prior knowledge of methods for solving them. For our purposes it will usually be sufficient to verify by substitution that a given function satisfies the differential equation.

1.5.1 Hydrostatics

Probably the simplest kind of situation is one where the only nonzero stress components have the character of pressure. This "hydrostatic" situation exists in a body of fluid which is motionless, or nearly so. The human body contains fluid-filled spaces for which this approximation is reasonable. For such a situation it is shown (Sect. 5.2) that f is given simply by $-\ \partial p/\partial x$; similar expressions apply for f_y and f_z . Under terrestial conditions the force of gravity must usually be considered. But for a fluid at rest in outer space where the force of gravity is negligible we can write (Cf. Eqs. (1.4)).

$$\partial p/\partial x = 0 \; ; \quad \partial p/\partial y = 0 \; ; \quad \partial p/\partial z = 0 \tag{1.6}$$

These three partial differential equations are satisfied simultaneously by the rather trivial "expression":

$$p = C \; , \tag{1.7}$$

where C is a constant independent of x, y and z. From this result we know that for a motionless fluid in a zero-gravity field the pressure is everywhere the same.

The more typical condition, where a gravitational field exists, is treated in Section 5.2 . Applications are to pressure distributions in the ocean, in the atmosphere and in tall animals or **trees** . Also, rather surprisingly, the same theory applies to the equilibrium distribution of small biological particles in suspension under the influence of gravity or other force fields. (Section 9.5).

1.5.2 Shear Fields; Parabolic Flow

Another kind of situation presents itself when a medium is subject to shearing stress. This kind of stress is experienced, for example, by a medium sandwiched between two surfaces, one of which slides relative the other. An elastic solid body subject to shear experiences a change in shape, but recovers immediately when the strain is removed. A liquid responds differently of course. A volume element of Newtonian liquid changes shape in a shearing action, but has no tendency to recover when stress is removed. A biological tissue is usually viscoelastic, with properties intermediate between those of an elastic solid and a Newtonian liquid. A volume element of this medium is deformed when shearing stress is applied and it tends to recover when the stress is removed; but the recovery is not immediate, and is not complete.

Fields of shear are discussed in Chap. 5 both for solids and liquids. Especially simple are arrangements in which the shearing takes place between parallel plates, one of which is fixed while the other is displaced sideways (i.e., in its own plane). Suppose for definiteness that the (very large) plates are perpendicular to the y direction, that the medium is a liquid, and that the moving plate travels in the x direction with constant velocity. Then after steady motion has been established the velocity u in the liquid will also be along x ; u will be constant with time but will vary linearly with y. The <u>velocity gradient</u> is defined as $\partial u/\partial y$ and is constant in this field of "simple shear".

Simple shearing flow is an idealization, but is approximately realized in <u>Couette flow</u>, a flow field set up in a fluid confined to a narrow space between concentric cylinders. Devices based on Couette flow are used to measure viscosity of biological fluids and other media.

Simple shearing flow, in which the velocity gradient is constant, is a useful idealization for gaining insight on properties of flow suspensions. For example, asymmetric particles, such as rods or discs, become aligned in shearing fields to an extent that depends on their shape and on the velocity gradient. For constant-gradient fields it has been possible to derive equations for predicting the degree of this alignment. Using these equations information on the shape of such biomolecules as fibrinogen and collagen are obtained from experimental measurements of flow alignment. (Chap. 9.)

When biological macromolecules or cells are suspended in a fluid and subjected to shearing flow, distortions are produced which may lead to rupture and other damage. These effects are discussed in Chap. 14 .

They are important in design of biomedical devices for assisting circulation of the blood, since these produce shear which might be damaging. Even in the normal animal circulation shearing effects are produced ; very possibly, these are significant in degrading elements of the blood which , in turn, are to be replaced by fresh elements.

In the circulatory system the flow pattern is not one of simple shear. In general, the pattern is complex, but a useful idealization is that of steady laminar flow through a tube of circular cross-section. Here the velocity gradient is not constant, but is zero at the center and has its greatest value at the wall; the velocity follows a parabolic distribution. In Chap. 5 a number of situations are discussed, involving parabolic flow occurring in channels. The analysis for a circular tube leads to Poiseuille's law, and is the basis for operation of the capillary-type viscosimeter which is in much use for making viscosity measurements.

Appeal to the theory for parabolic flow in a tube is also made in the "pore theory" for transport of ions and small molecules through biological membranes. (Chap. 11.)

1.5.3 Motion of Particles in Suspension.

Methods for determining the size and shape of biological macromolecules depend heavily on hydrodynamical methods. One important technique is that of centrifugation; here measurements are made of the rate at which macromolecules migrate or sediment, in a rotating vessel. Theory for the method is discussed in Chap. 6. Use is made of the idea that the velocity of migration for a particle sedimenting in a gravitational or centrifugal field is proportional to the net force on the particle.

A frictional coefficient f is defined such that f is the ratio of force to velocity. Expressions for f have been derived by Stokes and others for spheres, ellipsoids and rods. The analysis used in these derivations is much too long and complicated for inclusion in this book, but results are quoted in Chap. 6. It is because f is dependent on molecular shape and size that these quantities can be obtained from sedimentation experiments.

For a sphere of radius R immersed in a fluid of shear viscosity coefficient η , Stokes showed that the frictional coefficient is given by $6\pi\eta R$. This simple result is much used for various purposes. For example, a relatively simple approach to calculating the viscosity of a suspension of elongated particles is to consider each particle as equivalent to a pair of spheres connected with a thin (but rigid) rod; in other words, the particle is represented as a dumbbell. The Stokes expression is then used to calculate the force on each sphere resulting from relative motion between it and the surrounding fluid. (Chap. 6.) A similar approach is used in analyzing the orientation of elongated particles in a flow field. (Chap. 9.)

In Chap. 9 the topics of Brownian movement and diffusion are discussed, both being important for understanding the behavior of ions and molecules in biological media. Insight on the nature of these processes is gained from principles of thermal physics taken up in Chap. 8. It is shown that the diffusion coefficient D for a suspended particle is inversely proportional to the friction coefficient f, being given by KT/f ; here k is the Boltzman constant and T the Kelvin temperature. Diffusion experiments give important information on the size and shape of macromolecules. Also diffusion processes are important in transport of ions in and between cells.

Critical to the functioning of a biological cell is the transport of ions and molecules across its membrane. Existing theories for this transport are based on results of continuum mechanics, together with ideas from thermal physics. This topic is discussed in Chap. 11, following a brief presentation of relevant thermodynamics in Chap. 10.

1.5.4 Bioelasticity; Sound and Shear Waves; Environmental Effects

Principles of thermal physics are also required for treatment of bioelasticity in Chap. 12. Here the related concepts of entropy and probability find applications in understanding the elastic properties of flexible biomolecules, such as occur in arterial walls, bladder membranes and hair.

Much of our information on elastic and viscous coefficients comes from measurement of sound velocity and absorption. Theory and experimental results for acoustic measurements are presented in Chap. 13. Closely related, and also included, is a discussion of viscoelastic properties of biomolecules determined by use of oscillating shear.

Finally, in Chap. 14 we discuss mechanical and thermal aspects of our environment, and how these influence life processes and structures. Important here are effects of heat, cold, pressure, decompression, shearing stress and sound waves.

1.6 BIOLOGICALLY GENERATED FORCE: MUSCLE

Before proceeding with a detailed development of concepts in continuum mechanics we pause to reflect on the idea of "force". In macroscopic physics we consider forces as arising from gravitational, electrical and magnetic fields. At the atomic, nuclear and subnuclear levels other forces apply. All of these forces exist in biological media, of course. But in addition biological systems have unique means for exerting force. As force-generating tissue we think primarily of muscle, but must also include the cilia and flagella by means of which small organisms transport themselves. It is these biological "sources of force" that are taken up in this and the next section.

Animals move about and make changes in their environment by means of forces generated in specialized contractile tissue. Such a tissue is muscle, which constitutes a large fraction of the animal body, being found in the limbs, the heart, the intestines and other organs. On a microscopic scale there are contractile elements in the flagellae by means of which many kinds of sperm, bacteria and protozoa propel themselves through a liquid. Fundamental research on muscle contraction has shown that the action results from some kind of interaction between macromolecules. While these interactions are not understood in detail, they evidently are associated with forces between structures on an atomic scale, and are fueled by energy from biochemical reactions. Mechanisms for contraction seem to have no counterpart in nonliving systems.

Also unique to biology are the forces which move ions across membranes in the process known as _active transport_. Again the detailed mechanics of the action are yet to be discovered.

We shall present here a few basic facts about the structure and physiology of muscle. In doing so, we shall unfortunately need to omit description of the step-by-step development of the subject in many ingenious experiments, and instead present results as they are known today. For a more complete treatment texts and articles such as those listed at the end of this chapter should be consulted. The statements that follow apply especially to _striated_ muscle (so called because of its appearance) which includes skeletal and cardiac muscle. A whole muscle may have dimensions of the order of centimeters; contained within it is a (roughly) parallel array of smaller units, the _fibers_, together with an artery and vein as well as a nerve system.

A fiber may be many centimeters long and, say, 0.1 mm in diameter. It contains parallel subunits, the _fibrils_, as well as a system of _tubules_ (through which fluids and suspended ions and molecules are conveyed) and _mitochondria_ (subcellular units which act as factories for manufacture of various necessities of cell life).

The fibril, or myofibril, is a cylinder which may be about 1 μm in diameter and which runs from one end of the fiber to the other. As viewed in a suitable way with a microscope (especially if interference and polarization techniques are used) it has a periodic banded appearance as shown at the top of Fig. 1.1 . At locations marked Z are thin structures called Z discs. The section of myofibril lying between consecutive Z discs is the sarcomere. Two dark bands appear in each sarcomere; these are separated by a clear space, the H band. This H band and the two dark bands adjoining it comprise an A band. Between consecutive A bands lies an I band with a Z disc at its center.

Painstaking analysis using electron microscopy has led to an interpretation of the banded structure in terms of longitudinal fine protein filaments. In the lower left of Fig. 1.1 is represented part of the cross section of a myofibril. The large circles represent myosin filaments and the small ones actin filaments. These are in regular hexagonal array such that each filament of myosin is surrounded by six of actin, and each filament of actin is surrounded by three of myosin. In the living state the distance between myosin filaments is found (from Xray analysis) to be 455 A.

A longitudinal section (bounded by planes identified by dashed lines in the cross section) reveals a parallel array of alternating myosin and actin filaments. Each myosin filament is found to have a length m of about 1.65 μm and each actin filament a length a of 1.05 μm. These are arranged as indicated schematically in the lower right of Fig. 1.1. Attached to each side of the Z discs are the actin filaments, with the myosin interposed as shown. Comparing with the bandings defined at the top of Fig. 1.1 we see that the A band is the region occupied by myosin. The dark bands are regions where the myosin and actin overlap. In the I bands the actin exists alone; correspondingly, in the H bands the myosin exists alone.

From other studies it is known that an actin filament consists of two strings of protein globules, the strings being wound around each other to form a two-stranded rope. These globules, called G-actin molecules, are each about 55 A in diameter, and have a molecular weight

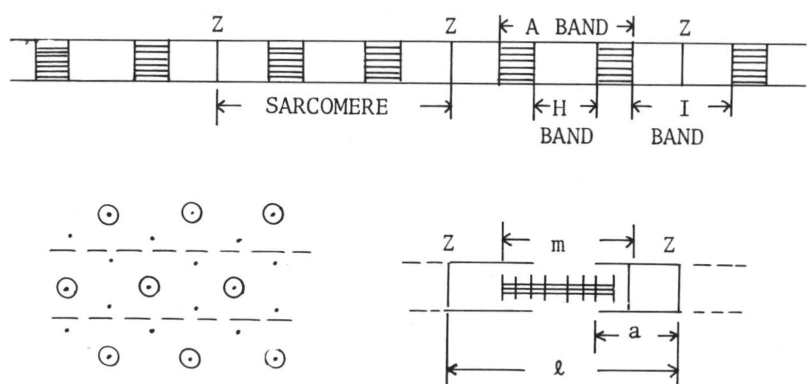

Fig. 1.1 Top: myofibril with Z discs and A-, H- and I- bands identified. Lower left: hexagonal array of filaments as revealed in a cross section; the large circles represent myosin and the small ones actin. Lower right: filaments as would hypothetically be seen in longitudinal section; the myosin length m is 1.65 μm ; the actin length $\underline{a}$ is 1.05 μm ; ℓ is variable.

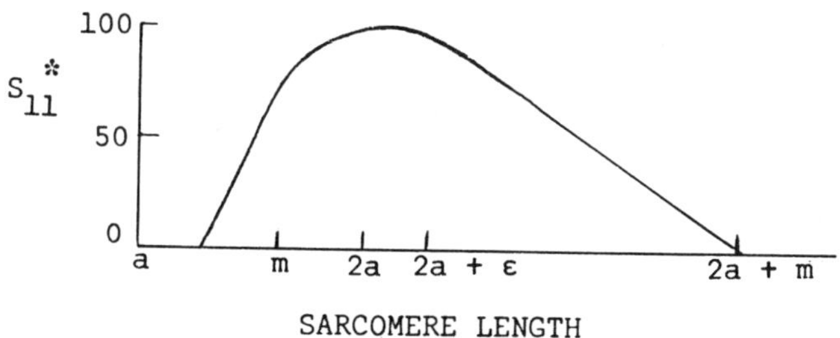

Fig. 1.2 Plot of longitudinal stress S_{11}^* versus sarcomere length ℓ . Here S_{11}^* is the constant value achieved in smooth tetanus under isometric conditions; its units are arbitrary. (Adapted from Gordon, Huxley and Julian, 1966).

of 60,000. A myosin filament is an aggregate of myosin molecules, each of which can be separated into two units; the intact molecule is about 1300 A long with diameter about 20 A along most of its length and 40 A at its head. In Fig. 1.1 , lower right, the small lateral projections from the myosin represent "cross-bridges"; spaced about 450 A apart, it is at these small structures that the force of contraction is thought to be developed.

A muscle may be stimulated electrically or chemically with the result that tension is developed in the muscle, or its dimensions change, or both. Isometric conditions are achieved if the muscle is stimulated while its ends are held fixed so that the muscle's length remains constant. Isotonic conditions hold if the muscle is allowed to contract while its tension is held constant.

For given conditions it is found that the force developed parallel to the axis of a muscle tends to be proportional to its cross-sectional area. In other words the force per unit area, or tensile stress, is essentially independent of area. Referring to Cartesian coordinates, if the axis of the muscle is along the x or 1 direction, we can represent the axial tensile stress as S_{11} in keeping with the notation of Chap. 2.

In experiments done isometrically with electric stimulation by voltage pulses, the stress developed depends on stimulating parameters such as the height V and duration τ of the pulse. After a pulse in which V and τ are sufficiently large the stress S_{11} increases to a peak value, then falls again, all within a time of the order of a second or less (depending on the type of muscle and the temperature, among other things).

The peak value of S_{11} is found to be independent of V and τ if these parameters are each sufficiently large; as might be expected the value of V required for this condition decreases as τ increases. If the pulses are repeated regularly at a high enough frequency the condition of smooth tetanus is achieved where S_{11} reaches a high value S_{11}^*, then is constant until fatigue occurs.

Much insight comes from some experiments in which S_{11}^* is determined as a function of the sarcomere length ℓ for isolated single fibers. See Fig. 1.2 . Results of the experiments are neatly explainable in terms of the filament model in the lower right of Fig. 1.1 .Thus when the fiber is stretched to a length such that ℓ is just (2a + m) the biologically generated stress is zero; in this fully extended state there is no overlap between myosin and actin filaments.

When the fiber length is fixed at successively smaller values we see from Fig. 1.2 that the available tensile stress increases nearly linearly until ℓ is slightly greater than 2a. When this condition (limit of the linear region) is reached the actin extends completely from one Z disc to another, except for a small interruption in the H zone near the unattached ends of the actin filaments. Also at this value of ℓ all cross bridges from the myosin can project laterally to a neighboring actin filament.

When ℓ is decreased still further S_{11}^* remains at a high value until ℓ is equal to 2a, then decreases monotonically; the decrease is especially rapid when ℓ becomes less than m , the normal length of a myosin filament. Perhaps the decrease results from disorientation of the myosin under the high state of contraction. The stress goes to zero as ℓ shrinks to approach the actin length a.

Thus the force developed by muscle appears to be associated with the cross bridges attached to the myosin. The linear region on the right of the plot in Fig. 1.2 corresponds to conditions where the force is proportional to the distance of overlap between actin and myosin, and hence proportional to the number of cross bridges formed between these two kinds of filaments. According to "sliding filament" theories these bridges somehow cause the actin to slide over the myosin during contraction. Energy is provided chemically; it is known that ATP (adenosine triphosphate) must be present in order for contraction to occur. The molecule ATP is a basic energy source in biology, not only for muscle contraction, but also for active ion transport across membranes, and for other processes. A challenging problem in biology is to learn just how forces are developed between "sliding filaments". What is the mechanics of the cross bridge action? How does ATP operate the machinery?

To pursue such questions, and for other reasons, we turn now to typical magnitudes of quantities involved in muscle contraction. It is found that for an optimum choice of ℓ a typical value of S_{11}^* is 1.5 kg/cm^2 . Using this value we can estimate the force developed in a single filament. As explained in connection with Fig. 1.1 the filaments are in a hexagonal array, the distance between myosin filaments being about 450 A in the living state. Hence there are nearly 6×10^{10} of these filaments per square centimeter of cross section in a fibril. In the H band it is reasonable to suppose that the myosin supports nearly all of the tensile stress developed in a fibril. Hence we find that the tension force T exerted by each myosin filament in the H band must be about 2.5×10^{-5} dyn. (Prob.1.1).

A fibril may contain 100 to 1500 such filaments and hence may support a tension of from 0.0025 dyn to 0.038 dyn.

From the calculated tension per filament we can draw illuminating conclusions about the force which must be exerted by each cross bridge, if the contraction force does indeed arise at these bridges. The bridges between a given actin-myosin pair are spaced about 450 A apart longitudinally, there being about 18 of these in a series for a given "half" of a myosin filament. Since a filament of myosin is surrounded by 6 of actin there is a possibility of 108 bridges altogether for the myosin half-filament under consideration; the average longitudinal spacing is about 75 A. Correspondingly the average space between bridges along the actin filament (since it has 3 close myosin neighbors) is 150 A. If all bridges were acting at once we could suppose the force of 2.5×10^{-5} dyn per myosin filament is distributed equally among the 108 bridges; each bridge would then contribute a force of 2.3×10^{-7} dyn. If only a fraction of the bridges are operating at a given time the force per bridge is, of course, greater than this.

It is especially interesting to consider a hypothesis referred to by H. E. Huxley (1960) and considered in some detail by A. F. Huxley (1957); by this hypothesis contraction occurs in a series of short steps, each caused by an elementary displacement of a bridge. If the displacement is x and the average force in the direction of the displacement is f, the work W done per cross bridge in a displacement is

$$W = f x . \tag{1.1}$$

Suppose W were the energy provided by one ATP molecule. When this is <u>dephosphorylated</u> (reduced to ADP, adenosine diphosphate, by splitting off one phosphate group) the energy released is about 2.8×10^{-13} ergs. Using this figure for W and 2.3×10^{-7} dyn for f we obtain from Eq. (1.1) that

$$x = 1.22 \times 10^{-6} \text{ cm} = 122 \text{ A} .$$

It is striking that this (in the spirit of a ballpark estimate) is reasonably close to the average spacing (150 A) between cross bridge attachment sites on an actin filament. The appealing idea arises therefore that an elementary step is fueled by one ATP molecule, and consists of a displacement (of the "actin end" of a cross bridge) from one attachment site to another on the actin filament. If this proves to be a valid concept, the challenge remains of learning the details involved in this basic step.

It has been estimated (Hill, 1949) that under favorable circumstances about 40% of the biochemical energy provided to the muscle is converted into mechanical work. This efficiency of 40% compares favourably with that of various manmade machines. Our muscles are truly "well designed".

1.7 BIOLOGICALLY GENERATED FORCES: CILIA AND FLAGELLA

Many animal and plant cells generate motion by means of hair-like appendages on their free surfaces. When these motile appendages are few in number and long in proportion to the size of the cell, it is customary to call them flagella (singular, flagellum); when they are numerous and relatively short they are called cilia (singular, cilium). In some instances these appendages have the function of transporting the cell itself; for example, mammalian spermatozoa swim by means of flagella as do many protozoans and bacteria. In other situations their function is to cause transport of fluids or small objects relative to a tissue surface; in this category are the cilia which line the respiratory tract of mammals, where they continuously convey foreign particles to the exterior. These structures are perhaps the smallest self-powered motile units that nature provides. Together with muscle they provide examples of biological machinery at a basic level. For an extensive treatise on cilia and flagella see Reference 1.7 (D. Fawcett).

Characteristically a cilium executes a kind of rowing motion, with a unique oar which bends in the upstroke but becomes straight and rigid in the downstroke (Sleigh, Reference 1.8). In a flagellum a wavelike motion takes place somewhat as in a vibrating string; the wavelength is typically less than the flagellar length. It is believed that these transverse motions or vibrations arise somehow from forces associated with a set of tubules which run the length of the vibrating unit. In the simplest examples (cilia, and sperm flagella for such invertebrates as sea urchins) there are just eleven of these fibers; in cross section they are arranged in the pleasing design seen in Fig. 1.3. Of the eleven fibers shown, two are in the center and nine in a circle of diameter about 0.15 μm .

Fig. 1.3 Schematic representation of cross section through a flagellum, emphasizing certain features: the "figure eight" appearance of the nine outer filaments; the arms; the radial "spokes". From Afzelius, Ref. 1.9.

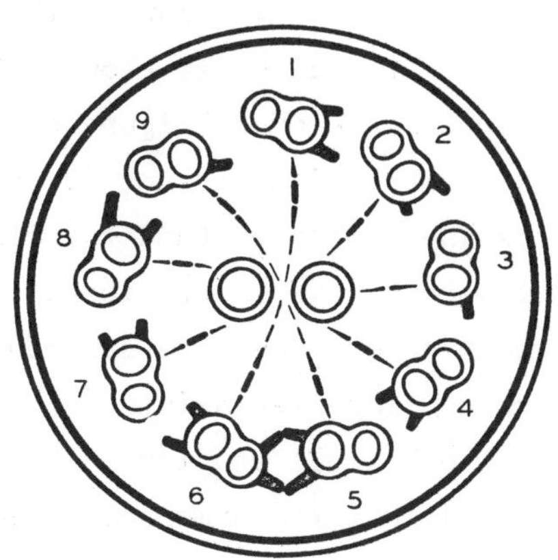

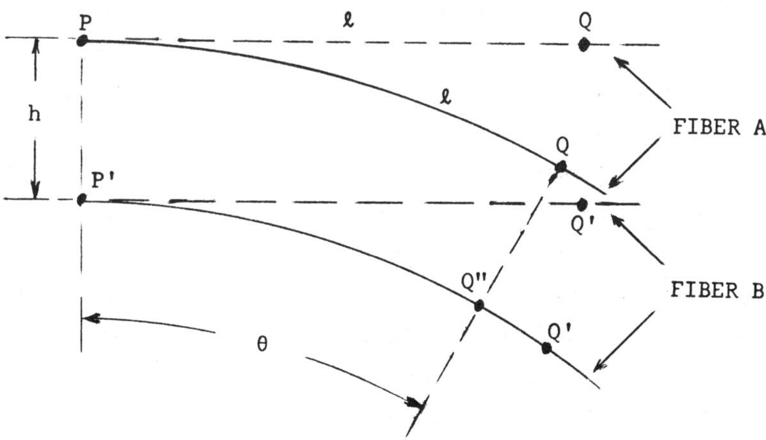

Fig. 1.4 Sketch for reference in discussing the sliding of one fiber over another when bending occurs. When Fibers A and B are straight Q' is the neighbor point to Q. When they are bent as shown Q' has "slid" to the right relative to Q.

Each of the two inner fibers is approximately circular in cross section, with diameter about 150 A. Each of the nine outer fibers has the appearance of a "figure eight" with dimensions about 150 A by 250 A. Considerable significance is attached to the "arms" which extend in the clockwise direction (and for fiber 6 also in the opposite direction) from each of the outer fibers (Afzelius, Reference 1.9). Also radial structures of some kind apparently extend to each of the outer fibers from the flagellar axis.

Biochemical studies (Gibbons and Rowe, References 1.10 and 1.11) indicate that the arms are molecules of a protein which has been given the name dynein. It is concluded that these molecules are somewhat oblong, with mean diameter about 100 A and molecular weight 600,000, and are spaced about 140 A apart along the length of a fiber. Each of the two central fibers is a tube while each of the 9 outer ones is a pair of tubes with a common wall. For any of these the wall is evidently constructed as a single layer of spherical building blocks, each a protein molecule about 40-50 A in diameter, with molecular weight about 40,000 (Ringo, Reference 1.12). The nature of the "mortar" if any is not known, but it must be of high quality since the fibers seem to be quite stiff.

How is vibrational movement produced in cilia and flagella? As with muscle the energy evidently comes from dephosphorylation of ATP, that is, its reduction to ADP. But here the reaction occurs at the dynein molecules while in muscle it occurs at the cross bridges. For this and other reasons it is believed that in cilia and flagella dynein molecules may play a role analogous to that of the cross bridges in muscle. Just as the cross bridges in muscle, powered by ATP, may cause sliding of actin relative to myosin, so dynein molecules in a cilium may cause sliding of any of the nine outer fibers relative to its nearest neighbor(s) (Brokaw, Reference 1.13).

In a flagellum the sliding action is closely associated with the curvature which varies with time and with position. In Fig. 1.4 let A and B identify two neighboring fibers, such as No. 3 and No. 4 in

Fig. 1.3; they are separated by a distance h. Let P and Q be two points on A separated by a distance ℓ . We shall refer to the point on B which is closest to P as the <u>neighbor point</u> to P, etc. When the flagellum is straight the fibers A and B are straight also; the neighbor points to P and Q are then P' and Q', respectively; the distance P' Q' is just ℓ.

Now suppose that in this region the flagellum is deformed as shown (that is, bent in a plane containing the filaments A and B) so that the filaments A and B form coplanar concentric circular arcs, with the arc PQ subtending an angle θ at the center of curvature. If P' is still neighbor point to P, then the neighbor point to Q is no longer Q', but a new point Q" , the arc length Q' Q" being just $h\theta$ (Prob. 1.2). The distance h between neighboring fibers is about 500 A ; if the arc subtended by PQ is 30° then Q'Q" is about 260 A.

Thus the point Q' on B has been displaced a distance 260 A relative to its former neighbor Q on the fiber A. Force is required to cause such displacements, especially because the fibers are enmeshed in a medium which resists the deformation caused by such a displacement. On the assumption that this force is supplied by the dynein arms Rikmenspoel (Reference 1.14) arrived at an estimate of 0.6×10^{-7} dyn for the peak force exerted by one dynein molecule. It may be significant that this is of the same order as the force associated with a myosin-actin cross bridge (Section 1.6). Assuming pairs of dynein molecules, i.e., pairs of "arms" , spaced 140 A along the fiber we have 1.4×10^{6} molecules per centimeter of length. All of these acting together might then produce a force along the fiber of 0.08 dynes per centimeter of length. The arms exert a tangential stress on the fiber surface which we might call S_{12} (as in Chap. 2, if the 1 direction is taken along the axis of the fiber and the 2 direction along its radius). This stress is not uniform but we can obtain a rough average value for it by considering the arms to act along a longitudinal strip of fiber surface whose width is the fiber diameter 150 A. We then obtain

$$S_{12} = \frac{\text{force}}{\text{area}} = \frac{0.08}{1.5 \times 10^{-6}} = 5 \times 10^{4} \text{ dyn/cm}^{2}.$$

PROBLEMS

Problem 1.1 Starting from the statement that myosin filaments are in a hexagonal array and 450 A apart calculate the number of filaments per unit area of muscle cross section. Show also that a stress of 1.5 kg/cm^2 distributed over these in the H band leads to a tension per filament of 2.5×10^{-5} dyn, as quoted in the text.

Problem 1.2 Referring to Fig. 1.4 show that the arc length Q'Q" is equal to $h\theta$.

CHAPTER 2
THE STRESS TENSOR

2.1 INTRODUCTION

In classical continuum mechanics, as in classical particle mechanics, we assume Newton's laws (Section 1.1) and proceed to discover their consequences for situations of interest. Typically we want to determine the deformation and/or flow of a medium in response to applied forces. It is helpful to select some definite region of the medium for our attention. Let the region be R, bounded by an imaginary surface S. (See Fig. 2.1). Our goal may be to determine the response of the medium in R to forces applied to this region.

Forces on the matter contained in R are partly of distant origin; gravity is such a force and in terrestrial experiments has its origin at the earth's center. Just as important, if not more so, are forces exerted by the matter immediately outside the (fictitious) surface S. These arise from interactions between molecules just outside S and their neighbors just inside, and are frequently called <u>contact forces</u> or <u>surface forces</u>. Our interest is frequently in the force per unit area, for which the term <u>stress</u> is used.

We shall see in the next section that six independent stress quantities must be specified for a complete description of the contact forces in a given surface region. In specifying these quantities it is in the spirit of continuum mechanics to ignore molecular details, even though the stresses are of molecular origin. Suppose we consider a stress component in the vicinity of a given point, say, P on the surface S. A continuum approximation is valid under two conditions: (i) our interest is in the average force over an area, say, ΔS (in the vicinity of P) whose characteristic dimension ε is large compared to the intermolecular spacing <u>a</u> and (ii) the stress is nearly uniform over ΔS. Condition (i) insures that the average is over a large number of molecules, and (ii) that the stress is characteristic of the region near P. These conditions might be stated compactly in the form of

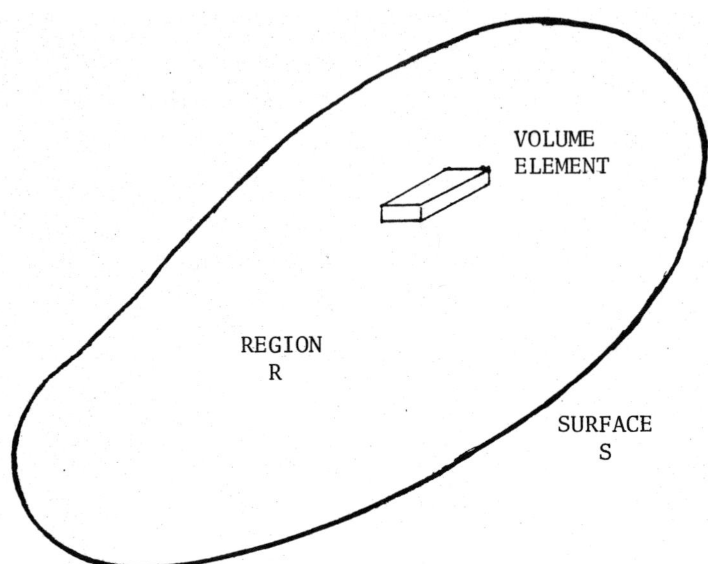

Fig 2.1 It is often convenient to speak of an arbitrary closed surface S which encloses a region R. Sometimes we consider a small volume element within R.

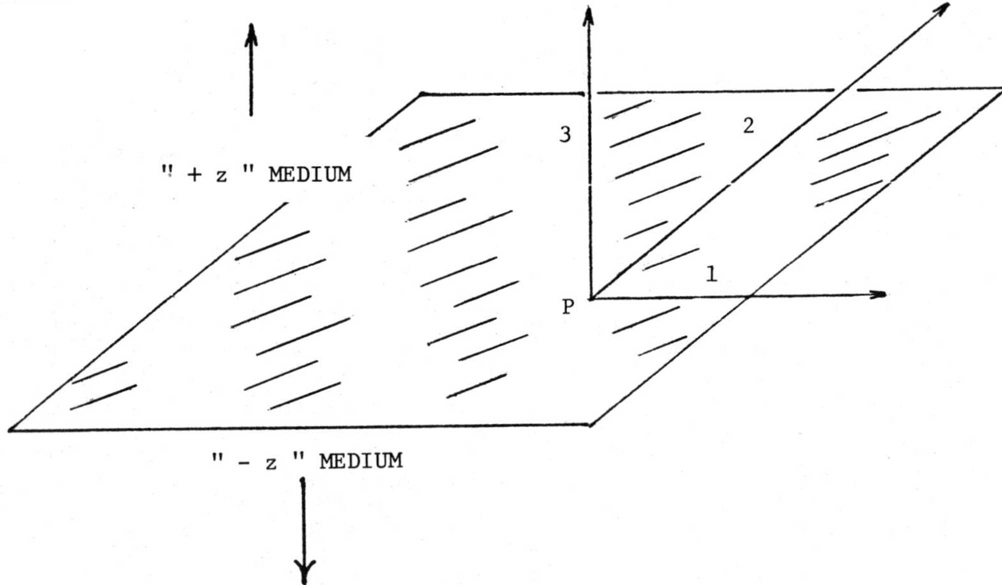

Fig. 2.2 In defining stresses at a point P it is necessary to refer to planes passing through P. In this figure the plane is such that its normal is along z or "3"; it separates the medium into two parts, one lying in the "+z" direction and the other in the "-z" direction.

Eq. (1.3) where ℓ is a typical nearest distance between maxima and minima of the stress on S . (Compare with the discussion relative to Eq. (1.3)).

2.2 STRESS COMPONENTS

In order to define the stresses more specifically we refer to a set of Cartesian coordinates (x,y,z). It will frequently be convenient to refer to the same axes as (x_1,x_2,x_2) or simply as (1,2,3); we shall alternate from one notation to another as the situation calls for it. Thus in Fig. 2.2 the "1" direction is that of the coordinate which we shall call either x or x_1 , the "2" direction is that of y or x_2 and the "3" direction is that of z or x_3 . Consider a point P in the interior of a continuous medium. Imagine that a surface cuts through the medium, passing through P, and for definiteness let this be the plane $z = c$, where c is constant. For economy of language we refer to the medium in the half-space $z > c$ as the "+z" medium and the medium defined by $z < c$ as the "-z" medium.

In general, as anticipated in earlier discussion, the "+z" medium exerts a contact force on the "-z" medium, across the plane $z = c$. By Newton's Third Law it then follows that the "-z" medium exerts an equal and opposite force on the "+z" medium, across the same plane. This action-reaction situation always presents itself in dealing with contact forces. (Although, of course, the Law also holds in isolated-particle situations, its consequences are often less obvious there since the origin of an "external field" may be far from the region of interest.) Because of this, we must exercise care; in giving the direction of a contact force or stress we must be sure to state which is the medium being acted upon.

Making use of these ideas and terminology we now define the stress quantities, starting with one which we symbolize as S_{33} . For simplicity we assume at present that the stresses are independent of

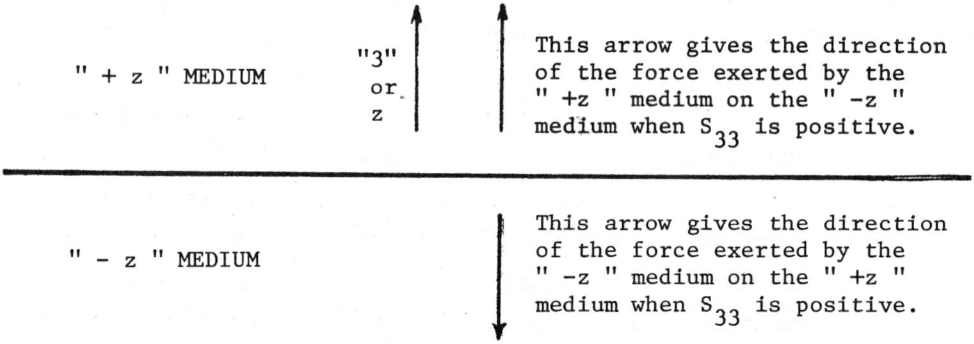

" + z " MEDIUM

"3" or z

This arrow gives the direction of the force exerted by the " +z " medium on the " -z " medium when S_{33} is positive.

" - z " MEDIUM

This arrow gives the direction of the force exerted by the " -z " medium on the " +z " medium when S_{33} is positive.

Fig. 2.3 **Any** of the **stress quantities** S_{ij} **represents force being exerted by** one **part** of **a medium** on **another part. In specifying the direction it is necessary to state which part is exerting the force and which part is being** acted **on. In this figure the convention is given for a normal or diagonal stress component.**

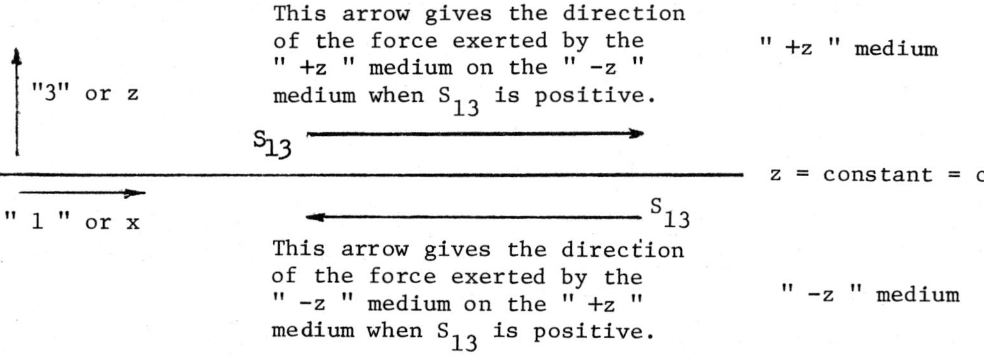

Fig. 2.4 **Here the sign convention is given for a typical** off-diagonal or **transverse stress component.**

x and y, but may depend on z. (We shall later relax this condition and let the stresses be functions of all coordinates.) By S_{33} at the plane z = c we shall mean the stress (contact force per unit area) in the "3" direction exerted by the " +z " medium on the " -z " medium; see upper arrow in Fig. 2.3. Invoking Newton's Third Law we see that S_{33} could equally well be defined as the stress in the " -3 " direction at z = c exerted by the " -z " medium on the " +z " medium. See lower arrow on Fig. 2.3.

Such a quantity as S_{33} is sometimes called a <u>normal</u> stress component, since its direction is normal to the surface in question. By contrast we now consider a quantity S_{13} of another kind. By S_{13} at the plane z = c we shall mean the stress in the "1" direction exerted by the " +z " medium on the " -z " medium across the plane z = c . See upper arrow in Fig. 2.4. By Newton's Third Law we could equally well define S_{13} as the stress in the " -1 " direction exerted by the " -z " medium on the " +z " medium.

Such a quantity as S_{13} is sometimes called a <u>transverse</u> stress component. A similar component would be S_{23} whose definition is analogous to that for S_{13} .

Clearly there are nine possible symbols S_{ij} . In this symbol the first subscript or index (i) gives the direction (except for sign) of the force or stress. The second index (j) gives the direction (except for sign) of the normal to the surface across which the stress is transmitted. Thus, as we saw in Fig. 2.4, for the stress component S_{13} the force is (positive or negative) along the " 1 " or x direction, and the reference surface has its normal along the (positive or negative) " 3 " or z direction. In a general field the components S_{ij} are functions of all three coordinates (x,y,z); the values of S_{ij} at a given point P must be defined in terms of the stresses as they exist at P. The nine stress components S_{ij} may be conveniently displayed in a 3 x 3 array:

$$S_{ij} = \begin{bmatrix} S_{11} & S_{12} & S_{13} \\ S_{21} & S_{22} & S_{23} \\ S_{31} & S_{32} & S_{33} \end{bmatrix} . \tag{2.1}$$

In such an array the terms S_{11} , S_{22} and S_{33} are the <u>diagonal</u> or <u>normal</u> stress components, while the others are the off-diagonal or <u>transverse</u> or <u>tangential</u> components. It can be shown (Prob. 2.1) that $S_{12} = S_{21}$, etc., so that one can always write the array in terms of six independent stress quantities; for example,

$$S_{ij} = \begin{bmatrix} S_{11} & S_{12} & S_{13} \\ S_{12} & S_{22} & S_{23} \\ S_{13} & S_{23} & S_{33} \end{bmatrix} . \tag{2.2}$$

As a special situation of interest suppose that a given body is subject to forces such that S_{33} is a positive constant S_o independent of x, y and z, while all other components S_{ij} are everywhere zero. The array then becomes

$$S_{ij} = \begin{bmatrix} 0 & 0 & 0 \\ 0 & 0 & 0 \\ 0 & 0 & S_o \end{bmatrix} \tag{2.3}$$

and represents a situation where a uniform tensile stress S_o exists along the z direction. Such an array gives an approximation to the stress field which exists in a stretched thin wire or thread, and probably in a contracted muscle filament. (But, in general, tensile stress applied in a given direction will be accompanied by nonzero stress components in other directions.) When Eq. (2.3) applies to a filament whose cross section has area A one frequently refers to the <u>tension</u> T^*, a quantity defined by

$$T^* = S_o A \quad . \tag{2.4}$$

We see that whereas a stress component S_{ij} (and, specifically, S_o) has units of force/area the tension T^* has units of force only.

In a liquid at rest under compression the off-diagonal or transverse components of stress are all zero, and the diagonal terms are all equal to (-p) where p is the <u>pressure</u>. Thus the array becomes

$$S_{ij} = \begin{bmatrix} -p & 0 & 0 \\ 0 & -p & 0 \\ 0 & 0 & -p \end{bmatrix} \tag{2.5}$$

Such an array is said to be <u>isotropic</u>. Even for a flowing liquid the stress array is isotropic if effects of viscosity can be neglected.

For a thin membrane under tension σ the stress array can be written

$$S_{ij} = \begin{bmatrix} S_o & 0 & 0 \\ 0 & S_o & 0 \\ 0 & 0 & S^* \end{bmatrix} \quad , \tag{2.6}$$

where the x_3 direction has been chosen perpendicular to the membrane surface. For a membrane of thickness h the quantity σ is defined by

$$\sigma = S_o h \quad . \tag{2.7}$$

We see that σ has units of force/length. The concept of membrane tension is an important one in biology.

2.3 THE STRESS TENSOR

In Section 2.2 our coordinates were Cartesian, by which we mean rectilinear coordinates that form an orthogonal set. The only other requirement on the set was that it be right handed. We could just as well have chosen another Cartesian set, obtained from the original one by a rigid rotation. In such a rotated set the 9 stress components would be defined relative to the "new" axes and would in general differ from the "old" components.

We will give the rule for transformation, but first we introduce appropriate symbols. Let the "new" axes be (1',2',3'), the "old" ones being (1,2,3) as before. Let i and j be two indices, either of which can represent any of the old axes (1,2,3). Correspondingly let m and n be two indices, either of which can represent any of the new axes (1',2',3'). We define the symbol a_{mi} as the cosine of the angle between the $m^{\underline{th}}$ new axis and the $i^{\underline{th}}$ old axis; for example a_{12} is the cosine of the angle between the (1') and the (2) axis. Similarly, a_{nj} gives the cosine of the angle between the $n^{\underline{th}}$ new axis and the $j^{\underline{th}}$ old one.

Using these symbols we can now state the transformation rule which applies to the stress quantities; justification for this rule is discussed in Section 2.4. Any component S'_{mn} in the new axes is related to the components S_{ij} in the old by the somewhat complicated equation

$$S'_{mn} = \sum_{i=1}^{3} \sum_{j=1}^{3} a_{mi}\, a_{nj}\, S_{ij} \,. \tag{2.8}$$

The rule in Eq. (2.8), applicable to Cartesian sets, is just that for a Cartesian tensor of rank two; hence the nine stress quantities are the components of such a tensor. We refer to the set of S_{ij}, or any set obtained from them by transforming according to Eq. (2.8), as the <u>stress tensor</u>.

We gain insight on the significance of Eq. (2.8) by applying it to specific examples, especially fairly simple ones. In the first place we restrict ourselves to rotations about one of the coordinate axes, such as the x_3 axis. Thus we consider two sets of Cartesian axes related as shown in Fig. 2.5. Here we find (Prob. 2.2) that the a_{ij} are given by

$$\begin{bmatrix} a_{11} & a_{12} & a_{13} \\ a_{21} & a_{22} & a_{23} \\ a_{31} & a_{32} & a_{33} \end{bmatrix} = \begin{bmatrix} \cos\phi & \sin\phi & 0 \\ -\sin\phi & \cos\phi & 0 \\ 0 & 0 & 1 \end{bmatrix} \tag{2.9}$$

We shall later consider biological effects of stress fields where, as viewed in the (x_1, x_2, x_3) coordinate system, the S_{ij} are of the form

$$S_{ij} = \begin{bmatrix} -p & A & 0 \\ A & -p & 0 \\ 0 & 0 & -p \end{bmatrix} \tag{2.10}$$

Here p is a hydrostatic pressure and A is a shear stress. It is convenient to think of S_{ij} as the sum of two simpler tensors as follows:

$$S_{ij} = G_{ij} + H_{ij} , \tag{2.11a}$$

where G_{ij} is an isotropic tensor

$$G_{ij} = \begin{bmatrix} -p & 0 & 0 \\ 0 & -p & 0 \\ 0 & 0 & -p \end{bmatrix} \tag{2.11b}$$

and

$$H_{ij} = \begin{bmatrix} 0 & A & 0 \\ A & 0 & 0 \\ 0 & 0 & 0 \end{bmatrix} \tag{2.11c}$$

Use has been made here of the rule for adding tensors; see Prob. 2.3. It follows from the tensor rules that S'_{mn}, which we shall call the transform of S_{ij}, is given by

$$S'_{mn} = G'_{mn} + H'_{mn} \tag{2.12}$$

where G'_{mn} and H'_{mn} are, respectively, the transforms of G_{ij} and H_{ij}. We see that G_{ij} is isotropic, being just the tensor of Eq. (2.5). It can be shown (Prob. 2.4) that such a tensor is unaltered by a rotation of coordinate axes; that is, an isotropic tensor is invariant with respect to such a transformation. Hence

$$G'_{mn} = \begin{bmatrix} -p & 0 & 0 \\ 0 & -p & 0 \\ 0 & 0 & -p \end{bmatrix} \tag{2.13}$$

To interpret this physically, consider a given point P of a medium where Eq. (2.11b) applies. The invariance of G_{ij} means that at (or very near) P the stress at any imaginary plane surface passing through P will be normal to the surface (regardless of the orientation of the plane) and will have the value (-p). If the pressure p is positive the stress is compressional, while negative p signifies that the medium is under tension.

By contrast, rotation has a considerable effect on H_{ij}. It can be shown (Prob. 2.5) that in the (x_1', x_2', x_3') coordinate system this tensor becomes

$$H'_{mn} = \begin{bmatrix} A \sin 2\phi & A \cos 2\phi & 0 \\ A \cos 2\phi & -A \sin 2\phi & 0 \\ 0 & 0 & 0 \end{bmatrix} \tag{2.14}$$

A test for consistency can be applied by letting $\phi = 0$; we do indeed find then that H'_{mn}, given by Eq. (2.14), reduces to H'_{ij} of Eq. (2.11c). We are probably not surprised that H'_{mn} and H_{ij} are again equal when $\phi = 180°$; this means that rotation through half a revolution does not change the stress components. However, a profound change is seen when $\phi = 45°$; this rotation yields

$$H'_{mn} = \begin{bmatrix} A & 0 & 0 \\ 0 & -A & 0 \\ 0 & 0 & 0 \end{bmatrix} \tag{2.15}$$

Thus, in this example, a 45° rotation of coordinates causes a tensor with only off-diagonal components (H_{ij} in Eq. (2.11c)) to become one with only diagonal components (H'_{mn} in Eq. (2.15)). A stress field appearing as pure shear in one system of coordinates appears as a combination of positive (tensile) and negative (compressional) normal stresses in another system.

A tensor such as that in Eq. (2.13) or (2.15) in which all off-diagonal terms are zero, is said to be in <u>diagonal</u> form. From a general theorem in tensor analysis any symmetrical tensor can be put in diagonal form by a suitable rotation of coordinates (Long, 1961).

Returning to the original stress tensor S_{ij} we can now write the result S'_{mn} of transforming according to the rotation defined by Eq. (2.9) and Fig. 2.5. Referring to Eq. (2.12) we find that G'_{mn} is given by Eq. (2.13) and H'_{mn} by Eq. (2.14). Hence we have

$$S'_{mn} = \begin{bmatrix} (A \sin 2\phi - p) & A \cos 2\phi & 0 \\ A \cos 2\phi & (-A \sin 2\phi - p) & 0 \\ 0 & 0 & -p \end{bmatrix} \qquad (2.16)$$

This result will find application when we consider the action of elastic and viscous shear fields on biological macromolecules, cells and other structures.

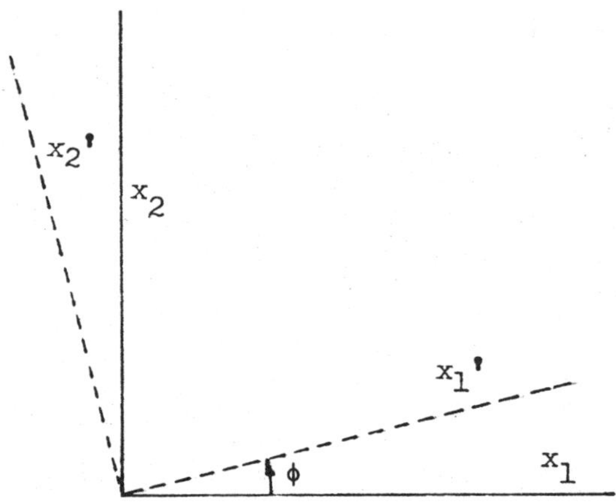

Fig. 2.5 Relations between new and old axes in a rotation about the "3" axis. The x_3 and x_3' axes are coincident and perpendicular to the figure.

2.4. VERIFICATION OF THE TENSOR RULE

In justifying the transformation in Eq. (2.8) we follow an approach typical of continuum mechanics in general. In essence we apply to a continuous medium (which may be in a solid, liquid, gaseous or other state) a basic principle in mechanics such as

"Newton's Second Law" (see Section 1.1), according to which the sum of forces on a body is equal to the time rate of change of its momentum. But unlike the situation in particle mechanics, there is no obvious well-defined body in the continuous medium to which the Second Law can be applied. Instead, as anticipated in Section 1.3, we rather arbitrarily select a small portion of the medium, frequently called a volume element or mass element, about which we make our statement.

In a general derivation of Eq. (2.8) all components of the stress are considered and a tetrahedron is chosen for the volume element (Long, 1961). However, for our purposes we get the idea fairly well by treating a simplified two-dimensional situation. Here we take the volume element to be a right-angled prism; see Fig. 2.6. Coordinate axes (1,2) and (1',2') are defined as indicated in the figure; the axes 3 and 3' are coincident, and directed toward the reader. Assume that the stresses exert no forces in the 3 or 3' direction; then all stress components involving a 3 or 3' axis are zero. Hence the tensors in the old and new axes are of the form

$$S_{ij} = \begin{bmatrix} S_{11} & S_{12} & 0 \\ S_{21} & S_{22} & 0 \\ 0 & 0 & 0 \end{bmatrix} \qquad (2.17)$$

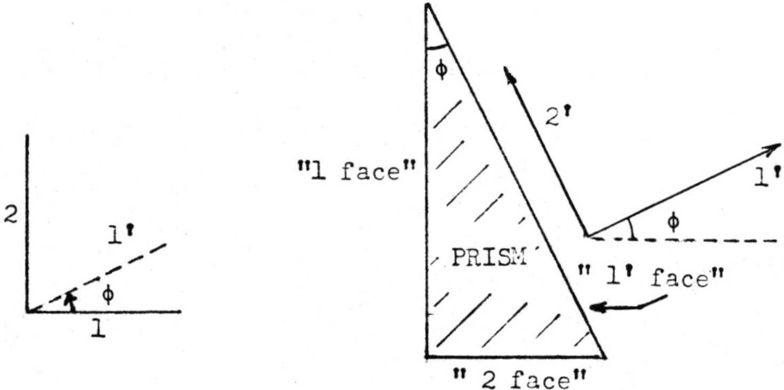

Fig. 2.6 Equilibrium of a prism under stresses on the three faces whose normals are along x_1 , x_2 and x_1' , respectively .

$$S'_{mn} = \begin{bmatrix} S'_{11} & S'_{12} & 0 \\ S'_{21} & S'_{22} & 0 \\ 0 & 0 & 0 \end{bmatrix} \tag{2.18}$$

Assume further that all stress components are constant, independent of space and time. Regarding the volume element itself, it is bounded by planes parallel to the plane of the figure, the distance between them being the prism thickness. For convenience we refer to the face whose normal lies along the 1 axis simply as the "1 face", etc. Then the oblique face is the 1' face; let its area be A'. One then readily finds (Prob. 2.7b) that the area A_1 of the 1 face and the area A_2 of the 2 face are given by

$$A_1 = a_{11}A'; \; A_2 = a_{12}\,A' \;, \tag{2.19}$$

where $a_{11} = \cos\phi$ and $a_{12} = \sin\phi$. In proceeding with our proof we now recall a basic principle for equilibrium, which follows from the three laws of Newton stated in Section 1.1:

> "If a body is in equilibrium the vector sum of forces on it must be zero".

We apply this principle by stating that, in particular, because the prism is in equilibrium the sum of forces along the 2' direction must be zero. Because of the nature of the stress field there are only three faces on which the stresses contribute to force in the 2' direction. These are the 1' , 1 and 2 faces; we let the force contributions along 2' from these faces be F_1', F_1 and F_2 , resp. The principle for equilibrium then becomes

$$F_1' + F_1 + F_2 = 0 .$$

It follows from definitions of the stress components that the outer medium exerts a force on the 1' face in the 2' direction given by

$$F_1' = A'S_{21}' .$$

We recall from Eq. (2.19) that the area of the 1 face is $a_{11}A'$. Hence the outer medium exerts a force $S_{11}a_{11}A'$ in the -1 direction, whose component along 2' is $-a_{21}S_{11}a_{11}A'$. Another force contribution from stress on the 1 face is $S_{21}a_{11}A'$ in the 2 direction, whose component along 2' is $-a_{22}S_{21}a_{11}A'$. Adding these contributions we obtain

$$F_1 = -a_{11}A'(a_{21}S_{11} + a_{22}S_{21}) .$$

Similarly it is readily shown that (Prob. 2.8)

$$F_2 = -a_{12}A'(a_{21}S_{12} + a_{22}S_{22}) .$$

Combining the above equations leads readily to the following expression for S_{21}' :

$$S_{21}' = \sum_{i=1}^{2} \sum_{j=1}^{2} a_{2i}a_{1j}S_{ij}' . \qquad (2.20)$$

This result is included in the more general Eq. (2.8); in our special two-dimensional situation i and j need be summed only over two values, 1 and 2. In the general derivation the stresses are arbitrary; each of the S_{ij} may vary with space coordinates and also with time.

2.5 PRESSURE AND TENSION IN A MODEL CELL

In research with the aim of learning basic principles of biology it is often useful to replace actual biological cells by relatively simple models. With models it is easier to repeat experiments under controlled conditions; because of this it is more feasible to study effects of varying a single parameter, say, A, while other parameters B, C and so on, are held fixed. Being simpler, models are also more susceptible than real cells to mathematical analysis in terms of physical principles, an activity which lies at the core of biological physics.

Of course, models or analogues must be treated with some caution. While a given analogue may indeed be helpful for exploring some aspects of cell behavior it may be useless or even misleading for other aspects.

For present purposes we choose as model for a biological cell a spherical liquid droplet of radius R surrounded by a film or membrane of thickness h. See Fig. 2.7 . Such a model has been much used; many choices have been made for the liquid and for the membrane in attempts to reproduce various cellular characteristics. Our aim here is to consider the tension in a cell membrane and its relation to intracellular pressure. For this purpose the detailed characteristics of the liquid and membrane are not very important except that the membrane must be homogeneous and isotropic. We assume that the stress tensor in the liquid is isotropic as given in Eq. (2.5) where p is

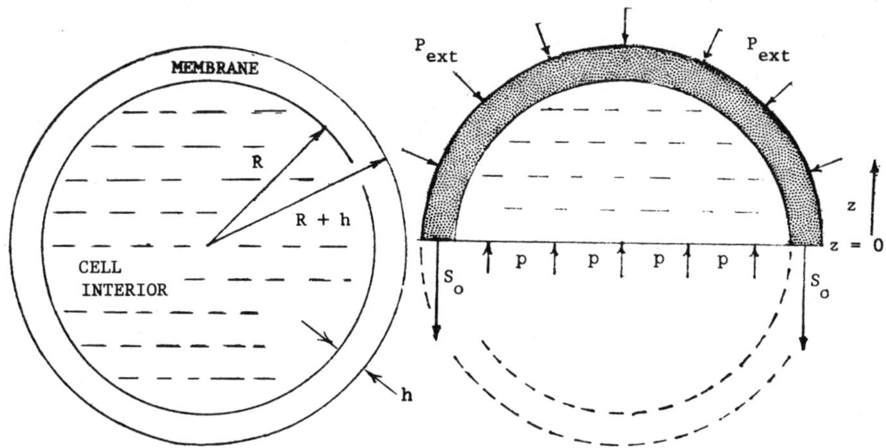

Fig. 2.7 Model for a biological cell. Characteristically $h \ll R$. We assume uniform stresses as follows: positive pressure p in the cell interior, positive pressure P_{ext} on the exterior, tensile stress S_o in the membrane. Arrows in the figure on the right show the direction of forces exerted via stresses on the body to which Newton's First Law is applied, namely, the upper half-cell.

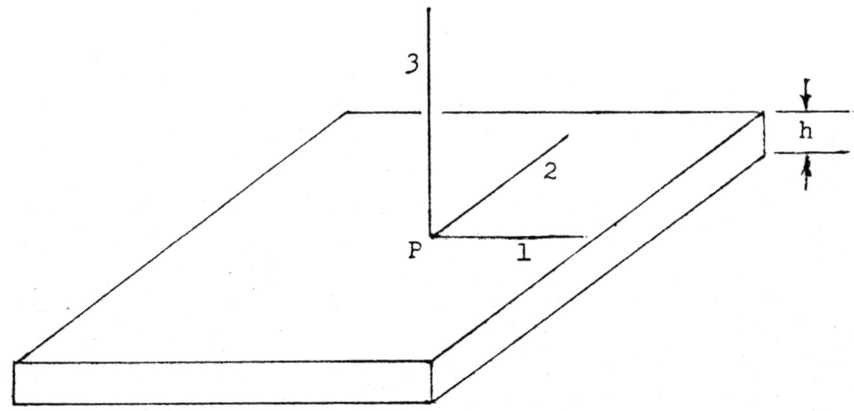

Fig. 2.8 Portion of membrane under tension. The membrane is bounded (locally) by planes whose normals are along the 3 or z direction.

constant throughout the interior region. Recall (Prob. 2.4) that this tensor is invariant with respect to rotation. Hence at any point P in the liquid the stress across any plane through P is then normal to the plane and given by -p . In a cell the stress is usually <u>compressional</u>, the pressure being equal to p .

We assume the membrane thickness h is small compared to the radius R. Near any definite point P in or on the cell membrane we can then regard the neighboring portion of the membrane as essentially flat, that is, as a part of a rectangular sheet; see Fig. 2.8 . It should be realized that h/R is typically much smaller than suggested in Fig. 2.7 . Taking P as the origin of Cartesian coordinates we let the 3 axis be normal to the sheet passing through P ; the 1 and 2 axes then lie in a plane parallel to the membrane surfaces. We assume the stress tensor has the form given in Eq. (2.6) ; using Eq. (2.7) this becomes

$$S_{ij} = \begin{vmatrix} S_o & 0 & 0 \\ 0 & S_o & 0 \\ 0 & 0 & S^* \end{vmatrix} = \begin{vmatrix} \sigma/h & 0 & 0 \\ 0 & \sigma/h & 0 \\ 0 & 0 & S^* \end{vmatrix} \qquad (2.21)$$

where σ is the membrane tension. We recall that this tensor is invariant for rotations about the z or 3 axis (Prob. 2.6) . Hence for any choice of direction for the 1 axis (so long as it is parallel to the membrane surface) we can state: S_o gives the force in the 1 direction across unit area of a 1 plane exerted by the membrane substance which lies in the "+1" direction on that in the "-1" direction. The area of membrane cut by the 1 plane is just h for unit length of the cut, that is, unit length in the 2 direction. Hence the force (along 1) per unit length (along 2) is hS_o or σ .

We obtain a relationship between the pressure p and the membrane tension σ by considering an application of the principle of equilibrium, somewhat as was done in Section 2.4 . There the principle was applied to a prismatic volume element representing part of a continuous medium. Here we select a hemispherical portion of the model cell. This portion is shown arbitrarily as the upper half of the cell in the right of Fig. 2.7 .

We distinguish three forces acting on this body. The outer medium exerts pressure P_{ext} on it which may, for example, be just the atmospheric pressure; this results in a downward force F_{ext} on the body given by AP_{ext} , where A is the area projected by the cell on a horizontal plane. (Prob. 2.9) . This projected area is a circle of radius (R + h) so we obtain

$$F_{ext} = \pi(R+h)^2 P_{ext} \, . \tag{2.22}$$

Another force exerted on the selected body (i.e., the upper half-cell) comes from pressure exerted by the liquid in the lower half-cell. By hypothesis the pressure is p everywhere in the liquid. At the lower boundary of the body, that is, at $z = 0$, the force F_p on the body associated with the stress p is upward and given by

$$F_p = \pi R^2 p \, , \tag{2.23}$$

since πR^2 is the area of a circular region over which the pressure is p . Thus F_p gives the force exerted by the medium below $z = 0$ on the medium above that plane.

The third force is also one which is exerted on the upper half-cell by the lower half-cell. This arises from tensile stress in the membrane. Let P be any point which is on or in the membrane and which is also on the equatorial plane $z = 0$. Considering the discussion in connection with Fig. 2.8 and the stress tensor of Eq. (2.21) we come to the following statement: The stress S_o gives the force per unit area exerted downward by the medium (membrane substance) below $z = 0$ on the medium above. Hence the total downward force F_t (on the "body") associated with the tensile stress is $S_o A'$

where A' is the area of membrane exposed by cutting through it with the plane z = 0 . This ring-shaped section will have the area

$$A' = \pi(R + h)^2 - \pi R^2 = \pi(2hR + h^2) .$$

Hence we have

$$F_t = S_o A' = \pi S_o(2hR + h^2) . \tag{2.24}$$

Since our "body", the upper half-cell, is in equilibrium under the forces considered, we realize (from the equilibrium principle, as in Section 2.4) that the total downward force on the body is zero, that is,

$$F_{ext} - F_p + F_t = 0 . \tag{2.25}$$

Combining equations given above and reorganizing we obtain (Prob. 2.10)

$$p = P_{ext} + \left(\frac{2h}{R} + \frac{h^2}{R^2}\right)(P_{ext} + S_o) . \tag{2.26}$$

We have already assumed that h << R and therefore should discard the term (h^2/R^2) on the right hand of Eq. (2.26); we have then

$$p = P_{ext} + \frac{2}{R}(\sigma + hP_{ext}) , \quad \sigma = hS_o ; \tag{2.27}$$

here σ has the units of force/length. Under some circumstances hP_{ext} is much smaller than σ ; when it is neglected we obtain simply

$$p = P_{ext} + \frac{2\sigma}{R} . \tag{2.28}$$

Some cells are far from spherical, and in fact are better represented as long thin cylinders. This is true for cells of nerve and muscle and of filamentous plant tissue. A more suitable model then has the shape of a long cylinder in which the liquid body has radius R and the membrane has thickness h . When h is sufficiently small it is readily shown here (Prob. 2.10) that

$$p = P_{ext} + \frac{\sigma}{R} . \tag{2.29}$$

For cell surfaces of other shapes results are less simple; σ may vary from point to point. At any ordinary point Q on the surface there are two principal directions, which are mutually perpendicular; along one of these the curvature is maximum and along the other it is minimum. Let the maximum and minimum radii of curvature be R_1 and R_2, respectively. Then it can be shown that Eq. (2.29) is replaced by

$$p = P_{ext} + \sigma\left(\frac{1}{R_1} + \frac{1}{R_2}\right) ; \qquad (2.30)$$

here σ is the tension at the point Q while p and P_{ext}, as before, represent the pressures inside and outside the cell, respectively. Equation (2.30) was first derived by Laplace (References 2.2 and 2.3). For a sphere R_1 and R_2 are equal (one might say that every direction is a principal direction on a sphere) and Eq. (2.30) reduces to Eq. (2.28). For a cylinder one of the radii of curvature is infinite and Eq. (2.30) reduces to Eq. (2.29).

It can be shown that the results in Eqs. (2.28) through (2.30) apply also to a body of fluid without a membrane, but surrounded by another fluid. In this situation σ represents the <u>surface tension</u> or <u>surface energy</u>, of the interface; σ gives the work required to form unit area of new surface and is a constant characteristic of a given pair of fluids. A few representative values of the surface tension may be found in Table A.1.

Equations (2.28) through (2.30) are useful in biology, partly for dealing with droplets and bubbles which may be present within cells and in body fluids. They are also used in analyzing behavior of cells, especially in response to environmental changes.

2.6 STRESSES IN LIVING SYSTEMS; UNITS

Concepts of force and stress are a vital part of human experience. Our vocabulary is full of such words as push, pull, attract, repel, tug and squeeze, in addition to force and stress themselves. Presumably language developed this way over the centuries because force-related words have been found necessary or, at least, helpful in describing what we see and do. But in everyday usage our language often lacks precision and is not suited for quantitative description. Hence we have taken pains to define terms and concepts in this chapter.

Stresses which occur in living systems may be generated internally, directly or indirectly; or they may be imposed by the environment. Certain tissues (especially, muscle) are able to generate stress directly; some characteristics of the basic process for stress generation are taken up in Sections 1.6 and 1.7. Other tissues which form walls of fluid-filled spaces are typically under stress from pressure of the contained fluid. In blood vessels, for example, this pressure arises ultimately from muscular activity (beating of the heart); hence the stress in blood vessels is generated internally, but indirectly. Similarly the tension in cells from osmotic pressure is an internally generated stress. On the other hand the structural units of our skeleton are typically under stress arising from our environment. In this class are (i) stresses which result from the force of gravity and (ii) pushes, pulls, blows and pressures which arise in encounters with surrounding matter (both animate and inaminate!).

In the next section and in later chapters we shall discuss forces, stresses and related quantities, as they appear in situations relevant to biology. To do this in a quantitative manner we must become familiar with appropriate units, especially those from the cgs and mks systems. In connection with filaments or thin rods, Eq. (2.4), we referred to the tension T . This quantity has simply the units of force, expressed as dynes (abbreviated as "dyn") in the cgs system and as newtons (abbreviated as "n") in the mks system.

In membranes it is convenient to speak of a membrane tension σ , as in Eq. (2.7). The units of σ are force/length, expressed in cgs units as dynes per centimeter (dyn/cm) and in mks units as newtons per meter (n/m).

Especially basic are the stress quantities S_{ij} . For these the units are force/area, expressed in the cgs system as dyn/cm^2 and in the mks system as n/m^2 . For pressure a variety of other units are also used, such as the atmosphere (atm).

Factors for converting from one system of units to another are listed in Table A.2 (in Appendix A); abbreviations are listed in Table A.3.

2.7 RESULTS FOR DROPLETS AND RED CELLS

We consider here experiments in which results of Section 2.5 are applied to tests on mechanical characteristics of cells and their membranes. There are various means by which changes can be made in the pressure difference Δp or $(p - P_{ext})$, where p is the pressure on the interior of a cell and P_{ext} the pressure outside. When the pressure difference Δp is altered, changes occur in the membrane tension σ as given, for example, by Eq. 2.28 . Changes in cell shape and size may occur as Δp increases; the cell may burst at some critical value of Δp.

A rather simple way of altering Δp for a cell is to adjust its osmotic environment as discussed in a later chapter. To illustrate the possibilities we consider a human red cell. This cell normally has the shape of a biconcave disc whose diameter is about 8 μm and whose width is somewhat greater than 2 μm. The cell retains this shape when placed in a simple solution of 0.9% Na Cl. See Fig. 2.9 (a) ; such a solution is said to provide an isotonic environment to the red cell.

When the Na Cℓ concentration is decreased the solution becomes more and more hypotonic to the cell. The pressure difference Δp then increases and the cell changes shape as shown in Fig. 2.9 (b) and (c). When the NaCℓ concentration becomes as low as 0.4 % to 0.5% the cell will burst releasing its hemoglobin contents into the surrounding medium. It has been calculated by Hoffman (Reference 2.5) that the value of Δp at which the membrane bursts is approximately 6500 dyn/cm^2; under conditions just before bursting the cell is spherical with radius about 3 μm. From Eq. (2.28) it then follows that the tension σ in the membrane when bursting occurs is about 1 dyn/cm. A similar result was obtained by Katchalsky, et al. (Reference 2.6). For a membrane of thickness h we see from Eq. (2.7) that the stress is just σ/h; if h is 100 A for the red cell membrane we obtain about 10^6 dyn/cm^2 as the breaking stress for this membrane. This value may be compared to a breaking stress of 1.2×10^8 dyn/cm^2 for lead, 4×10^9 dyn/cm^2 for copper and 4×10^{10} dyn/cm^2 for tungsten.

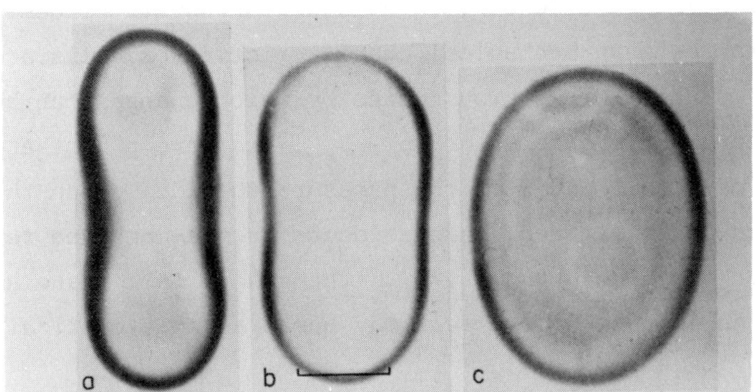

Fig. 2.9 Equilibrium shape of a red cell in isotonic (a) and in increasingly hypotonic ((b) and (c)) environment. From Rand and Burton, Reference 2.4.

Another method for altering the pressure difference Δp involves mechanical manipulation. Mitchison and Swann (Reference 2.7) devised an approach which makes use of a micropipette, a small glass capillary with inner bore no more than a few microns in diameter. They arranged to bring the tip of the micropipette in contact with part of a cell. Suction was then applied, thus reducing P_{ext} and increasing Δp locally; the resulting cell deformation was observed. This method was also used by Rand and Burton (Reference 2.4) in their studies on the red cell which we now discuss. As shown in Fig. 2.10 a portion of the cell is drawn into the pipette whose bore radius is R_p. The portion drawn into the tube is partly bounded by a surface of relatively small radius of curvature r; the main part of the cell which remains outside is assumed to be spherical with radius R_c. The pressure inside the pipette is P_1 while the outer pressure is P_2; inside the cell the pressure is P_3. It is assumed that a tension σ is established at the cell surface, and that the tension is the same in the small portion of surface in the capillary as in the main outer portion. Under these assumptions it follows from Section 2.5 that

$$P_3 - P_2 = 2\sigma/R_c \; ; \quad P_3 - P_1 = 2\sigma/r \; . \tag{2.31}$$

As Rand and Burton used the technique, the pressure difference $(P_2 - P_1)$ is always adjusted so that the portion drawn into the capillary has a hemispherical shape, i.e., so that r is equal to R_p. When this is true Eqs. (2.31) combine to yield

$$P_2 - P_1 = 2\sigma(R_p^{-1} - R_c^{-1}) \; . \tag{2.32}$$

Confidence in Eq. (2.32) was gained by applying it to experiments on fluid drops or droplets, for which the interfacial tension σ is known. Thus Fig. 2.11 shows a plot of $(P_2 - P_1)$ versus $(1/R_p)$ for a very large drop of isobutyl alcohol in water; the radius R_c was of the order of centimeters. These data were obtained by use of a series of micropipettes of various radii R_p. Experimental results, represented by the plotted points, are fitted very well by a straight line as shown. Such a linear relationship is, of course, to be expected from the theory according to Eq. (2.32). From the latter equation the slope of the line gives 2σ; the investigators obtained 2.42 dyn/cm for σ, in reasonable agreement with values obtained by other methods on the interfacial tension between

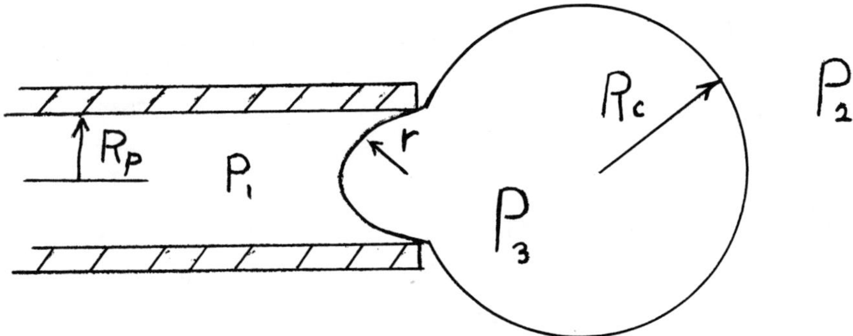

Fig. 2.10 Schematic of arrangement used by Mitchison and Swann (Reference 2.7) and also by Rand and Burton (Reference 2.4) for studies on mechanical properties of cells.

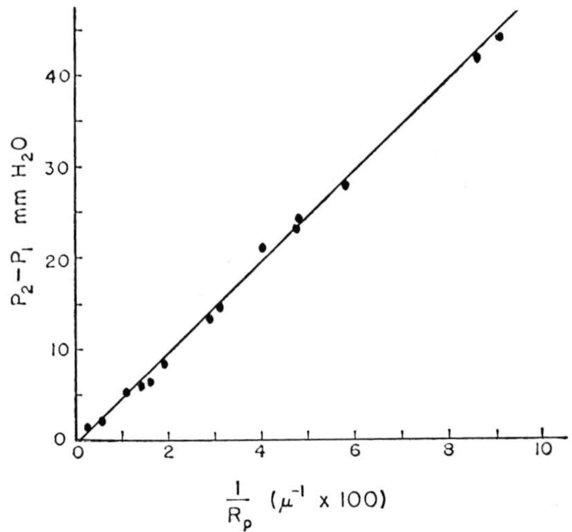

Fig. 2.11 Critical pressure (P_2 - P_1) <u>versus</u> ($1/R_p$) for a large drop of isobutyl alcohol in water. From Rand and Burton, Reference 2.4.

water and isobutyl alcohol. According to Eq. (2.32) the pressure difference $(P_2 - P_1)$ should not be zero when $(1/R_p)$ is zero, but should have the value $(-2\sigma/R_c)$. But the magnitude of this term is small when the drop is large, as was the situation for the data of Fig. 2.11; hence it is reasonable that the intercept should hardly be perceptible here.

Figure 2.12 shows results of an analogous experiment done with red cells. These are either in isotonic or hypotonic solution; hence they are either normal in shape as in Fig. 2.9(a), or are swollen as in Fig. 2.9 (b) and (c). Again, as with the drop of Fig. 2.11, the data are fitted well by a straight line; but here there is a signific intercept when $(1/R_p)$ is zero. These are remarkable results, especially in view of the difficulty in obtaining them; data shown here are for values of R_p varying from 1 μ downward to 0.3 μ , the latter being smaller than the shortest wavelength of visible light .

What do we learn about the red cell from the results of Fig. 2.12? At the least we have, of course, the information contained in the individual data points; from these we know for each R_p what value of $(P_2 - P_1)$ is required to pull part of the cell surface into a hemispherical cap of radius R_p . Such information, which seems very specialized, takes on added significance when considered together with other observations of Rand and Burton; they found in their experiments that a critical value of $(P_2 - P_1)$ was required to form a hemispherical cap in the micropipette (except for very swollen cells). When this value was reached not only did the cap form but it was also immediately drawn into the pipette a short distance, followed by part of the remaining cell, as suggested in Fig. 2.13 . Presumably if R_p were large enough the entire cell would be drawn into the pipette, as if it were a liquid droplet (Prob.2.12). Such studies help in understanding the behavior of red cells in the normal circulation of the human body, where they must deform greatly to pass through the small capillaries.

It is characteristic of the simple liquid droplet that its value of σ does not depend on its size or shape (unless the radius of curvature is

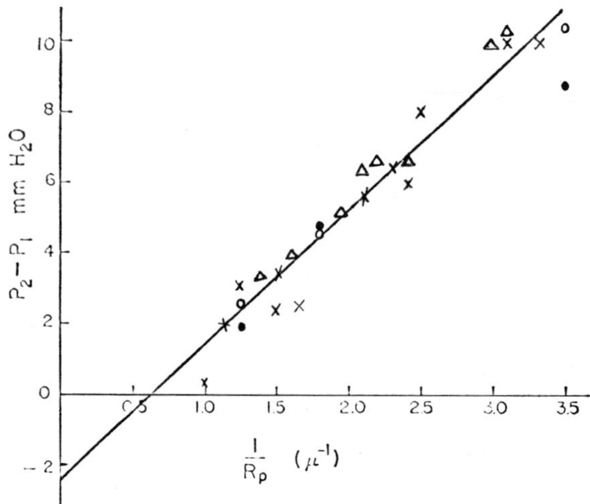

Fig. 2.12 Plot analogous to that in Fig. 2.11 but for a red cell. From Rand and Burton, Reference 2.4.

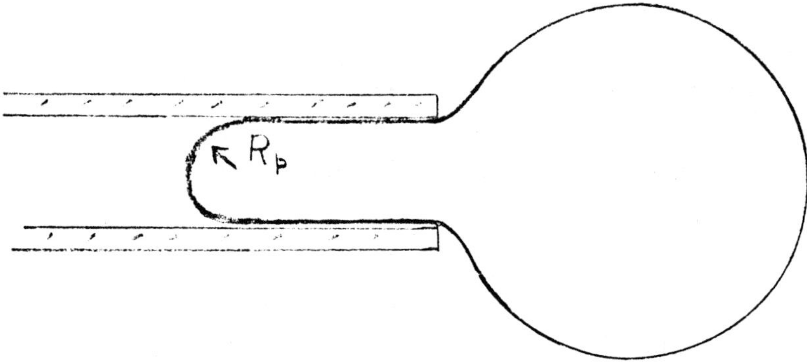

Fig. 2.13 Instability phenomenon for a droplet. Once a hemispherical portion of radius R_p is pulled into the pipette the entire droplet enters.

very small). Here the quantity σ , called the interfacial tension or interfacial energy, results from a tendency for the liquid molecules in the droplet to "prefer the interior to the surface". In more technical language the potential energy of such a liquid molecule is lower in the interior (where it is surrounded by like molecules) than when it is at the surface (where it is only partially so surrounded). the droplet therefore tends to minimize its surface area, as if it were under tension. The observed linear relation in Fig. 2.11 attests to the constancy of σ relative to R_p for the drop studied there.

From Fig. 2.12 we conclude that for a red cell also, under some circumstances, the quantity σ required for Eq. 2.32 is essentially independent of R_p. Specifically Rand and Burton obtained for σ the (very low) value of 0.019 dyn/cm. Evidently the cell behaves in some ways like a liquid drop under the conditions of Fig.2.12. This observation in itself suggests information about the cell membrane. Perhaps under the conditions of Fig. 2.12 the membrane is rather relaxed and contributes very little tension. Then for the medium or cytoplasm inside the hemispherical cap it is reasonable that the effective value of σ arises in the same way as for a fluid drop. Thus molecules of the cytoplasm "prefer" to be in the interior rather than at the periphery and there is a tendency for the cytoplasm to minimize its surface area.

However, further information about the red cell membrane is derived from the fact that at the critical value of $(P_2 - P_1)$ the cell does not flow entirely into the pipette, as a simple droplet would. Rand and Burton explained this by assuming an increase in σ, noting that as a cell proceeds to enter the micropipette (Fig.2.13) its surface area increases. Actually the area begins to increase as the hemispherical cap is being formed, but apparently not enough to cause much membrane stretching. (It is as if the membrane were initially in a "folded" state; a small increase in gross area of the membrane then causes it to unfold without much actual stretching.) Evidently, when the cap begins to move down the pipette the membrane area increases rapidly and it is then that σ increases significantly, because the membrane is being stretched.

In other studies the authors carried out similar experiments with cells in hypertonic solutions, that is solutions with NaCl concentrations greater than the isotonic value. Again a linear relation was found between $(P_2 - P_1)$ and $(1/R_p)$; the constant σ was in this case only 0.0065 dynes/cm, about one-third the value in hypotonic solution.

PROBLEMS

2.1 The aim of this problem is to give a simplified proof that the stress tensor is symmetric, i.e. that $S_{ij} = S_{ji}$. See Fig. 2.1; assume the region R contains matter at equilibrium under stresses S_{ij} which are constant in space (i.e., independent of x, y and z). Suppose the volume element within R is a rectangular parallelopiped whose sides are the planes $x = 0$, $x = a$, $y = 0$, $y = b$, $z = 0$ and $z = c$. Draw a labelled sketch of this volume element. Obtain an expression for the torque L about the z axis as a result of stresses acting on the various faces of the element. Show that L_z is zero if and only if $S_{12} = S_{21}$.

(Since the torque must be zero in an equilibrium situation, the last step constitutes a proof that $S_{12} = S_{21}$. Similar arguments lead to the conclusion that $S_{13} = S_{31}$ and $S_{23} = S_{32}$, and hence a justification for writing S_{ij} in the symmetrical form of Eq. 2.2. It is also possible, though the required reasoning is more extended, to show that Eq. 2.2 is justified even when the S_{ij} are non-uniform, and even if equilibrium does not exist. (Long, 1961).

2.2 Suppose two sets of Cartesian coordinates are related as shown in Fig. 2.5, the axes x_3 and x_3' being identical and directed out of the figure. Show that the direction cosines are as given in Eq. (2.9).

2.3 By definition the sum of two tensors G_{ij} and H_{ij} is an array of nine quantities S_{ij}, where each quantity S_{ij} is the sum of the respective G_{ij} and H_{ij} components. Thus

$$S_{11} = G_{11} + H_{11} \; ; \quad S_{12} = G_{12} + H_{12} \; ; \quad \text{etc.}$$

Show that the array S_{ij} is a tensor. Thus show that in a rotation of coordinates S_{ij} transforms according to the rule in Eq. (2.8), it being given that G_{ij} and H_{ij} do so.

2.4 Given that in the (x_1 , x_2 , x_3) coordinate system the stress tensor is isotropic, that is, has the form of G_{ij} in Eq. (2.11b). Show that in a rotation about the x_3 axis the stress tensor is unchanged regardless of the value of ϕ . In doing this you will have justified Eq. (2.13) and will have shown, for a special situation, that an isotropic tensor is invariant for rotation.

2.5 Given that in the (x_1 , x_2 , x_3) coordinate system the stress tensor has the form of H_{ij} in Eq. (2.11c). Show that in a rotation about the x_3 axis the tensor transforms to H'_{mn} as given in Eq. (2.14).

2.6 Show that S_{ij} given in Eq. (2.6) is invariant for rotation about the z axis.

2.7 Considering the prism of Fig. 2.6 verify the area relationships stated in Eq. (2.19).

2.8 Complete missing steps in verifying the two-dimensional tensor transformation equation, Eq. (2.20). In particular, derive the expression given for F_2 .

2.9 Verify the expression

$$F_{ext} = AP_{ext} ,$$

which was required in deriving Eq. (2.22). One rather simple way to do this is to apply the principle of equilibrium to a fictitious hemispherical object immersed in fluid where the pressure is everywhere equal to P_{ext} .

2.10 Combine equations as indicated and obtain the expression for $(p-P_{ext})$ given in Eq. (2.26).

2.11 Derive Eq. (2.29), which applies to a model cell of cylindrical shape, by means of reasoning similar to that used in deriving Eq. (2.28). Ignore end effects, assuming a long cylinder.

2.12 Justify the statement in the caption for Fig. 2.13, namely, that (for a droplet) "Once a hemispherical portion of radius R_p is pulled into the pipette the entire droplet enters."

CHAPTER 3

STRAIN AND RATE OF STRAIN

We continue here with the development of terminology and concepts in continuum mechanics. Living organisms typically contain fluids or suspended objects that are in motion. Also macromolecules, cells and organs are continually changing shape, either through natural processes or in response to environment. In this chapter we define and discuss geometric and kinematic quantities useful for describing motions and deformations. Especially important are tensors for strain and rate-of-strain. In succeeding chapters we learn how to relate these quantities to stresses and external force fields.

3.1 DISPLACEMENTS.

Stresses acting on a body may cause it (1) to accelerate as a whole, (2) to rotate about an axis, or (3) to experience change in shape and/or size. It is the third possibility which most interests us here. The body may be a discrete solid object, or it may be an arbitrary part of a continuous medium.

We consider the body to be divided up into a large number of small portions which we shall call "particles". Each particle is small enough so that for many purposes we can regard all of it as being at a single point, say P. It is convenient, in fact, to speak of "the particle P ", meaning a particle at P (or, more accurately, a particle which contains the point P).

To describe deformation or strain of a body one refers to positions of representative particles. One must first identify an "initial" or "undeformed" state of the body (by what is sometimes a rather arbitrary choice). This initial state is then compared with another state in which the body is said to be "deformed" or "under strain". Description of the strain in the second state is achieved, in essence, by comparing positions of representative particles (assumed identifiable) in the body in the two states.

Our interest will be primarily in changes of <u>relative</u> positions for different particles in the body. If in a given change of state all particles are displaced without alteration of any inter-particle distances, the change does not involve strain. Such a change represents a rigid-body transformation; by this is meant a translation, rotation, or both, of the body as a whole.

For specifying positions and displacements of particles we choose a set of Cartesian axes numbered (1, 2, 3). Let P be a particle whose coordinates with respect to these axes in the initial state of the body are (a_1, a_2, a_3), and whose coordinates with respect to the same axes in the deformed states are (x_1, x_2, x_3). Let us designate the changes in coordinates along the three axes as (ξ_1, ξ_2, ξ_3); these components are given by

$$\begin{aligned} \xi_1 &= x_1 - a_1 \\ \xi_2 &= x_2 - a_2 \\ \xi_3 &= x_3 - a_3 \ . \end{aligned} \tag{3.1}$$

We may imagine a vector (Fig. 3.1) called the <u>displacement vector</u>, from the original position (a_1,a_2,a_3) to the displaced position (x_1,x_2,x_3) of the point P. The quantities (ξ_1,ξ_2,ξ_3) in Eqs. (3.1) are just the components of the displacement vector; we shall call them, alternatively, the <u>displacement components</u>, the <u>components of displacement</u>, or, simple the <u>displacements</u>. In a given state of strain the ξ_i vary from particle to particle. In expressing these variations mathematically we need a convenient means of identifying each of the very many particles which comprise the body. For this purpose the original coordinates serve very well, since each particle has its own distinct set of a_i. Even more convenient are the coordinates x_i of the particle in a given state of strain in spite of the fact that these, unlike the a_i, are dependent on the state.

As measures of the interparticle variation of the ξ_i we use such derivatives as $\partial\xi_1/\partial a_2$ or $\partial\xi_1/\partial x_2$ and, more generally the derivatives $\partial\xi_i/\partial a_j$ or $\partial\xi_i/\partial x_j$. Usually the deformations of interest are small such that, for example, the change in a dimension of an object is small compared to the original dimension. In a small deformation the derivatives $\partial\xi_i/\partial a_j$ and $\partial\xi_i/\partial x_j$ are each in magnitude much less than unity. It can also be shown (Prob. 3.1) that under these conditions the derivative of a displacement ξ_i with respect to a given original coordinate a_j is approximately equal to the derivative of the same displacement ξ_i with respect to the corresponding strained coordinate x_j ; that is,

$$\partial\xi_i/\partial a_j \cong \partial\xi_i/\partial x_j \ . \tag{3.2}$$

We shall henceforth for convenience refer primarily to the x_j .

Since i and j can each take on any of the values 1, 2 or 3 we see that there are nine of the $\partial\xi_i/\partial x_j$. We find also, as for the stress quantities S_{ij} discussed in Chap. 2, that the $\partial\xi_i/\partial x_j$ are components of a Cartesian tensor of rank two (Prob. 3.2).

3.2 ROTATION AND STRAIN TENSORS

We now consider more specifically how the displacement field can be described. In general any of the displacements ξ_i may be a function of all three of the coordinates x_i . In the vicinity of a given point 0 a very useful expression can be given for ξ_i in terms of a Taylor series. At a neighboring point P (See Fig. 3.2) whose strained coordinates x_i differ from those at 0 by $(\Delta x_1, \Delta x_2, \Delta x_3)$ we have

$$(\xi_i)_P = \xi_{io} + (\partial\xi_i/\partial x_1)_o \, \Delta x_1 + (\partial\xi_i/\partial x_2)_o \, \Delta x_2 + (\partial\xi_i/\partial x_3)_o \, \Delta x_3 + - - \tag{3.3}$$

On the right hand of Eq. (3.3) the subscript "O" on a quantity such as appears on ξ_{io}, $\partial\xi_i/\partial x)_o$, etc., means that the quantity is to be evaluated at the point O. When the Δx_i are sufficiently small the terms shown on the right hand side of Eq. (3.3) give an accurate representation of ξ_i at P ; for all other terms are of the order of $(\Delta x_1)^2$, etc., and higher. We see that the nine derivatives $\partial\xi_i/\partial x_j$ are useful quantities. When these are known at any arbitrary point O, we can use them in Eq. (3.3) to determine for any displacement component ξ_i the difference between its value at O and that at any nearby point P. If all $\partial\xi_i/\partial x_j$ are zero the ξ_i are the same at all points and the motion is one of pure translation.

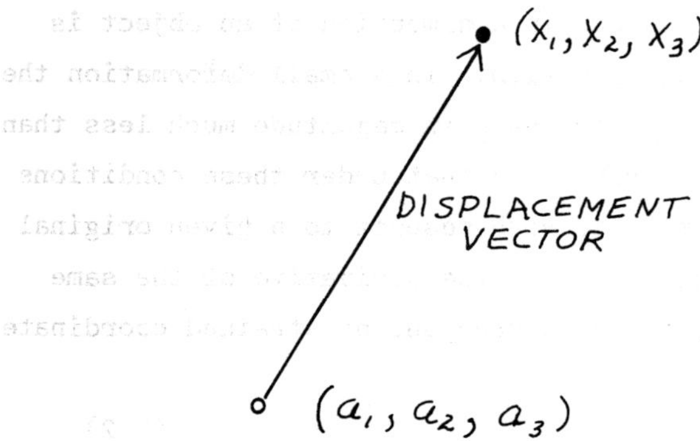

Fig. 3.1 The displacement vector for a given particle at a given time t extends from its original position (a_1 , a_2 , a_3) to its position (x_1 , x_2 , x_3) at the time t.

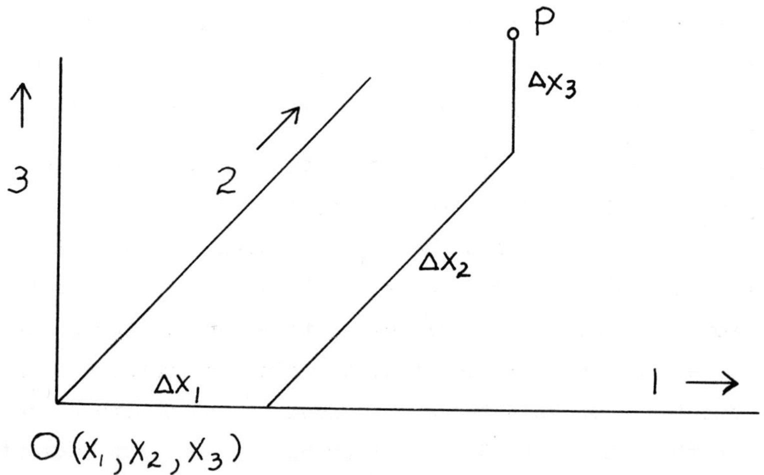

Fig. 3.2 A given point O and a neighboring point P whose coordinates differ from those at O by Δx_1 , Δx_2 , Δx_3 , respectively.

However, the $\partial\xi_i/\partial x_j$ are not as generally useful as other quantities, to which these are closely related, and which we now consider. These new quantities present themselves when we note, for example, that

$$\partial\xi_2/\partial x_1 = \tfrac{1}{2}(\partial\xi_2/\partial x_1 - \partial\xi_1/\partial x_2) + \tfrac{1}{2}(\partial\xi_2/\partial x_1 + \partial\xi_1/\partial x_2) \ .$$

In general one may write

$$\partial\xi_i/\partial x_j = \rho_{ij} + \varepsilon_{ij} \tag{3.4a}$$

where

$$\rho_{ij} = \tfrac{1}{2}(\partial\xi_i/\partial x_j - \partial\xi_j/\partial x_i)$$

$$\varepsilon_{ij} = \tfrac{1}{2}(\partial\xi_i/\partial x_j + \partial\xi_j/\partial x_i) \ . \tag{3.4b}$$

Hence Eq. (3.3) might be rewritten in terms of the ρ_{ij} and ε_{ij} rather than in terms of the $\partial\xi_i/\partial x_j$. It is readily shown (Prob. 3.3) that ρ_{ij} and ε_{ij} are both Cartesian tensors of rank two. Clearly ε_{ij} is a symmetric tensor, that is, $\varepsilon_{ij} = \varepsilon_{ji}$. On the other hand ρ_{ij} is <u>antisymmetric</u>, by which we mean that $\rho_{ij} = -\rho_{ji}$. An advantage of using ρ_{ij} and ε_{ij} is that important features of the displacement field can be expressed directly in terms of them.

Thus the ρ_{ij} are components of the so-called <u>rotation tensor</u>, and describe the rotational features of the displacement field. For example, the value of ρ_{21} at a given point 0 can be shown (Prob. 3.4) to give the angle of rotation about the "3" axis experienced by a volume element in the vicinity of 0 as a result of the deformation. A necessary and sufficient condition for the absence of rotation is that all the ρ_{ij} are zero.

On the other hand the ε_{ij} are components of the <u>strain tensor</u>, and represent changes in size and/or shape. A necessary and sufficient condition for rigid body motion (translation and/or rotation) is that all ε_{ij} are everywhere zero.

We have now defined the various quantities a_i , x_i , ξ_i , ρ_{ij} and ε_{ij} which are useful in describing displacements, rotations and strain. To give meaning to these quantities, and to gain experience in using them, we take up a few examples. For some of these we first specify the displacement field, that is, we give expressions for the ξ_i . We then evaluate the ρ_{ij} and ε_{ij} quantities and write the rotation and strain tensors as follows:

$$\rho_{ij} = \begin{bmatrix} 0 & \rho_{12} & \rho_{13} \\ -\rho_{12} & 0 & \rho_{23} \\ -\rho_{13} & -\rho_{23} & 0 \end{bmatrix} ; \qquad \varepsilon_{ij} = \begin{bmatrix} \varepsilon_{11} & \varepsilon_{12} & \varepsilon_{13} \\ \varepsilon_{12} & \varepsilon_{22} & \varepsilon_{23} \\ \varepsilon_{13} & \varepsilon_{23} & \varepsilon_{33} \end{bmatrix} \quad (3.5)$$

In writing Eq. (3.5) it has been recognized that $\rho_{ij} = -\rho_{ji}$ and that $\varepsilon_{ij} = \varepsilon_{ji}$.

3.3 CONSTANT LONGITUDINAL STRAIN IN ONE DIMENSION

First suppose the ξ_i are constants, that is, the same at all points of the body:

$$\xi_1 = b_1 \ ; \quad \xi_2 = b_2 \ ; \quad \xi_3 = b_3 \ . \qquad (3.6)$$

We see from Eqs. (3.4) that all ρ_{ij} and ε_{ij} are zero here. In this trivial example the displacement field, represented by Eqs. (3.6), is a rigid body translation.

Next consider the displacements

$$\xi_1 = b_1 x_1 \ ; \quad \xi_2 = b_2 \ , \quad \xi_3 = b_3 \ . \qquad (3.7)$$

For this field the rotation and strain tensors are given by

$$\rho_{ij} = \begin{bmatrix} 0 & 0 & 0 \\ 0 & 0 & 0 \\ 0 & 0 & 0 \end{bmatrix} \quad ; \qquad \varepsilon_{ij} = \begin{bmatrix} b_1 & 0 & 0 \\ 0 & 0 & 0 \\ 0 & 0 & 0 \end{bmatrix} \quad . \tag{3.8}$$

Since all ρ_{ij} are zero we know that there is no rotation. All ε_{ij} are zero except ε_{11} which has the value b_1 . We shall now show that the displacement field specified by Eqs. (3.7) or (3.8) represents uniform expansion along the x_1 direction. Turning to Fig. (3.3) we see a pattern of displacement vectors representing the field of Eqs. (3.7) or (3.8) in an arbitrary body. This is the pattern applicable to any section through the body by a plane whose normal is perpendicular to the x_1 axis. All vectors are in the "1" or x_1 direction. The magnitude of the displacement depends only on x_1 ; at any particular value of x_1 all displacement vectors are identical, independent of x_2 and x_3 .

For more detailed examination of the situation we consider a selected portion of the body, that which in the strained state lies between the planes of particles at $x_1 = A$ and $x_1 = B$, where $B > A$. The distance L between the planes in this(strained) state is just (B-A). By Eq. (3.7), since $b_1 = \varepsilon_{11}$, we can write the x_1-displacements ξ_1 of the two planes as

$$(\xi_1)_A = \varepsilon_{11} A \ ; \qquad (\xi_1)_B = \varepsilon_{11} B \ . \tag{3.9}$$

Suppose the distance between the planes of particles in the unstrained state is L_o . Let ΔL be $(L - L_o)$, the increase of inter-plane distance brought about by the strain; of course ΔL may be either positive or negative. Then from the meaning of displacement, and from Eqs. (3.9) we see that

$$\Delta L = (\xi_1)_B - (\xi_1)_A = \varepsilon_{11}(B - A) = \varepsilon_{11} L \ . \tag{3.10}$$

Remembering that L is equal to $(L_o + \Delta L)$ we find that we can also write

$$\Delta L = \frac{\varepsilon_{11} L_o}{1 - \varepsilon_{11}} \quad . \tag{3.11}$$

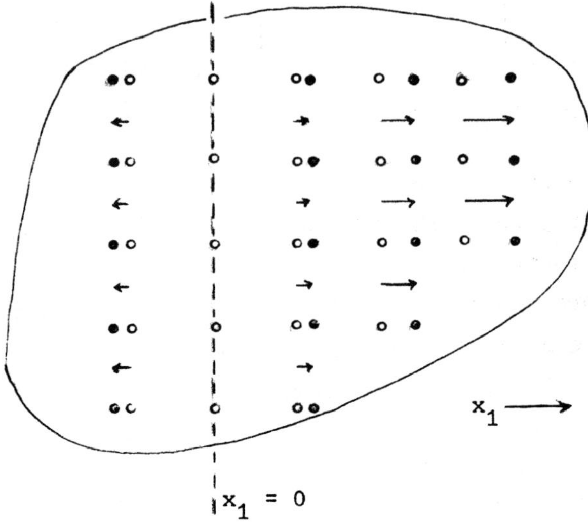

Fig. 3.3 The displacement field described by Eqs. (3.7). Open circles show initial positions of various particles and filled circles the positions after displacement.

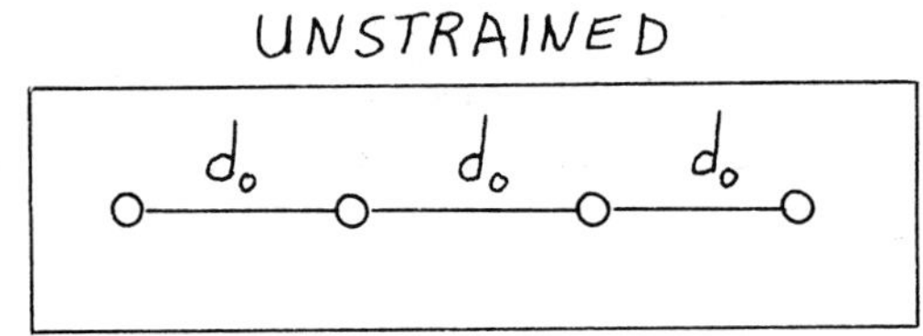

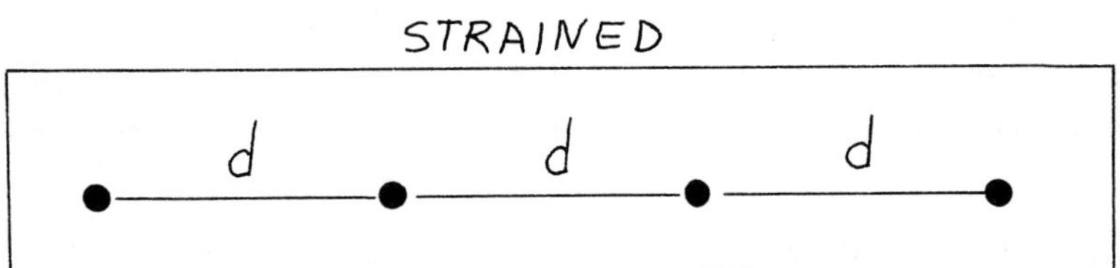

Fig. 3.4 The displacement field of Eq. 3.7 causes increased spacing between particles aligned with the x axis.

It is commonly assumed that $\varepsilon_{11} << 1$; when this is true we make little error (Prob. 3.5) by neglecting ε_{11} relative to unity in the denominator of Eq. (3.11). The right hand side of the equation then becomes simply $\varepsilon_{11} L_o$. We have found then that in a displacement field represented by Eqs. (3.7) or (3.8) with $b_1 << 1$, the increase ΔL in distance between any two parallel planes with normal along x_1 is given adequately by either of the equations

$$\Delta L = \varepsilon_{11} L \; ; \qquad \Delta L \cong \varepsilon_{11} L_o \; . \tag{3.12}$$

We see that the strain component ε_{11} is here just the fractional increase in length (or increase in length per unit length) along x_1 brought about by the strain. When ε_{11} is positive the deformation is an elongation or stretch; if negative, ε_{11} represents a shortening or compression. It is readily verified that the strained and unstrained lengths L and L_o are related by

$$L \cong L_o (1 + \varepsilon_{11}) \quad \text{when } \varepsilon_{11} << 1 \; . \tag{3.13}$$

For a line of particles which in the unstrained state are equally spaced along x_1 with separation d_o between adjacent particles, the spacings are also equal in a strained state but with separation d (see Fig. 3.4) given by

$$d \cong d_o (1 + \varepsilon_{11}) \; . \tag{3.14}$$

We now consider the effect of the deformation represented by Eqs. (3.7) or (3.8) on the volume of a body. For a pencil-shaped body with length parallel to x_1 , let its unstrained length and volume be L_o and V_o respectively, and let its strained length and volume be L and V; then L and L_o are related as in Eq. (3.13). For a pencil whose cross-sectional area is constant along its length and does not change in the deformation the volume is proportional to the length and we see that

$$\frac{V}{V_o} = \frac{L}{L_o} \cong 1 + \varepsilon_{11} \; , \tag{3.15a}$$

or

$$\varepsilon_{11} \cong \frac{V - V_o}{V_o} \; . \tag{3.15b}$$

It follows likewise that ε_{11} may be interpreted as the fractional change in volume in the deformation.

For a body of arbitrary shape and size which is everywhere subject to the deformation represented by Eqs. (3.7) we may, as is common in continuum mechanics, imagine its volume divided into a large number of volume elements of suitable size and shape. Here it is convenient to make the typical volume element of pencil-like shape with length along x_1 . (See Fig. 3.5). Equation (3.15) then applies to the element; its volume changes in the ratio $(1 + \varepsilon_{11})$ as a result of the deformation. Since for all elements the volume varies by the same ratio it is clear that the total volume of the body also varies in this ratio. Hence Eq. (3.15) again applies, where now V and V_o are the strained and unstrained volume, respectively, of an entire body of arbitrary shape.

If the material of the body is of uniform density ρ_o in the unstrained state its density after the kind of deformation we have been discussing will still be uniform, but with a new value ρ . Since the total mass of the body is unaffected by the deformation we know that $(\rho_o V_o)$ must be equal to (ρV). Hence

$$\frac{\rho_o}{\rho} = \frac{V}{V_o} \cong 1 + \varepsilon_{11} \; . \tag{3.16}$$

Assuming as before that $\varepsilon_{11} << 1$, we find (Prob. 3.6) that we can also write, approximately,

$$\frac{\rho}{\rho_o} \cong 1 - \varepsilon_{11} \; ; \quad \varepsilon_{11} \cong - \frac{\rho - \rho_o}{\rho_o} \; . \tag{3.17}$$

It follows likewise that ε_{11} may be interpreted as the negative of the fractional increase in density resulting from the deformation.

3.4 GENERAL LONGITUDINAL STRAIN IN ONE DIMENSION

When the displacement field is as given in Eqs. (3.7) or (3.8) the strain component ε_{11} is constant; the fractional increase in length $(\Delta L/L_o)$ of a line of particles along x_1 is then the same regardless of position and regardless of the original length L_o . Let us now consider the somewhat less restricted situation where the displacements are given by

$$\xi_1 = f(x) \; ; \quad \xi_2 = b_2 \; ; \quad \xi_3 = b_3 \; . \tag{3.18}$$

In Eqs. (3.18) the expression f(x) represents any suitably well-behaved function of x. The displacement vectors represented by Eqs. (3.18) are similar to those corresponding to Eqs. (3.7) (portrayed in Fig. (3.3)) in that, first, the direction is always parallel to x_1 and, second, the magnitude varies only with x_1 . Also the tensor components ρ_{ij} and ε_{ij} are the same in the two situations with one important exception: in the fields given by Eqs. (3.18) the strain component ε_{11} is not (in general) constant, but varies with x_1 . To evaluate ε_{11} at a given value of x_1 , say at $x_1 = \underline{a}$, one uses the relation (see Eqs. (3.4))

$$(\varepsilon_{11})_a = \left(\frac{\partial \xi_1}{\partial x_1}\right)_a = \left(\frac{\partial f}{\partial x}\right)_a \; . \tag{3.19}$$

Unless $f(x_1)$ has the form $b_1 x_1$ (or $b_1 x_1 + c_1$, where c_1 is constant) as in Eqs. (3.7), its derivative $\partial f/\partial x_1$ will vary with x_1 . In Fig. (3.6) is shown a plot of a possible function $f(x_1)$. The value of $\partial f/\partial x_1$ at $x_1 = a$ is represented on such a plot as the slope ($\tan \alpha$) of the tangent to the curve at $x = \underline{a}$. Clearly the slope of the tangent varies with the choice of $\underline{a}$. However, in physical situations (including those applicable to biology) the function f(x) can be expected to be "suitably well-behaved"; the mark of good conduct here is that neither of the derivatives $\partial f/\partial x$ or $\partial^2 f/\partial x^2$ is ever infinite. Under these good-behavior conditions there is always a range of x_1 in the vicinity of $x_1 = \underline{a}$ where the slope of the curve differs from that at $x_1 = \underline{a}$ by less than any prescribed amount (however small). For example, in Fig. 3.6 a region of x_1 is suggested by shading where the slope varies by, perhaps, less than 5% .

We now apply these conclusions to the physical situation where a body is subject to the displacement field given by Eq. (3.18).

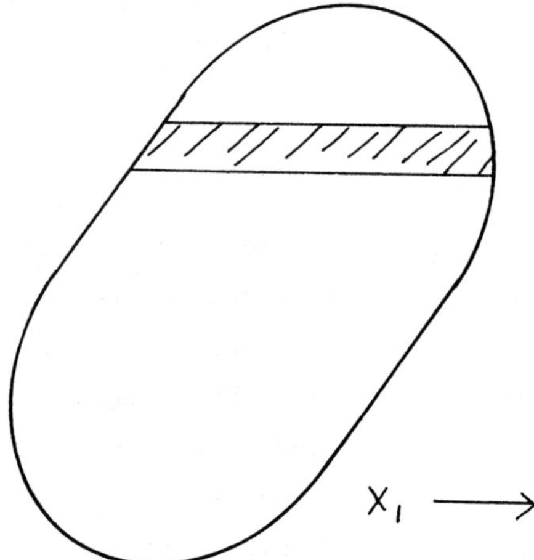

Fig. 3.5 A pencil-like volume element.

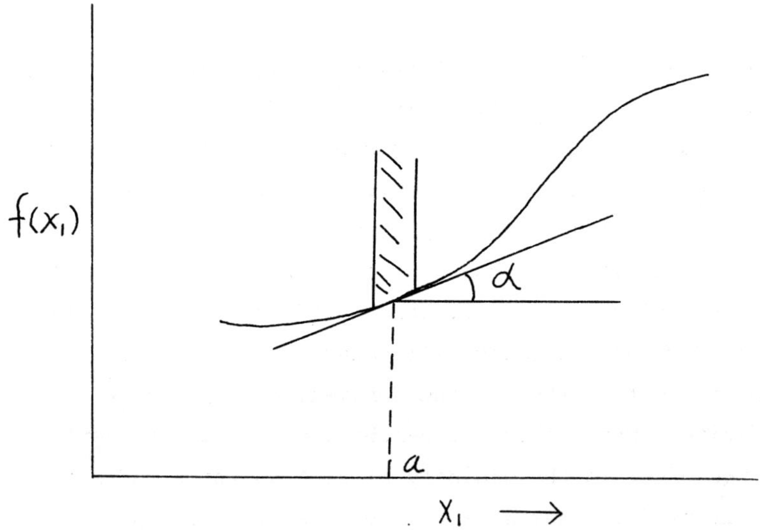

Fig. 3.6 For a well-behaved function $f(x_1)$ near any point $x_1 = a$ there is a small range of x_1 within which $\partial f/\partial x_1$ is nearly constant.

Throughout that (typically small) region of the body which lies sufficiently near $x_1 = \underline{a}$ (Cf the region indicated by shading in Fig. 3.6) the strain component ε_{11} is nearly constant and equal to $(\partial f/\partial x)_a$. It follows that in this region, or element of volume, the strain component ε_{11} can be interpreted as in connection with the constant-strain field, Eqs. (3.7). Thus $(\varepsilon_{11})_a$ gives, in a volume element near $x_1 = \underline{a}$, the fractional increase in length for any line of particles along x_1 lying in this region. From previous results it follows also that $(\varepsilon_{11})_a$ represents the fractional increase in volume, and the negative of the fractional increase in density for a volume element near $x_1 = \underline{a}$, in this deformation.

3.5 LONGITUDINAL STRAIN IN THREE DIMENSIONS

3.5.1 The Diagonal Strain Tensor

In the situations represented by Eqs. (3.7) and (3.18) where all ρ_{ij} and ε_{ij} are zero except ε_{11} , the deformation consists of a stretch or compression along x_1 . Related to this situation is the more general one where there is still no rotation, but where any of the ε_{11} , ε_{22} , and ε_{33} may be nonzero. The rotation and strain tensors are then of the form:

$$\rho_{ij} = 0 \; ; \qquad \varepsilon_{ij} = \begin{bmatrix} \varepsilon_{11} & 0 & 0 \\ 0 & \varepsilon_{22} & 0 \\ 0 & 0 & \varepsilon_{33} \end{bmatrix} . \qquad (3.20)$$

In the strain tensor here all components are zero, except along its diagonal (from upper left to lower right); such a tensor is said to be of <u>diagonal</u> form. By simple extension of previous discussion we see that when the strain tensor is diagonal one has, in general, the following deformations existing simultaneously:

(i) A stretch or compression along the "1" axis specified by ε_{11}

(ii) A stretch or compression along the "2" axis specified by ε_{22}

(iii) A stretch or compression along the "3" axis specified by ε_{33}

Thus ε_{22} gives the fractional increase in length (positive or negative) for a line of particles along x_2, and similarly for ε_{11} and ε_{33}. In general the strain components are functions of the x_i, but for some purposes may be regarded as essentially constant in a small volume element in the vicinity of any point of interest (as in the analogous situation considered in Section 3.4).

3.5.2 Constant Strain

We consider now the fractional change in volume or density of a body when subject to the deformation represented by Eqs. (3.20). For this purpose it is convenient to consider the body to be a rectangular parallelopiped whose sides are parallel to the coordinate axes. As a further restriction we suppose at first that the strain components are constant throughout the body, as if the displacements were given by

$$\xi_1 = b_1 x_1 \; ; \qquad \xi_2 = b_2 x_2 \; ; \qquad \xi_3 = b_3 x_3 \; ; \tag{3.21}$$

then ε_{11}, ε_{22}, ε_{33} are given by b_1, b_2, b_3, respectively. Suppose the lengths of the respective sides are (L_{10}, L_{20}, L_{30}) in the unstrained and (L_1, L_2, L_3) in the strained state. From previous analysis (Cf Eq. (3.13) we know that, approximately,

$$\begin{aligned} L_1 &= L_{10}(1 + \varepsilon_{11}) \\ L_2 &= L_{20}\,(1 + \varepsilon_{22}) \\ L_3 &= L_{30}\,(1 + \varepsilon_{33}) \,. \end{aligned} \tag{3.22}$$

Hence the strained volume V is given approximately by

$$V = L_1 L_2 L_3 = V_o\,(1 + \varepsilon_{11})(1 + \varepsilon_{22})(1 + \varepsilon_{33}) \,, \tag{3.23}$$

where V_o is the unstrained volume($L_{1o}L_{2o}L_{3o}$). When, as has been assumed, the magnitude of each strain component is much less than unity, we make little error in writing simply (Prob. 3.6)

$$V/V_o = (1 + \theta) \,, \tag{3.24a}$$

where

$$\theta = \varepsilon_{11} + \varepsilon_{22} + \varepsilon_{33} \ . \qquad (3.24b)$$

The quantity θ is called the dilatation. It follows from Eqs. (3.24a) that θ gives the fractional increase of volume, and from Eq. (3.16) that θ also gives the negative of the fractional increase in density in the deformation.

3.5.3 General Longitudinal Strain

In generalizing the above results we still consider a parallelopiped but do not require it to be a distinct body in the sense that it is physically isolated from other matter. Instead we now assume it to be a portion of a much larger body, and to be defined by imaginary boundaries in an arbitrary manner. We assume further that the parallelopiped is small enough to be regarded as the "volume element" so often referred to in continuum mechanics.

Suppose the displacements are given throughout the body by

$$\xi_1 = f_1(x_1) \ ; \quad \xi_2 = f_2(x_2) \ ; \quad \xi_3 = f_3(x_3) \ , \qquad (3.25)$$

where the functions $f_1(x_1)$, etc. are well-behaved in the same sense as $f(x_1)$ in Sect. 3.4. For these displacements the rotation tensor is zero and the stress tensor is diagonal; in other words the tensors have the form already given in Eq. (3.20). In general the components ε_{11} , etc., are not constant throughout the body in this situation. However, as in Sect. 3.4 , the components may be assumed essentially constant throughout a volume element, say, in the vicinity of an arbitrary point P. It follows then from previous results that the fractional change in volume of the element and (approximately) the negative of the fractional change in density are given by the dilatation θ , evaluated at P. Specifically

$$(\theta)_P = (\varepsilon_{11} + \varepsilon_{22} + \varepsilon_{33})_P = (\partial\xi_1/\partial x_1 + \partial\xi_2/\partial x_2 + \partial\xi_3/\partial x_3)_P \ , \qquad (3.26)$$

where the subscript P on a quantity signifies that the quantity is to be evaluated at P.

3.6 SIMPLE SHEAR

In previous examples the displacement along a given direction was required to be either a constant, or a function of the corresponding coordinate only. We now consider a displacement field of the form

$$\xi_1 = bx_2 \; ; \quad \xi_2 = 0 \; ; \quad \xi_3 = 0 \; . \tag{3.27}$$

For this field the rotation and strain tensors are:

$$\rho_{ij} = \begin{bmatrix} 0 & \frac{1}{2}b & 0 \\ -\frac{1}{2}b & 0 & 0 \\ 0 & 0 & 0 \end{bmatrix} ; \qquad \varepsilon_{ij} = \begin{bmatrix} 0 & \frac{1}{2}b & 0 \\ \frac{1}{2}b & 0 & 0 \\ 0 & 0 & 0 \end{bmatrix} . \tag{3.28}$$

In Fig. 3.7 the deformation corresponding to Eqs. (3.27) or (3.28) is depicted by line segments representing displacements of various particles. Open circles represent positions of particles in the unstrained state, and closed ones, the strained state. All displacements are along x_1, with magnitude and direction depending on x_2. A line of particles which is vertical in the unstrained state is tilted by the deformation through an angle given by $\tan^{-1} b$, or essentially b in a small deformation. Since both tensors in Eqs. (3.28) have nonzero elements the displacement field in Eqs. (3.27) represents both rotation and strain. This displacement field is sometimes called a field of simple shear.

One may obtain a rotation-free field by adding to Eqs. (3.27) the respective displacements $(0, bx_1, 0)$ so that the new field is

$$\xi_1 = bx_2 \; ; \quad \xi_2 = bx_1 \; ; \quad \xi_3 = 0 \; . \tag{3.29}$$

The tensors are now

$$\rho_{ij} = 0 \; ; \qquad \varepsilon_{ij} = \begin{bmatrix} 0 & b & 0 \\ b & 0 & 0 \\ 0 & 0 & 0 \end{bmatrix} . \tag{3.30}$$

In Fig. 3.8 the deformation corresponding to Eqs. (3.29) and (3.30) is depicted for a body initially of square cross-section two of whose boundaries are in the planes $x_1 = 0$ and $x_2 = 0$, respectively,

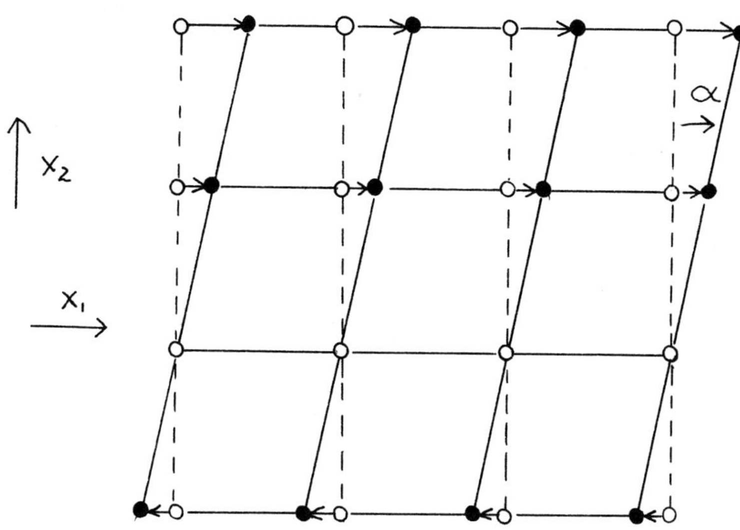

Fig. 3.7 The displacement field (simple shear) given by Eqs. (3.27). For each particle represented an open circle gives its position before, and a filled circle after, the displacement.

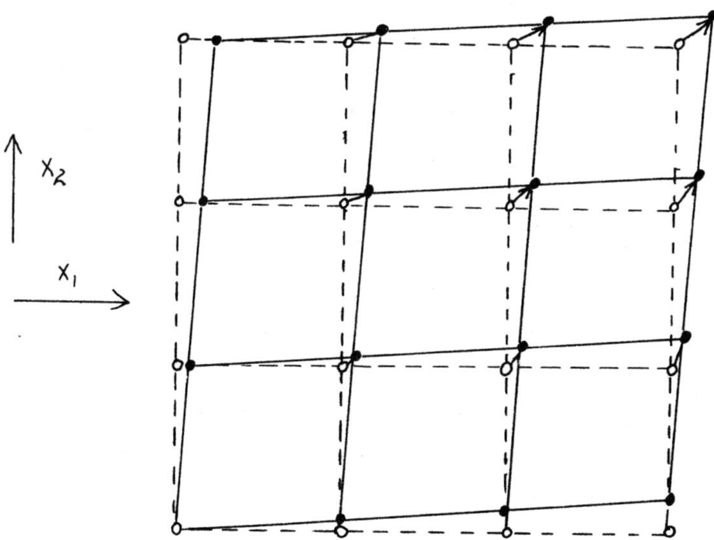

Fig. 3.8 The displacement field (pure shear) given by Eqs. (3.29). Code for circles same as in previous figure.

in the unstrained state. The cross-section changes shape after strain as shown in the figure. This deformation is known as pure shear.

If we add to Eqs. (3.27) the respective displacements $(0, -bx_1, 0)$ we obtain the displacements

$$\xi_1 = bx_2 \; ; \quad \xi_2 = -bx_1 \, , \quad \xi_3 = 0 \, . \tag{3.31}$$

In this case the tensors are

$$\rho_{ij} = \begin{bmatrix} 0 & b & 0 \\ -b & 0 & 0 \\ 0 & 0 & 0 \end{bmatrix} \; ; \qquad \varepsilon_{ij} = 0 \, . \tag{3.32}$$

Evidently the displacements in Eqs. (3.31) represent a strain-free rotation, i.e., rotation as of a rigid body. See Fig. 3.9.

3.7 RATE OF STRAIN AND ROTATION

In general the displacements ξ_i are functions of the time t. Let u_i be velocities defined by

$$u_i = \partial \xi_i \, / \, \partial t \, . \tag{3.33}$$

We let the components of a "rate-of-rotation" or angular velocity tensor be $\dot{\rho}_{ij}$, and the components of a "rate-of-strain" tensor be $\dot{\varepsilon}_{ij}$, such that

$$\begin{aligned} \dot{\rho}_{ij} &= \tfrac{1}{2} \left(\partial u_i / \partial x_j - \partial u_j / \partial x_i \right) \\ \dot{\varepsilon}_{ij} &= \tfrac{1}{2} \left(\partial u_i / \partial x_j + \partial u_j / \partial x_i \right) . \end{aligned} \tag{3.34}$$

Interpretations of these tensor components may be given in analogy to those for ρ_{ij} and ε_{ij}. Thus each nonzero, that is, off-diagonal component of $\dot{\rho}_{ij}$ represents angular velocity about one of the coordinate axes. Also, the sum of the diagonal terms of the rate-of-strain tensor gives $\partial\theta/\partial t$, the time derivative of the dilatation.

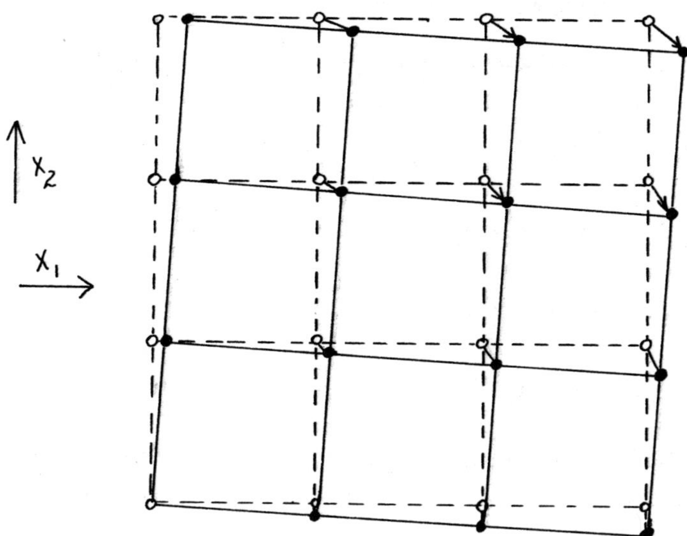

Fig. 3.9 The displacement field (pure rotation) given by Eqs. (3.31). Code for circles same as in Fig. 3.7.

PROBLEMS FOR CHAPTER 3

3.1 Justify Eq. (3.2), assuming $|\partial\xi_i/\partial x_j|$ and $|\partial\xi_i/\partial a_j|$ are each much less than unity.

In doing this, regard the x_i as given in terms of the a_j so that

$$x_1 = x_1\ (a_1, a_2, a_3)$$

$$x_2 = x_2\ (a_1, a_2, a_3)$$

$$x_3 = x_3\ (a_1, a_2, a_3)\ .$$

Similarly, regard the ξ_i as given <u>either</u> as functions of the a_j or as functions of the x_j . From the chain rule one may then obtain, for example,

$$\frac{\partial\xi_1}{\partial a_2} = \frac{\partial\xi_1}{\partial x_1}\ \frac{\partial x_1}{\partial a_2} + \frac{\partial\xi_1}{\partial x_2}\,\frac{\partial x_2}{\partial a_2} + \frac{\partial\xi_1}{\partial x_3}\ \frac{\partial x_3}{\partial a_2} \qquad .$$

From this proceed to the desired proof.

3.2 Justify the statement that (in a Cartesian coordinate tranformation) the quantities $\partial\xi_i/\partial x_j$ transform according to the rules for a Cartesian second-order tensor. In doing this let the new coordinates be (x_1' , x_2', x_3') and make use of the transformation relations (which are not hard to derive):

$$x_1' = a_{11}x_1 + a_{12}x_2 + a_{13}x_3$$

$$x_2' = a_{21}x_1 + a_{22}x_2 + a_{23}x_3$$

$$x_3' = a_{31}x_1 + a_{32}x_2 + a_{33}x_3\ .$$

Here the a_{ij} are the direction consines for the transformations. The inverse transformation is

$$x_1 = a_{11}x_1' + a_{21}x_2' + a_{31}x_3'$$

$$x_2 = a_{12}x_1' + a_{22}x_2' + a_{32}x_3'$$

$$x_3 = a_{13}x_1' + a_{23}x_2' + a_{33}x_3'\ .$$

The displacement vector has components (ξ_1,ξ_2,ξ_3) in the (1,2,3) system; let its components be $(\xi_1',\xi_2',\xi_3',)$ in the (1',2',3') system. Then one finds that the transformation rule for the ξ_1 is the same as for the x_i; i.e.,

$$\xi_1' = a_{11}\xi_1 + a_{12}\xi_2 + a_{13}\xi_3$$

$$\xi_2' = a_{21}\xi_1 + a_{22}\xi_2 + a_{23}\xi_3$$

$$\xi_3' = a_{31}\xi_1 + a_{32}\xi_2 + a_{33}\xi_3 \ .$$

Consider now as an example the quantity $\partial\xi_1'/\partial x_2'$. By the chain rule we have

$$\frac{\partial\xi_1'}{\partial x_2'} = \frac{\partial\xi_1'}{\partial x_1}\frac{\partial x_1}{\partial x_2'} + \frac{\partial\xi_1'}{\partial x_2}\frac{\partial x_2}{\partial x_2'} + \frac{\partial\xi_1'}{\partial x_3}\frac{\partial x_3}{\partial x_2'} \ .$$

Using expressions for ξ_1' and the x_i given above you may now proceed to obtain an expression for $\partial\xi_1'/\partial x_2'$, which is a special example of the rule,

$$\frac{\partial\xi_i'}{\partial x_j'} = \sum_\ell \sum_m a_{i\ell}\, a_{jm} \frac{\partial\xi_\ell}{\partial x_m} \ ;$$

the latter is the transformation rule for a second-order Cartesian tensor (Cf Eq. 2.8).

3.3 Given the result of Prob. 3.2 show that ρ_{ij} and ε_{ij} (defined in Eqs.(3.4b)) are Cartesian second-order tensors.

3.4 Show that if a body rotates through a small angle ϕ about the "3" axis this angle is given by ρ_{21} ; the angle is regarded as positive if it is clockwise when viewed along the positive "3" direction. Do this by first showing that in this rotation the displacements of a representative point are given by

$$\xi_1 = -\phi x_2$$
$$\xi_2 = \phi x_1$$
$$\xi_3 = 0 \ ,$$

where (x_1,x_2) are the coordinates of the point relative to the axis of rotation. Then evaluate ρ_{21} .

3.5 Suppose $\varepsilon_{11} = 10^{-4}$. Estimate the percentage error in approximating ΔL in Eq. (3.11) by $\varepsilon_{11} L_o$.

3.6 (a) Justify the approximation in Eq. (3.17) by showing that if $\varepsilon_{11} << 1$ then

$$(1 + \varepsilon_{11})^{-1} \cong 1 - \varepsilon_{11} .$$

(b) Show that under these conditions it is also true that

$$(\rho - \rho_o) \cong -\varepsilon_{11} \rho .$$

(c) Justify the approximation in Eqs. (3.24) by showing that under the assumed conditions

$$(1 + \varepsilon_{11})(1 + \varepsilon_{22})(1 + \varepsilon_{33}) \cong 1 + \varepsilon_{11} + \varepsilon_{22} + \varepsilon_{33} .$$

(d) Given Eqs. (3.24), show that θ gives the fractional increase of volume and the negative of the fractional increase of density in a small deformation.

CHAPTER 4

CONSTITUTIVE EQUATIONS FOR SIMPLE MEDIA

4.1 GENERAL CONSIDERATIONS

In previous chapters we became acquainted with the concepts of stress, strain and rate-of-strain, and specifically with their respective components S_{ij}, ε_{ij} and $\dot{\varepsilon}_{ij}$. For many media it is possible to write equations, called constitutive equations, which relate these quantities. Constitutive equations depend very much on the nature of the medium, and for biological media are rather complex. But fortunately considerable insight on the difficult biological situations can be gained by analysis of simpler "physical" ones. (This approach to difficult problems by way of simpler ones is effective generally and might be said to form the basis of biological physics.)

We consider first of all the linear relations that have proved very useful in dealing with media which are at equilibrium; when these are motionless the $\dot{\varepsilon}_{ij}$ are all zero. These linear relations give expression to the familiar Hooke's Law of elementary physics which states that "Stress is proportional to strain." In terms of the S_{ij} and ε_{ij} we may rephrase the statement to yield "Any stress component S_{ij} can be written as a linear combination of the strain components ε_{ij}". Mathematically Hooke's Law takes the form of six equations, one for each of the six independent stress components; for any of the S_{ij} the equation is

$$S_{ij} = a_{ij11}\,\varepsilon_{11} + a_{ij12}\,\varepsilon_{12} + a_{ij13}\,\varepsilon_{13} + a_{ij22}\,\varepsilon_{22} + a_{ij23}\varepsilon_{23} + a_{ij33}\,\varepsilon_{33}\,, \tag{4.1}$$

where a_{ij11}, etc., are constants. For crystalline solids one must indeed work with lengthy equations of this kind. Only a little comfort derives from the fact that not all of the 36 constants a_{ijkl} are independent; 21 are needed in the general case.

But the constitutive relations reduce considerably when the medium is isotropic, i.e., when the elastic properties of the medium are the same in all directions. (Liquids and gases are usually isotropic, as are also many noncrystalline or polycrystalline solids.) For such media it can be shown that only <u>two</u> of the constants a_{ijkl} are independent. If we call these λ and μ the constitutive relation for an isotropic medium can be written in the compact form

$$S_{ij} = \lambda\theta\delta_{ij} + 2\mu\varepsilon_{ij} \; . \tag{4.2}$$

Here θ is the dilatation, given by the sum of the diagonal terms in the strain tensor (Eq. 3.24b); as shown in Chap. 3 the dilatation θ gives the fractional increase of volume of an element of the medium. The quantity δ_{ij} which appears in Eq. (4.2) is the "Kronecker delta"; it has the following convenient properties:

$$\delta_{ij} = 0 \text{ for } i \neq j \; ; \qquad \delta_{ij} = 1 \quad \text{for } i = j \; . \tag{4.3}$$

Thus Eq. (4.2) represents the six equations

$$\begin{aligned} S_{11} &= \lambda\theta + 2\mu\varepsilon_{11} \\ S_{22} &= \lambda\theta + 2\mu\varepsilon_{22} \\ S_{33} &= \lambda\theta + 2\mu\varepsilon_{33} \\ S_{12} &= 2\mu\varepsilon_{12} \\ S_{13} &= 2\mu\varepsilon_{13} \\ S_{23} &= 2\mu\varepsilon_{23} \; . \end{aligned} \tag{4.4}$$

Since the ε_{ij} are dimensionless we see that λ and μ have the same dimensions as S_{ij} .

4.2 PURE SHEAR

In these equations the constant μ is known as the shear elastic coefficient. Its significance is best seen by considering a displacement field representing pure shear; thus when Eqs. (3.29) or (3.30) and Fig. 3.8 apply the strain tensor has the form.

$$\varepsilon_{ij} = \begin{bmatrix} 0 & \varepsilon_{12} & 0 \\ \varepsilon_{12} & 0 & 0 \\ 0 & 0 & 0 \end{bmatrix} ; \tag{4.5}$$

Here all strain components are zero except ε_{12} . According to Eqs. (4.4) the stress tensor then has the form

$$S_{ij} = \begin{bmatrix} 0 & 2\mu\varepsilon_{12} & 0 \\ 2\mu\varepsilon_{12} & 0 & 0 \\ 0 & 0 & 0 \end{bmatrix} ; \tag{4.6}$$

Thus Eqs. (4.4) reduce to statements that all stress components are zero except S_{12} and that

$$S_{12} = 2\mu\varepsilon_{12} . \tag{4.7}$$

In this situation 2μ gives the ratio of stress to strain; for a medium characterized by a large value of μ a relatively large stress is required to produce a given amount of shearing strain. Values of μ for certain materials are given in Appendix A . Such "stiff" solids as steel and glass have high values of μ while plastics and gels have low values. For liquids and gases we usually take μ to be zero .

4.3 HYDROSTATICS

When we set $\mu = 0$ we get from Eq. (4.2) that

$$S_{ij} = \lambda\theta\delta_{ij} . \tag{4.8}$$

The stress array is then isotropic, given by

$$S_{ij} = \begin{bmatrix} \lambda\theta & 0 & 0 \\ 0 & \lambda\theta & 0 \\ 0 & 0 & \lambda\theta \end{bmatrix}. \tag{4.9}$$

Comparing Eq. (4.9) with Eq. (2.5) we see that the diagonal stress terms S_{11}, S_{22}, S_{33} can be written either as $\lambda\theta$ or as $(-p)$; hence the hydrostatic pressure p can be written

$$p = - \lambda\theta. \tag{4.10}$$

In applying Eq. (4.10) to a fluid it is important to realize that in this equation p does not usually represent the <u>total</u> hydrostatic pressure. Typically one starts with a situation where the pressure is nonzero; then p represents an <u>increase</u> in pressure over its initial value and θ represents the fractional change in volume relative to the initial situation. With this in mind one can relate the constant λ to the bulk modulus of elasticity B which is commonly tabulated in handbooks. One defines B by the relationship

$$\Delta p = -B \frac{\Delta V}{V_o} ; \tag{4.11}$$

here Δp is the increase in pressure required to cause the volume of a (portion of) fluid to increase by ΔV from an initial value V_o. The negative sign means, of course, that a positive increase in volume requires a negative increment (a decrease) of pressure. Identifying p and θ in Eq. (4.10) with, respectively, Δp and $(\Delta V/V_o)$ in Eq. (4.11) we see that the equations agree if

$$\lambda = B . \tag{4.12}$$

The <u>compressibility</u> β is frequently referred to:this quantity is just the reciprocal of B. Equation (4.12) applies only to fluids, i.e., media for which $\mu = 0$. When μ is not zero one can give a generalized meaning for pressure and obtain the relation (Prob. 4.1)

$$B = \lambda + (2/3)\mu \ . \tag{4.13}$$

For an ideal gas subject to a small adiabatic compression it can be shown (Prob. 4.2) that

$$B = \gamma p_o \ , \tag{4.14}$$

where γ is the ratio of specific heats for the gas and p_o is the equilibrium hydrostatic pressure. For an ideal gas subject to a small isothermal compression one has (Prob. 4.3) simply

$$B = p_o \ . \tag{4.15}$$

For most media expressions for B cannot be derived from first principles; instead B must be determined experimentally. In Appendix A values of B are given for a number of materials.

4.4 "SIMPLE" EXTENSION

Here we treat stretching of wires or other bodies, a subject commonly taken up in elementary physics courses as a simple one-dimensional problem. But we shall see that the problem cannot be made fully one-dimensional unless μ is zero. For suppose all components of the strain tensor ε_{ij} are zero except ε_{11} ; then we find that the stress tensor will have three nonzero components S_{11} , S_{22} and S_{33} . On the other hand suppose all S_{ij} are zero except S_{11} ; it then turns out that all three diagonal components of ε_{ij} are nonzero. We shall now justify these statements, considering first the case where the strain tensor has the form in Eq. (3.8) so that

$$\varepsilon_{ij} = \begin{bmatrix} \varepsilon_{11} & 0 & 0 \\ 0 & 0 & 0 \\ 0 & 0 & 0 \end{bmatrix} . \tag{4.16}$$

Then by Eqs. (4.2) or (4.4) the stress array becomes

$$S_{ij} = \begin{vmatrix} (\lambda + 2\mu)\varepsilon_{11} & 0 & 0 \\ 0 & \lambda\varepsilon_{11} & 0 \\ 0 & 0 & \lambda\varepsilon_{11} \end{vmatrix} . \tag{4.17}$$

We see that a simple stretch along the "1" direction has associated with it normal stress components along all three coordinate axes. This result is applicable to elongation or compression of a disc-like body whose lateral dimesnions, i.e., those along the "2" and "3" directions, are large compared to those along the direction of stretch. Under these conditions little contraction or stretch occurs laterally.

We now suppose that the stress tensor has only one component so that, specifically,

$$S_{ij} = \begin{bmatrix} S_{11} & 0 & 0 \\ 0 & 0 & 0 \\ 0 & 0 & 0 \end{bmatrix} . \tag{4.18}$$

The corresponding components of the strain tensor are not evident immediately but must be obtained by solving the simultaneous equations which result from Eqs. (4.2) or (4.4):

$$\begin{aligned} S_{11} &= \lambda(\varepsilon_{11} + \varepsilon_{22} + \varepsilon_{33}) + 2\mu\varepsilon_{11} \\ 0 &= \lambda(\varepsilon_{11} + \varepsilon_{22} + \varepsilon_{33}) + 2\mu\varepsilon_{22} \\ 0 &= \lambda(\varepsilon_{11} + \varepsilon_{22} + \varepsilon_{33}) + 2\mu\varepsilon_{33} \\ 0 &= 2\mu\varepsilon_{12} = 2\mu\varepsilon_{13} = 2\mu\varepsilon_{23} . \end{aligned} \tag{4.19}$$

Thus the off-diagonal components of ε_{ij} are zero. From the first three equations we find quickly that $\varepsilon_{22} = \varepsilon_{33}$ and then (Prob. 4.4) that

$$S_{11} = E\varepsilon_{11} ; \quad \varepsilon_{22} = \varepsilon_{33} = -\nu\varepsilon_{11} ; \tag{4.20}$$

where

$$\begin{aligned} E &= (3\mu\lambda + 2\mu^2)/(\lambda + \mu) \\ \nu &= \tfrac{1}{2}\lambda/(\lambda + \mu) . \end{aligned} \tag{4.21}$$

We see that when Eq. (4.18) applies a stretch along "1" is accompanied by contractions along "2" and "3" . This result applies to the stretching of a thin wire or filament, where it is found that little lateral stress is developed; thus S_{22} and S_{33} are small.

It may seem reasonable that S_{22} and S_{33} should be small since (except for effects of atmospheric pressure, usually negligible) the lateral stress must be zero at the surface of the wire, and in a thin wire the surface is not far from any interior point. An alternative explanation would be that the thin filament can easily contract thus relieving any lateral stress which might otherwise develop.

In Eq. (4.21) the constant E is the familiar Young's Modulus; it gives the ratio of stress (S_{11}) to strain (ε_{11}) in an experiment involving simple stretching of a wire or filament. The constant ν is known as Poisson's ratio; it gives the magnitude of the ratio of lateral to longitudinal strain for the stretched wire or filament. Values of E and ν for various materials are tabulated in handbooks; representative values are given in Appendix A.

4.5 FLOWING MEDIA

In a solid subject to stresses the equilibrium state is characterized by a field of static strain. Specifically, for an isotropic medium, the equilibrium static strain quantities are related to the stresses by Eqs. (4.2) or (4.4). In a liquid or gas the same is true for some situations, such as that of a fluid in a motionless vessel subject to gravity. But in other situations steady motions will be set up in the fluid as a result of stresses. In general, application of steady stresses may give rise to a field of static strain as well as to a steady velocity field (or field of rate-of-strain). For a <u>linear</u> <u>isotropic</u> <u>viscoelastic</u> <u>medium</u> one commonly used generalization of Hooke's Law leads to the tensor relation

$$S_{ij} = \lambda\theta\delta_{ij} + 2\mu\varepsilon_{ij} + \lambda'\dot{\theta}\delta_{ij} + 2\eta\dot{\varepsilon}_{ij} , \qquad (4.22)$$

where λ' and η are constants and where, as in Section 3.7, with

$$u_i = \partial\xi_i/\partial t ,$$

$$\begin{aligned} \dot{\varepsilon}_{ij} &= \partial\varepsilon_{ij}/\partial t = \tfrac{1}{2}(\partial u_i/\partial x_j + \partial u_j/\partial x_i) \\ \dot{\theta} &= \partial\theta/\partial t = \partial u_1/\partial x_1 + \partial u_2/\partial x_2 + \partial u_3/\partial x_3 . \end{aligned} \qquad (4.23)$$

Restricting our attention to steady flow situations (where θ at a given point does not vary with time) we set $\dot{\theta} = 0$. Taking the medium to be a liquid we also let $\mu = 0$ and obtain

$$S_{ij} = \lambda\theta\delta_{ij} + 2\eta\dot{\varepsilon}_{ij} \; . \tag{4.24}$$

One sometimes replaces $\lambda\theta$ by $-p$ as in Eq. (4.10); the component equations then become

$$\begin{aligned} S_{11} &= -p + 2\eta\, \partial u_1 / \partial x_1 \\ S_{22} &= -p + 2\eta\, \partial u_2 / \partial x_2 \\ S_{33} &= -p + 2\eta\, \partial u_3 / \partial x_3 \\ S_{12} &= \eta(\partial u_1 / \partial x_2 + \partial u_2 / \partial x_1) \\ S_{13} &= \eta(\partial u_1 / \partial x_3 + \partial u_3 / \partial x_1) \\ S_{23} &= \eta(\partial u_2 / \partial x_3 + \partial u_3 / \partial x_2) \; . \end{aligned} \tag{4.25}$$

Here p is the hydrostatic pressure; it is the normal stress which would exist in the absence of flow for given θ. The constant η is the coefficient of shear viscosity; clearly it is analogous to μ , the coefficient of shear elasticity. In Appendix A values of η are given for various media.

4.6 SIMPLE SHEARING FLOW

Suppose a velocity field is set up in a fluid with components given by

$$u_1 = Gx_2 \; , \quad u_2 = u_3 = 0 \; . \tag{4.26}$$

We shall call this motion <u>simple shearing flow</u>, by analogy to the displacement field, Section 3.6 , called "simple shear". The motion represented by Eq. (4.26) is such as would occur in a thin liquid film which fills the space between two parallel plates one of which moves in its own plane while the other is fixed. (See Fig. 4.1). The motion is also similar to <u>Couette flow</u> , taken up in Section 5.8 .

There is an obvious analogy between the velocity field in Eq. (4.26) and the displacement field of Eq. (3.27). One may compare the rotation and strain tensors of Eq. (3.28) with the tensors $\dot{\rho}_{ij}$ and

$\dot{\varepsilon}_{ij}$ which apply to Eq. (4.26) , namely,

$$\dot{\rho}_{ij} = \begin{bmatrix} 0 & \tfrac{1}{2}G & 0 \\ -\tfrac{1}{2}G & 0 & 0 \\ 0 & 0 & 0 \end{bmatrix} ; \qquad \dot{\varepsilon}_{ij} = \begin{bmatrix} 0 & \tfrac{1}{2}G & 0 \\ \tfrac{1}{2}G & 0 & 0 \\ 0 & 0 & 0 \end{bmatrix} . \qquad (4.27)$$

Thus in simple shearing flow fluid elements are continually subject to both rotation and shear. The angular velocity of a fluid element about the z axis is given by $\dot{\rho}_{21}$ (Prob. 4.6) and hence by $-\tfrac{1}{2}G$; this means that the magnitude of the angular velocity is $\tfrac{1}{2}G$ and the sense is as shown in Fig. 4.1 . It is instructive to note that the velocity field of Eq. (4.26) may be thought of as a superposition of two fields R and S , defined below:

$$\begin{aligned} &\text{FIELD R:} \quad u_1 = \tfrac{1}{2}\, Gx_2 , \quad u_2 = -\tfrac{1}{2}\, Gx_1 , \quad u_3 = 0 \\ &\text{FIELD S:} \quad u_1 = \tfrac{1}{2}\, Gx_2 , \quad u_2 = \tfrac{1}{2}\, Gx_1 , \quad u_3 = 0 \end{aligned} \qquad (4.28)$$

The field R is easily shown to represent pure rotation (since $\dot{\varepsilon}_{ij} = 0$) while the field S represents pure shear (since $\dot{\rho}_{ij} = 0$).

From Eq. (4.24) the stress components for simple shearing flow, Eq. (4.26) are (assuming θ nonzero)

$$S_{ij} = \begin{bmatrix} \lambda\theta & \eta G & 0 \\ \eta G & \lambda\theta & 0 \\ 0 & 0 & \lambda\theta \end{bmatrix} . \qquad (4.29)$$

Again, as in Eqs. (4.25), the quantity $\lambda\theta$ might be replaced by $-p$. If the coordinate system is rotated through an angle ϕ about the z axis the tensor will vary with ϕ . (Prob. 4.5) .

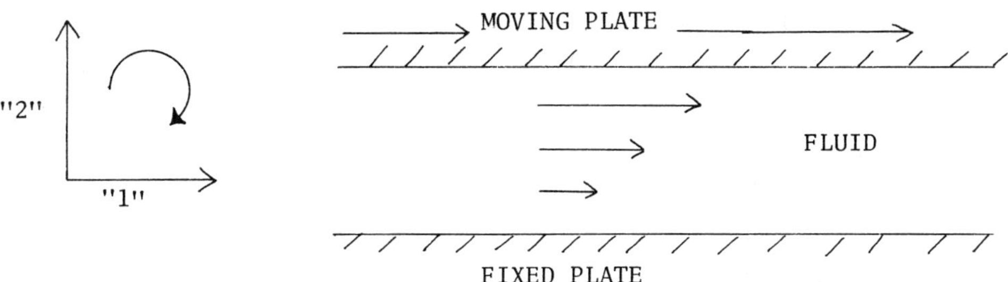

Fig. 4.1 The simple shearing flow field described by Eqs. (4.26).

PROBLEMS FOR CHAPTER 4

4.1 One sometimes defines a mean pressure $\overline{p}$ by the relation

$$\overline{p} = -(1/3)(S_{11} + S_{22} + S_{33}) .$$

Show from Eq. (4.2) or (4.4) that if also B is defined by

$$\overline{p} = -B\theta$$

one then obtains the expression in Eq. (4.13) for B.

4.2 For an ideal gas whose volume v is varied adiabatically the pressure p varies in such a way that

$$pv^{\gamma} = p_o v_o^{\gamma} ,$$

where p_o and v_o are initial values of the pressure and volume. Let p be $(p_o + \Delta p)$ and let v be $(v_o + \Delta v)$; show that when $\Delta p \ll p_o$ and $\Delta v \ll v_o$ one obtains

$$\Delta p = -\gamma P_o (\Delta v/v_o) .$$

(You may use the series expansion for $(1 + x)^n$; see Appendix B). Referring to Eq. (4.11) we see that this leads to $B = \gamma P_o$ as stated in Eq. (4.14).

4.3 Refer to Prob. 4.2 ; if the volume varies isothermally one has

$$pv = p_o v_o .$$

Show then that $B = p_o$ as stated in Eq. (4.15) .

4.4 Show that Eqs. (4.19) for a thin wire lead to Eqs. (4.20) and (4.21) .

4.5 (a) Using the transformation rule of Eq. (2.8) give the stress array for simple shearing flow which results from Eq. (4.29) when axes are rotated an angle ϕ about the "3" axis. It will be helpful to refer to Probs. 2.4 and 2.5 of Chap. 2 . It will also be helpful to think of the stress tensor in Eq. (4.29) as the superposition of two tensors, whose transformations can be considered separately.

(b) Give the special results which apply when $\phi = 45^\circ$.

4.6 Show by analogy with Prob. 3.4 that the quantity $\dot{\rho}_{21}$ gives the angular velocity of any volume element of fluid about the z axis.

4.7 When a strip of soft biological tissue is stretched it contracts laterally in such a way that its volume remains nearly constant. Show that for such material the Poisson ratio ν is approximately 0.5 .

4.8 (a) From Eqs. (4.13) and (4.21) show that

$$E = \frac{9\,B\mu}{3B + \mu}$$

and

$$\nu = \tfrac{1}{2}\,\frac{3B - 2\mu}{3B + \mu} \quad .$$

(b) Show that if $\mu \ll B$ approximate expressions for E and ν are

$$E \cong 3\mu\left(1 - \frac{\mu}{3B}\right) \cong 3\mu$$

$$\nu = \tfrac{1}{2}\left(1 - \frac{\mu}{B}\right) \cong 0.5 \ .$$

4.9 From Eqs. (4.21) show that

$$\lambda = \frac{2\mu\nu}{1 - 2\nu}$$

and

$$E = 2\mu(1 + \nu) \ .$$

CHAPTER 5

FIELDS OF STRESS, STRAIN AND FLOW

Here we shall take up the problem of calculating distributions of the S_{ij}, ε_{ij}, u_i, etc., in situations of interest. In doing this the procedure is to develop or find a suitable differential equation for each problem and then arrive at a solution which satisfies reasonable boundary conditions. To minimize mathematical complications we shall usually confine our attention to fields of stress, strain and flow which can be described in just two dimensions and sometimes in only one. A number of concepts important to biology can be studied in this way.

5.1 VOLUME ELEMENTS AND THE VOLUME FORCE

In general each of the components ξ_i , u_i , S_{ij} , etc., is a field quantity; that is, each varies continuously in space and time. For S_{12} , for example, this idea is expressed symbolically by the equation

$$S_{12} = S_{12}\,(x\ ,\ y\ ,\ z\ ,\ t)\ .$$

To find how S_{12} or any other quantity is distributed we follow an approach typical of continuum mechanics. We consider a region of matter which may be in the state of a solid, liquid or gas or, as is typically true of biological media, in a state which is not really any of these. Out of this region we select a small portion which we call an <u>element</u> (or <u>volume element</u> or <u>mass element</u>). We determine the force on this element and, following Newton's Second Law, set this equal to its rate of change of momentum. The relevant quantities characteristically vary in space, and hence are not constant throughout the element. Definite results, valid near a given point at a given time, are obtained from expressions which apply when dimensions of the element become very small or "approach zero".

For the volume element it is convenient to pick a rectangular parallelopiped or box with edges parallel to the (1 , 2 , 3) axes. As in Fig. 5.1 we choose one corner as the point P with coordinates (x_a , y_a , z_a), and let the sides be of length $(\Delta x , \Delta y , \Delta z)$.

We now consider a situation where S_{11} depends only on x while all other S_{ij} are zero. Let ΔF_x be the x component of force on this element exerted by the outer medium by way of the stress S_{11}. Only two faces are acted on by S_{11}, both perpendicular to the 1 direction: one of these passes through P and lies in the plane $x = x_a$; the other lies in the plane $x = x_a + \Delta x$ and is shown shaded in Fig. 5.1. Let $S_{11}(x_a)$ be the value of S_{11} at the first face and $S_{11}(x_a + \Delta x)$ that at the second. Both faces have the area $\Delta y \, \Delta z$. Then, recalling the sign convention on S_{11}, we see that the outer medium exerts on the element the forces:

$-\Delta y \Delta z \, S_{11} (x_a)$ across the face at $x = x_a$

$\Delta y \Delta z \, S_{11} (x_a + \Delta x)$ across the face at $x = x_a + \Delta x$.

Hence the total force on the element is

$$\Delta F_x = \Delta y \, \Delta z \, [S_{11}(x_a + \Delta x) - S_{11}(x_a)] \ .$$

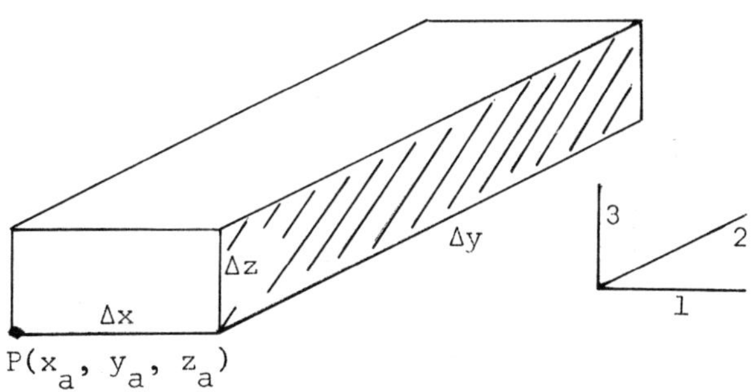

Fig. 5.1 Defining sketch for the volume element.

Since the element has the volume Δx Δy Δz the force per unit volume is

$$\frac{\Delta F_x}{\Delta x\ \Delta y\ \Delta z} = \frac{[S_{11}(x_a + \Delta x) - S_{11}(x_a)]}{\Delta x} . \tag{5.1}$$

Recalling from calculus the definition of a derivative we see that the right hand side of Eq. (5.1) is closely related to the value of $\partial S_{11} / \partial x$ at $x = x_a$. Specifically we have, by definition,

$$\left(\frac{\partial S_{11}}{\partial x}\right)_a = \lim_{\Delta x \to 0} \frac{S_{11}(x_a + x) - S_{11}(x_a)}{\Delta x} . \tag{5.2}$$

Hence if we define f_{xa} as the limit of the left hand of Eq. (5.1) as Δx approaches zero we obtain from Eqs. (5.1 and (5.2)

$$f_{xa} = \left(\frac{\partial S_{11}}{\partial x}\right)_a . \tag{5.3}$$

We refer to f_{xa} as the x component of the force per unit volume at $x = x_a$, and let f_x have the same meaning at any x. The quantity f_x (or f_{xa}) is sometimes called the <u>volume force.</u> Since x_a is an arbitrary value of x we can omit the subscript "a" on both sides of Eq. (5.3) and write simply

$$f_x = \frac{\partial S_{11}}{\partial x} , \tag{5.4}$$

with the understanding that both the volume force f_x and the stress derivative $\partial S_{11} / \partial x$ are to be evaluated at the same point.

Clearly f_x depends on spatial variation of the stress. If S_{11} is constant the volume force is zero no matter how large S_{11} may be.

While Eq. (5.4) was derived by assuming S_{11} depends only on x and all other S_{ij} are zero it is valid under much more general circumstances. Although the proof required is more lengthy, it is not very hard to show that Eq. (5.4) applies locally, that is, near the point at which $\partial S_{11} / \partial x$ is evaluated, even if S_{11} depends on all three coordinates. Also if other S_{ij} are nonzero additional terms will, in general have to be added to the expression for f_x but $\partial S_{11} / \partial x$ still gives the correct contribution from the stress component S_{11}. In addition it can be shown that Eq. (5.4) applies to a volume element of any shape; it need not be the box pictured in Fig. 5.1.

In proceeding from Eq. (5.1) to Eq. (5.4) we followed procedures and used the language of mathematical calculus; these are suitable for a medium which is truly continuous.* For such a medium the ratio on the right of Eq. (5.1) has a definite value no matter how small Δx, Δy and Δz become. However a physical medium (or, perhaps we should say, a real medium) consists of atoms which, in turn, are composed of electrons and nuclei. When Δx, etc., become small enough to be comparable to atomic dimensions the right hand of Eq. (5.1) has no clear meaning; for example, how would we define S_{11} at points "inside" an atom? If we did arrive at a reasonable definition for S_{11} we probably would find that the quantity varies considerably from one part of an atom to another.

To avoid complications it should be understood that Eq. (5.1) applies strictly only to volume elements whose dimensions are large compared to atomic dimensions. Field quantities such as S_{11} are then defined as averages over a relatively large number of atoms. In "physical calculus" the symbol "$\Delta x \to 0$" might be replaced by "$\Delta x \to \varepsilon$", where ε is small compared to typical macroscopic lengths (such as

---

* A continuous medium (a convenient fiction) is defined as one whose macroscopic properties such as density, temperature and elastic moduli, apply to any volume element no matter how small. R. B. Lindsay has illustrated the concept by quoting a political remark of former New York Governor Al Smith: "No matter how thin you slice it - it's still boloney".

the width of a channel) but large compared to atomic dimensions. In Eq. (5.4) f_x is then the average force per unit volume in an element whose dimensions are of the order of ε ; likewise $\partial S_{11} / \partial x$ is defined by the ratio on the right of Eq. (5.1) but with Δx of the order of ε . In some situations it may not be possible to find a length ε satisfying the conditions described above; in this circumstance the differential equations of continuum mechanics would be of doubtful value. Fortunately there are many situations where the conditions are satisfied very well and continuum mechanics provides a valid, powerful and convenient approach.

We now consider another stress field, namely, one in which S_{12} depends only on y, while all other S_{ij} are zero (except S_{21} which is, as usual, equal to S_{12}). We shall leave it as an exercise (Prob. 5.1) to show that for this field the volume force in the x direction is given by

$$f_x = \frac{\partial S_{12}}{\partial y} \quad . \tag{5.5}$$

Actually it can be shown that Eq. (5.5) applies locally even if S_{12} varies with x and z , and that the right hand of Eq. (5.5) gives the correct contribution from S_{12} regardless of whatever values the other S_{ij} may have.

It is clear by analogy that the contribution to f_x from S_{13} is $\partial S_{13} / \partial z$. In a general stress field where all S_{ij} may be nonzero the total expression for f_x is the sum of contributions from S_{11} , S_{12} and S_{13} , namely

$$f_x = \frac{\partial S_{11}}{\partial x} + \frac{\partial S_{12}}{\partial y} + \frac{\partial S_{13}}{\partial z} \quad . \tag{5.6}$$

It is sometimes convenient to revert to terminology where the coordinates are (x_1 , x_2 , x_3) and the corresponding components of the volume force are (f_1 , f_2 , f_3). An expression for any component f_i is then

$$f_i = \frac{\partial S_{i1}}{\partial x_1} + \frac{\partial S_{i2}}{\partial x_2} + \frac{\partial S_{i3}}{\partial x_3} \quad . \tag{5.7}$$

Equation (5.6) is readily seen to be a special case of Eq. (5.7) when $i = 1$.

We now make use of these expressions for the volume force to discuss situations involving stress, strain and/or flow. We obtain differential equations by applying Newton's laws of mechanics to an arbitrary volume element of the medium. Thus if the medium is at rest, or in motion such that a volume element travels at constant velocity, the total force on the element must be zero. If the element accelerates the total force on it must be equal to the product of its mass and its acceleration. Having obtained the differential equation for a problem the next task is to arrive at a solution which satisfies reasonable boundary conditions. This solution then gives the distribution-in-space of the field quantity of interest, say, a component of stress, or displacement, or flow.

5.2 PRESSURE DISTRIBUTION IN A MOTIONLESS FLUID

As our first example we consider the stress distribution that exists in a quiescent body of liquid, say, in a lake or reservoir. The results are useful also as an approximation to the situation in fluid systems of a mammal, or of a plant. In a liquid the shear elastic coefficient μ is zero and, since the liquid is at rest, all the $\dot{\varepsilon}_{ij}$ components are zero. The stress-strain relations then become simply (See Section 4.3)

$$S_{ij} = \lambda\theta\delta_{ij} = B\theta\delta_{ij} = -\ p\ \delta_{ij} \ , \tag{5.8}$$

Thus S_{ij} is an isotropic tensor given by Eq. (4.9) which we can also write as

$$S_{ij} = \begin{bmatrix} -p & 0 & 0 \\ 0 & -p & 0 \\ 0 & 0 & -p \end{bmatrix} \tag{5.9}$$

As stated in Section 4.3 this isotropic tensor S_{ij} is unchanged by rotation of coordinates. Hence at any point P in a quiet body of liquid the stress is normal to every plane through P. Choosing the 1 axis upward we consider the stress component S_{11} which acts normal to a horizontal plane. We have from Eq. (5.8) that

$$S_{11} = -p = B\theta \ . \tag{5.10}$$

We have already seen that the stress components S_{12} and S_{13} are zero so that the volume force from stresses is, from Eq. (5.6), simply

$$f_x = \frac{\partial S_{11}}{\partial x} = -\frac{\partial p}{\partial x} \ . \tag{5.11}$$

On an element whose volume is dV the force along x from stresses is f_x dV. The volume element is also acted upon by gravity. We call gravity an <u>external force</u>, by contrast to the stresses which arise from <u>contact forces</u>. The source of the gravitational field is effectively the center of the earth and is evidently "external" to most bodies of liquid we would consider. For an element of small volume dV and density ρ the gravitational force along the (upward) x direction is $-\rho g$ dV ; here g is the gravitational constant and is about 980 cm sec^{-2} on the earth's surface. Since each volume element is in equilibrium the sum of forces on it must be zero and we have

$$-\frac{\partial p}{\partial x} dV - \rho g \, dV = 0$$

or

$$\frac{\partial p}{\partial x} = -\rho g \ . \tag{5.12}$$

This is the differential equation we sought. We now proceed to find a solution for our problem, an easy matter if ρ and g are assumed constant. By a solution of Eq. (5.12) we mean an expression for p in terms of x such that $\partial p/\partial x$ is equal to $-\rho g$. Clearly any solution must be of the type

$$p = -\rho g x + c \,, \tag{5.13}$$

where c is a constant. The constant c is determined by a <u>boundary condition</u>; for a lake a reasonable boundary condition would be that p is equal to the atmospheric pressure P_{atm} at the top of the lake, say, at x = 0. Thus

$$p = P_{atm} \quad \text{at} \quad x = 0 \,. \tag{5.14}$$

Applying this condition to Eq. (5.13) we obtain

$$P_{atm} = c \,. \tag{5.15}$$

Hence the constant c is determined and Eq. (5.13) becomes

$$p = -\rho g x + P_{atm} \,. \tag{5.16}$$

The expression for p in Eq. (5.16) is the distribution for the problem we had posed; it satisfies the differential equation in Eq. (5.12) and the boundary condition in Eq. (5.14). Since x is positive upward and is zero at the lake surface, we see that in the lake itself x is always negative. To find the pressure p at a depth d below the surface we set x equal to -d in Eq. (5.16), and find that p at depth d is greater by $\rho g d$ than at the surface. For water we find that the depth required to increase p by one atmosphere is about 10.3 m or 32 ft.

When the compressibility of the medium is important the density can not be assumed constant. It can sometimes be set equal to kp, where k is a constant. Then Eq. (5.12) becomes

$$\frac{\partial p}{\partial x} = -kgp \,. \tag{5.17}$$

It can be shown by substitution that a solution of Eq. (5.17) is

$$p = A\, e^{-kgx} , \tag{5.18}$$

where A is an arbitrary constant. For an ideal gas at temperature T the constant k proves to be (Prob. 5.3)

$$k = \frac{M}{R_g T} , \tag{5.19}$$

where M is the molecular weight of the gas and R_g is the gas constant, 8.3×10^7 ergs mol^{-1} per degree Kelvin. Equation (5.18) can be used to describe the variation of pressure with height in the atmosphere, whenever the assumption of constant temperature is valid. We use a modified form of the equation later in discussing the distribution of biological particles in suspension. (Section 9.5).

5.3 SHEARING OF A SOLID-LIKE MEDIUM

Consider a slab of material such as gelatin or biological tissue (here idealized to be a solid) confined between planes at y = 0 and y = h. See Fig. 5.2. This situation has some resemblance to that seen by the synovial substance which fills the space between opposing surfaces in the joint of, say, a knee, hip or arm. Suppose the upper surface is displaced in the x direction a distance X while the lower surface remains fixed. We wish to determine the

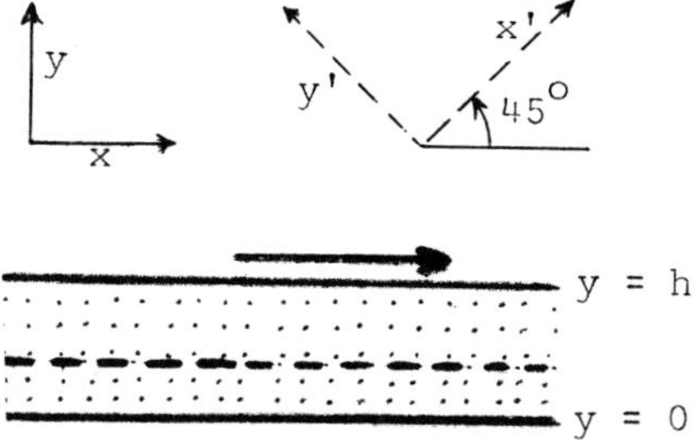

Fig. 5.2 A layer of medium between planes at y = 0 and y = h is subjected to shear and rotation by displacing its surface at y = h in the x direction relative to the surface at y = 0. Matter above any plane of constant y, such as that indicated by the heavy dashed line, exerts stress to the right on the matter below the plane.

displacement and stress at all values of y in the slab. It is reasonable to assume (if the y dimension of the slab is small compared to the dimensions along x and z) that the displacements are primarily in the x direction and vary only with y (except near the edges). Then the displacements are of the form

$$\xi_1 = f(y) \; ; \quad \xi_2 = \xi_3 = 0 \; , \tag{5.20}$$

where f(y) is a function of y to be determined. (Here we ignore effects of hydrostatic pressure.) From Eqs. (5.20) and the definition of ε_{ij} , Eq. (3.4b) , we see that

$$\varepsilon_{12} = \varepsilon_{21} = \frac{1}{2}\frac{\partial \xi_1}{\partial y} \; , \tag{5.21}$$

while all other ε_{ij} are zero. Hence if the stress-strain relation of Eq. (4.2) is valid we find that

$$S_{12} = S_{21} = 2\mu\varepsilon_{12} = \mu \frac{\partial \xi}{\partial y} \; , \tag{5.22}$$

and all other S_{ij} are zero. In Eq. (5.22) and in the rest of this Section we drop the subscript on ξ_1 and simply write ξ . From Eq. (5.6) the volume force along x from stresses is

$$f_x = \frac{\partial S_{12}}{\partial y} = \mu \frac{\partial^2 \xi}{\partial y^2} \; . \tag{5.23}$$

Since there are no external forces involved here f_x in Eq. (5.23) gives the total force per unit volume acting on any volume element in the solid. Since also the element is in equilibrium the total force must be zero and hence f_x given by Eq. (5.23) must be zero. Our differential equation for ξ is therefore simply

$$\frac{\partial^2 \xi}{\partial y^2} = 0 \; . \tag{5.24}$$

In order for Eq. (5.24) to be true it must be that $\partial\xi/\partial y$ is a constant; letting the constant be G^* we have

$$\frac{\partial \xi}{\partial y} = G^* \; . \tag{5.25}$$

We call G^* the displacement gradient. In order for Eq. (5.25) to be true it must be that

$$\xi = G^* y + A \ , \tag{5.26}$$

where A is another constant. The expression for ξ in Eq. (5.26) is a solution of Eq. (5.24), as may be verified by substitution. The constants G^* and A are arbitrary so far, but may be determined by requiring ξ to satisfy the boundary conditions.

$$\begin{aligned} \xi &= 0 \quad \text{at} \quad y = 0 \\ \xi &= X \quad \text{at} \quad y = h \ . \end{aligned} \tag{5.27}$$

We find readily that Eqs. (5.27) applied to Eq. (5.26) yield the values

$$A = 0 \ ; \qquad G^* = \frac{X}{h} \ . \tag{5.28}$$

Hence, finally, the function which satisfies both the differential equation in Eq. (5.24) and the boundary conditions in Eqs. (5.27) is simply

$$\xi = G^* y = \frac{X}{h} y \ . \tag{5.29}$$

Equation (5.29) (where ξ represents ξ_1) combined with statements that ξ_2 and ξ_3 are zero is equivalent to Eqs. (3.27) the displacement pattern corresponds to simple shear and is pictured in Fig. 3.7 . As explained in Section 3.6 the motion is a combination of shear and rotation; the components of ρ_{ij} and ε_{ij} are the same as in Eqs. (3.28) if b is replaced by G^* . Assuming the stress-strain relation given by Eq. (4.2) the stress tensor is given by

$$S_{ij} = \begin{bmatrix} 0 & \mu G^* & 0 \\ \mu G^* & 0 & 0 \\ 0 & 0 & 0 \end{bmatrix} . \tag{5.30}$$

We see that S_{12} is constant, independent of y , and is given by

$$S_{12} = \mu G^* \ . \tag{5.31}$$

This means that for any horizontal plane of constant y passing through the slab (See dashed line in Fig. 5.2) the matter above the plane exerts a stress μG^* on that below. This applies also at $y = h$; the medium above the slab (which might be the "upper" bone of a joint) must exert a stress to the right equal to μG^*.

Similarly the medium below any plane must exert a stress μG^* to the left on the medium above. Extending this reasoning to the plane $y = 0$ we see that the medium below the slab (say, the lower bone of a joint) must exert a stress μG^* on the slab in the -x direction.

This situation seems to provide a simple definition for μ, the shear coefficient of elasticity; this coefficient is just the ratio of the stress S_{12} to the displacement gradient G^* in the strain field under consideration.

As was shown in Section 2.3 the stress tensor, given by Eq. (5.30) in the (x, y, z) coordinate system, will typically have different components in a rotated system. Of special interest is the form of the tensor in coordinates (x', y', z) arrived at by rotating the reference axes 45° about the z axis, as shown in Fig. 5.2; here the tensor has the diagonal form

$$S_{ij}' = \begin{pmatrix} \mu G^* & 0 & 0 \\ 0 & -\mu G^* & 0 \\ 0 & 0 & 0 \end{pmatrix}, \qquad (5.32)$$

which may be compared with Eq. (2.15). This shows that the deformation under consideration is equivalent to a tensile stress μG^* along the x' direction and a compressional stress of the same magnitude along the y' direction.

5.4 SIMPLE SHEARING FLOW

Here we take up a situation analogous to that of Section 5.3. We consider a layer of fluid confined between parallel planes such that one plane, at $y = 0$, is fixed while the other, at $y = h$, moves in the x direction with velocity U. Again see Fig. 5.2. The flow which results is similar to the Couette flow which occurs in a rotating-cylinder viscometer, a device for measuring viscosity.

More important, acquaintance with the features of this simple flow pattern is helpful in understanding more complex patterns. If the dimensions of the layer along x and z are large compared to that along y we may assume the velocity of the fluid is primarily along x, and varies only with y. Then by analogy to Eq. (5.20) we have

$$u_1 = f(y) \; ; \qquad u_2 = u_3 = 0 \; ; \tag{5.33}$$

here f(y) is to be determined. In the rest of this Chapter we shall usually omit the subscript on u_1 and simply write u. From Eqs. (5.33) and the definition of $\dot{\varepsilon}_{ij}$, Eq. (3.34), we see that

$$\dot{\varepsilon}_{12} = \dot{\varepsilon}_{21} = \frac{1}{2}\frac{\partial u}{\partial y} , \tag{5.34}$$

while all other $\dot{\varepsilon}_{ij}$ are zero. Hence if the stress-strain relation of Eq. (4.24) is valid we find that

$$S_{12} = S_{21} = 2\eta\dot{\varepsilon}_{12} = \eta \frac{\partial u}{\partial y} . \tag{5.35}$$

From Eq. (5.6) the volume force along x from stresses is

$$f_x = \frac{\partial S_{12}}{\partial y} = \eta \frac{\partial^2 u}{\partial y^2} . \tag{5.36}$$

If the fluid is at atmospheric pressure P_{atm} the diagonal stress components S_{11} , S_{22} , S_{33} are also present; by Eq. (4.24) each of these is equal to $\lambda\theta$ or $-p$, where (ignoring changes in pressure with depth) p is approximately equal to P_{atm} . Since S_{11} , etc., are thus assumed constant it follows from Eq. (5.6) that they do not contribute to f_x . When effects of external forces can be ignored it follows as in Sect. 5.3 that f_x must be zero for an equilibrium state; hence from Eq. (5.36)

$$\frac{\partial^2 u}{\partial y^2} = 0 . \tag{5.37}$$

By comparison with Eq. (5.26) we can write a solution to Eq. (5.37) as

$$u = Gy + A , \tag{5.38}$$

where G and A are constants to be determined from the boundary conditions, namely,

$$u = 0 \quad \text{at} \quad y = 0 ,$$
$$u = U \quad \text{at} \quad y = h . \tag{5.39}$$

We find readily that for Eqs. (5.39) to hold we must have

$$A = 0 ; \quad G = \frac{U}{h} . \tag{5.40}$$

Hence the solution of the differential equation, Eq. (5.37), which satisfies the boundary conditions in Eqs. (5.39) is

$$u = Gy = \frac{U}{h} y , \tag{5.41}$$

exactly analogous to Eq. (5.29) . The quantity G is the <u>velocity gradient</u> $\partial u/\partial y$; it is constant in this field of simple shearing flow, already discussed in Section 4.6 . Equation (5.41) represents a field of shear and rotation; the tensors $\dot{\rho}_{ij}$ and $\dot{\varepsilon}_{ij}$ are given in Eqs. (4.27) and the stress components S_{ij} in Eq. (4.29) . By arguments similar to those in Section 5.3 we conclude that maintenance of this flow field requires that a stress S_{12} of magnitude ηG in the positive x direction be maintained on the fluid (say, by a moving plate) at $y = h$; also an equal and opposite stress is required at $y = 0$.

Again by analogy to Section 5.3 we can use this situation to define η , the coefficient of shear viscosity; η gives the ratio of stress S_{12} to the velocity gradient G in the flow field under consideration.

In a rotation of coordinates about the z axis the stress components will change as described in Section 2.3 . Thus in (x' , y' , z) axes rotated 45° about z, as shown in Fig. 5.2, the stress tensor is

$$S_{mn}' = \begin{pmatrix} -p+\eta G & 0 & 0 \\ 0 & -p-\eta G & 0 \\ 0 & 0 & -p \end{pmatrix} . \tag{5.42}$$

We see that the flow field contributes tensile stress along the x' direction and compressional stress along the y' direction.

5.5 FLOW BETWEEN PARALLEL WALLS

5.5.1 Horizontal Flow

In biology the concept of viscous fluid flow is important in connection with the motion of blood and other body fluids, in the various channels of the body. This concept has also been invoked in explaining the passage of water across membranes. In these and other applications the flow through a tube of circular cross section is a useful model. We shall take this up in a later section, but first treat a problem which is simpler in some respects, that of flow in a rectangular channel, one of whose cross-sectional dimensions is much smaller than the other. To find the flow pattern for this problem we can proceed as in the previous Sections 5.3 and 5.4, making use of results in Section 5.1.

We consider flow in the x direction, which we first take to be horizontal, between motionless walls at $y = h$ and $y = -h$. The velocity components are

$$u = u_1 = f(y) \ ; \quad u_2 = u_3 = 0 \ . \tag{5.43}$$

As in Section 5.4 the viscous stress S_{12} associated with the flow, in a fluid whose stress-strain relation is given by Eq. (4.24), is $\eta\partial u/\partial y$; also S_{13} is zero and the stress S_{11} is $\lambda\theta$ or $-p$. From Eq. (5.6) we then see that

$$f_x = -\frac{\partial p}{\partial x} + \eta\frac{\partial^2 u}{\partial y^2} \ . \tag{5.44}$$

Since gravity does not affect horizontal flow (and other external forces are assumed zero) the expression for f_x in Eq. (5.44) gives the total force on any volume element in the fluid, per unit volume. Since also each element moves in a straight line with constant speed its acceleration is zero and the total force on it must be zero. Letting $f_x = 0$ Eq. (5.44) yields

$$-\frac{\partial p}{\partial x} + \eta\frac{\partial^2 u}{\partial y^2} = 0 \ . \tag{5.45}$$

Our task now is to find expressions for p and u which will satisfy Eq. (5.45). While there are systematic ways to attack such problems it will serve our purposes here simply to observe that a suitable pair of functions is:

$$p = -ax + b \ ; \qquad u = \frac{1}{2} cy^2 + d \ , \tag{5.46}$$

where a, b, c, d are constants. In testing the validity of these functions it is not necessary to know how they are arrived at; they might, for example, have been discovered by trial and error. One can easily verify the fact that they <u>are</u> solutions by substituting them into Eq. (5.45). In addition we shall see that the expressions in Eq. (5.46) can be made to satisfy reasonable physical requirements, given as boundary conditions; for this problem these are:

$$\begin{aligned} u &= 0 \quad \text{at} \quad y = h \\ u &= 0 \quad \text{at} \quad y = -h \\ p &= p_1 \quad \text{at} \quad x = x_1 \\ p &= p_2 \quad \text{at} \quad x = x_2 \ . \end{aligned} \tag{5.47}$$

By mathematical "uniqueness theorems" one can often show that functions like those in Eq. (5.46) which satisfy a differential equation (such as Eq. (5.45))and also suitable boundary conditions (such as Eqs. (5.46)) are the <u>only</u> such solutions.

In Eqs. (5.46) the constant c is determined in terms of <u>a</u> by requiring that the functions satisfy Eq. (5.45); specifically, we find that

$$c = -\frac{a}{\eta}. \tag{5.48}$$

From the boundary conditions on u in Eqs. (5.48) we obtain the constant d:

$$d = -\frac{1}{2} ch^2 \ . \tag{5.49}$$

From the boundary conditions on the pressure p in Eqs. (5.47) we find, letting a = G' , that

$$G' = \frac{p_1 - p_2}{x_2 - x_1} ; \quad b = p_1 + G' x_1 . \tag{5.50}$$

Hence for two points at x_1 and x_2 in the channel the constant G' gives the magnitude of the ratio of the pressure difference between these points to their separation along x. An alternative expression for G' can be obtained by differentiating with respect to x the expression for p in Eq. (5.46) , obtaining

$$G' = -\partial p/\partial x . \tag{5.51}$$

In the channel flow under consideration we see that the pressure gradient $\partial p/\partial x$ is constant. The pressure gradient can be established by various means, for example, by means of pumps or other devices which apply the definite pressures p_1 and p_2 at x_1 and x_2 . Combining Eqs. (5.46), (5.48), (5.49) and (5.50) we obtain

$$p = p_1 + G'(x_1 - x) ; \quad u = \frac{G'}{2\eta} (h^2 - y^2) ; \tag{5.52}$$

Here G' is the magnitude of the pressure gradient. Plots of these functions are shown in Fig. 5.3. We see that p varies linearly with x, while u varies parabolically with y.

5.5.2 Downward Flow

Here we consider the same problem as in Section 5.5.1 except that this time we suppose the x axis is directed downward (toward the earth). The expression for f_x in Eq. (5.44) is still correct in principle (although we shall later conclude that $\partial p / \partial x$ is zero). But in this orientation gravity provides an external force ρg per unit volume in the x direction. Since the sum of forces must be zero we obtain

$$\rho g - \frac{\partial p}{\partial x} + \eta \frac{\partial^2 u}{\partial x^2} = 0 , \tag{5.53}$$

instead of Eq. (5.45). The boundary conditions on velocity are the same as in Eq. (5.47). However in this orientation we assume the ends of the channel at x_1 and x_2 are both open to the atmosphere. Hence p_2 and p_1 are both equal to the atmospheric pressure and $\partial p/\partial x$, being constant, must be zero. The boundary conditions are

$$\begin{aligned} u &= 0 \quad \text{at} \quad y = h \\ u &= 0 \quad \text{at} \quad y = -h . \end{aligned} \tag{5.54}$$

Following procedures similar to those in Section 5.5.1 we find that the solutions of Eq. (5.53) which satisfy the boundary conditions in Eq. (5.54) are

$$u = \left(\frac{\rho g}{2\eta}\right)(h^2 - y^2) \; ; \quad p = P_{atm} . \tag{5.55}$$

The expression for u in Eq. (5.55) is the same as that in Eq. (5.52) except that ρg replaces G' ; that is, the gravitational force per unit volume now replaces the volume force associated with compressional stresses. The flow is now "driven" by gravity (an external force field) rather than by a pressure gradient.

5.6 FLOW IN A TUBE

5.6.1 Horizontal Flow

We now consider flow through a tube of circular cross section; this situation is a useful model for a topic of much importance in biology and medicine. Let the tube have inside radius R and indefinite length. We suppose its axis of symmetry coincides with the x axis and let r measure distance radially out from this axis. See Fig. 5.4. To obtain a differential equation for this flow we proceed somewhat differently than in previous sections of this Chapter. Instead of directing our attention to an indefinitely small volume element, it is convenient here (especially because of the symmetry) to consider a representative body of fluid of finite size and require that the forces on it be zero.

For this purpose we choose a cylindrical body of fluid of radius r_e centered on the axis and bounded by planes at $x = x_1$ and

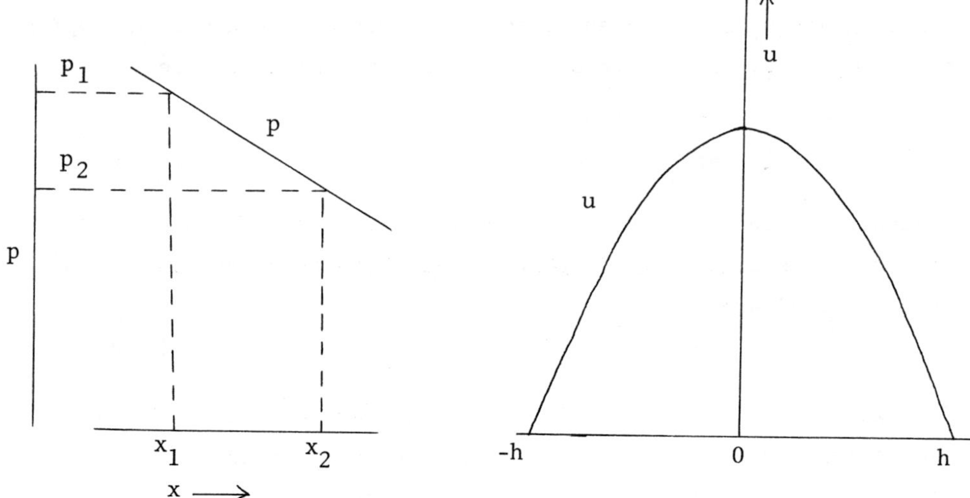

Fig. 5.3 Distribution of pressure p and velocity u in horizontal channel; from Eqs. (5.52).

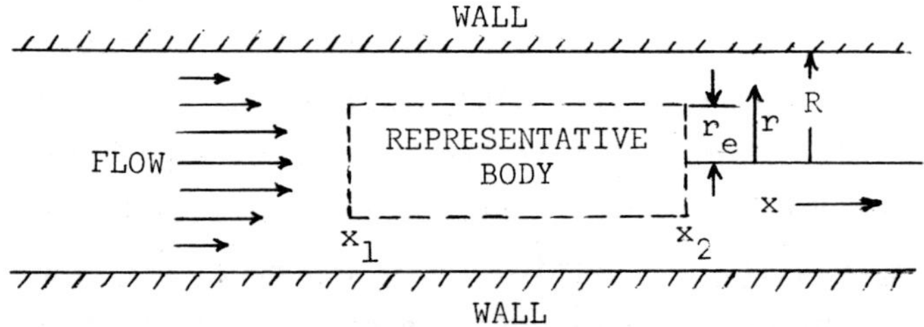

Fig. 5.4 Defining sketch for flow through a tube of circular cross section. View is in a plane through the axis of symmetry.

$x = x_2$; see Fig. 5.4. We assume the flow (in the region of interest) is everywhere along the x direction and call its velocity u. Further we assume the flow is symmetrical about the axis, so that u is a function only of r. We take the axis to be horizontal and consider contributions to the force on the representative body in the x direction. Part of the force, which we call F_x', comes from the stress S_{11} acting on the faces at x_1 and x_2 ; from Eq. (4.24) we see that since $\partial u/\partial x$ is zero, S_{11} is just $\lambda\theta$ or $-p$. Letting p_1 be the pressure at x_1 and p_2 that at x_2 we see that

$$F_x' = (p_1 - p_2)\pi r_e^2 \ . \tag{5.56}$$

Viscous stresses also exert a force on the cylindrical body. The fluid just outside this body will pull it in the x direction with a force S_{12} per unit area of the cylindrical surface. This statement needs some justification, since it is true that the quantity S_{12} was defined in Chap. 2 for a Cartesian coordinate system. It may nevertheless be used here if we define a Cartesian system which applies locally near any point P of interest on the cylindrical surface. Specifically we choose the local y axis normal to the surface at P (thus coincident with the radial or r direction) and the z axis perpendicular to those for x and y. It is then consistent with definitions of the S_{ij} in Chap. 2 to define S_{12} near the arbitrary point P as the "force per unit area of the local cylindrical surface exerted in the x direction by the medium outside the surface on the medium within".

From Eq. (4.24), again taking r as locally equivalent to y, we see that near the point P referred to earlier S_{12} is given by $\eta\partial u/\partial r$ the latter to be evaluated at $r = r_e$. Since the cylindrical surface bounding the body of interest has circumference $2\pi r_e$ and length (x_2-x_1) we see that the force F_x'' associated with the viscous stress S_{12} is

$$F_x'' = 2\pi r_e(x_2 - x_1)\eta \left(\frac{\partial u}{\partial r}\right)_e , \tag{5.57}$$

where the subscript "e" on $(\partial u/\partial r)$ means that the quantity is to be evaluated at $r = r_e$.

We are now ready to apply the condition that the sum of forces on the representative body of fluid (contained at a moment of interest in the cylinder of radius r_e and length (x_2-x_1)) must be zero. This condition holds since each volume element making up the body moves in a straight line with constant speed, and hence the total momentum of the body remains constant (even though the body changes shape as the flow occurs).

Hence

$$F_x' + F_x'' = 0 . \tag{5.58}$$

Combining the last three equations we obtain,

$$G' + \frac{2\eta}{r}\frac{\partial u}{\partial r} = 0 , \quad G' = \frac{p_1 - p_2}{x_2 - x_1} , \tag{5.59}$$

where it is understood that both r and $\partial u/\partial r$ are to be evaluated at r_e; but since r_e is arbitrary it need no longer be distinguished from r. Here G' is the magnitude of the pressure gradient, which may be thought of as driving the motion. Equation (5.59) is the desired differential equation for the velocity u. As a boundary condition we require that

$$u = 0 \quad \text{at } r = R \tag{5.60}$$

It is easily verified that Eqs. (5.59) and (5.60) are satisfied by

$$u = \frac{G'}{4\eta}(R^2 - r^2) . \tag{5.61}$$

We find that the velocity varies parabolically with the transverse coordinate r; compare the expression for u in Eqs. (5.61) with an analogous expression in Eq. (5.52). The volume of liquid which passes through a given cross section of the tube in time t_o is given by

$$V = 2\pi t_o \int_0^R ur\, dr , \tag{5.62}$$

since $2\pi r\ dr$ is the area of an elemental ring and the velocity u gives the volume passing through unit area per unit time. One finds (Prob. 5.7) that

$$\frac{V}{t_o} = \frac{\pi G' R^4}{8\eta} \quad , \tag{5.63}$$

where G' is the magnitude of the pressure gradient along the tube. For a tube of length L between the ends of which the pressure drop is Δp we see that G' is just $\Delta p/L$. Equation (5.63), or its equivalent, is often called Poiseuille's law, after the nineteenth century French physiologist Jean L. M. Poiseuille .

5.6.2 Downward Flow

If we suppose the x axis is directed downward we must consider another contribution to the x component of force on the representative body, to which we directed our attention in Section 5.6.1 . See Fig. 5.4. This contribution F_x''' gives the "weight" or gravitational force and has the value

$$F_x''' = \rho g \pi r_e^2 \ (x_2 - x_1) \ ; \tag{5.64}$$

here ρg gives the gravitational force per unit volume, while $\pi r_e^{\ 2}$ gives the cross-sectional area, and $(x_2 - x_1)$ the length of the representative body. Setting the sum of forces on the fluid body equal to zero we have that

$$F_x' + F_x'' + F_x''' = 0 \ , \tag{5.65}$$

or, combining the last two equations with Eqs. (5.56) and (5.57)

$$G' + \frac{2\eta}{r}\frac{\partial u}{\partial r} + \rho g = 0 \ , \tag{5.66}$$

where G', the magnitude of the pressure gradient, is zero as explained in Section 5.5.2. As a boundary condition we assume

$$u = 0 \quad \text{at} \quad r = R \ , \tag{5.67}$$

which is identical with Eq. (5.60). The solution of Eq. (5.66) which satisfies Eq. (5.67) is

$$u = \frac{\rho g}{4\eta} (R^2 - r^2) . \tag{5.68}$$

The expression for u in the above equation is the same as for horizontal flow through a tube, Eqs. (5.61), except that ρg replaces G'.

To find the volume V which passes in time t_o we obtain, by analogy with Eq. (5.63) that

$$\frac{V}{t_o} = \frac{\pi g R^4}{8\nu} \tag{5.69}$$

where ν stands for the ratio (η/ρ) and is called the kinematic viscosity coefficient, the constant η being called the dynamic viscosity coefficient. Solving Eq. (5.69) for ν we obtain

$$\nu = Ct_o \text{ where } C = \frac{\pi g R^4}{8V} . \tag{5.70}$$

Equation (5.70) is a working equation for capillary-type viscometers. In this instrument a liquid whose viscosity is to be measured is caused to fall through a vertical capillary under the influence of gravity. An observer determines the time t_o required for a given volume to flow. According to Eq. (5.70) the viscosity coefficient ν is proportional to the time t_o, the proportionality constant being the calibration constant given by C. We see that C is independent of properties of the liquid and is critically dependent on the capillary radius R. In principle one could calculate the calibration constant according to the expression given in Eq. (5.70). In practice this does not give a sufficiently accurate value for C, principally because of inaccuracies in the measurements of R. Instead one usually determines C by measuring t_o for a liquid of accurately known viscosity.

5.7 PRESSURE DISTRIBUTION IN A ROTATING VESSEL

The centrifuge is much used in biophysical experiments, either for separation of small biological particles or for obtaining information on the nature of the particles. We learn much about the basic principles involved in centrifugation by considering a simple situation, that of a cylindrical bucket with vertical axis, filled with liquid and set into rotation about its axis. When equilibrium is established the liquid rotates with the bucket as if it were a solid body. We let the origin of Cartesian coordinates (x, y, z) be on the axis of rotation. We choose the z direction along the axis of rotation; the x and y directions are then along radii outward from this axis. We consider a volume element instantaneously on the x axis, at a distance r from the origin. See Fig. 5.5. Any such element experiences centripetal acceleration (towards the center) with magnitude $\omega^2 r$ where ω is the angular velocity.

By Newton's Second Law this means that a force acts on the element in the x direction. This force must arise from stresses and is obtained from the general expression for f_x in Eq. (5.6). In using the latter equation we assume the stress-strain relation of Eq. (4.24) applies, recognizing that the velocities u_i to be used in this expression can be taken relative to the bucket, and hence by hypothesis all u_i are zero. Then Eq. (4.24) reduces to Eq. (5.8) and f_x is given simply by Eq. (5.11). Letting f_x be equal to the product of ρ (the mass per unit volume of the element) and the acceleration we find that $-\partial p/\partial x$ is equal to $-\rho\omega^2 r$ or

$$\frac{\partial p}{\partial r} = \rho\omega^2 r , \qquad (5.71)$$

since the left hand side is the same as "$\partial p/\partial x$ evaluated at $x = r$". A possible boundary condition comes from assuming that the center of the bucket is in equilibrium with atmospheric pressure P_{atm}, that is,

$$p = P_{atm} \quad \text{at } r = 0 \tag{5.72}$$

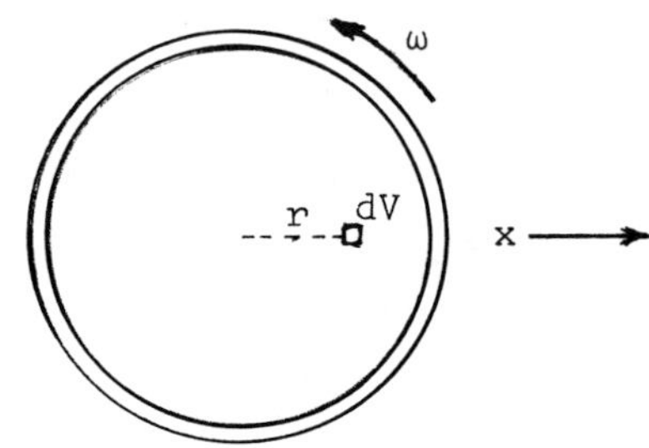

Fig. 5.5 A volume element of fluid in a rotating bucket.

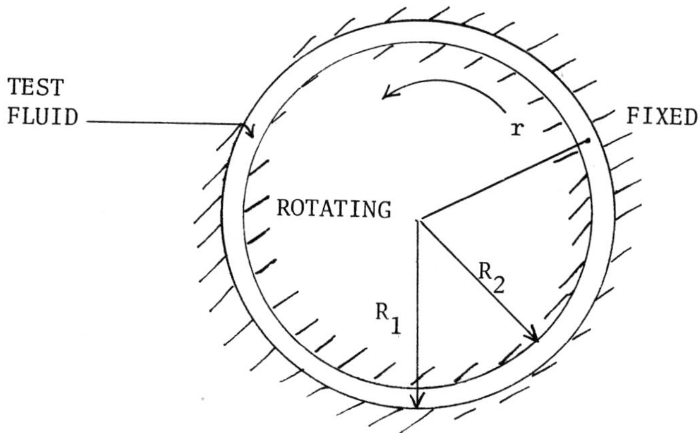

Fig. 5.6 Arrangement for generating Couette flow. A test fluid lies in the space indicated, between an outer fixed cylinder of radius R_2 and an inner rotating one of radius R_1. The radial distance to an arbitrary point in the fluid is r.

The solution of Eq. (5.71) which satisfies Eq. (5.72) is

$$p = \frac{1}{2}\rho\omega^2 r^2 + P_{atm} \quad . \qquad (5.73)$$

We see that the pressure increases with the square of distance from the axis. In a high speed centrifuge p reaches very high values near the outer wall of the bucket (Prob. 5.8) .

Migration of particles in a centrifuge will be discussed in Chap. 6.

5.8 COUETTE FLOW

We refer to the Couette arrangement shown in Fig. 5.6 and assume the flow lines are circles (i) concentric with the cylinder and (ii) lying in planes perpendicular to the axis. Imagine a cylindrical surface $r = a$ passing through the test fluid, the latter being a ("Newtonian") fluid with constitutive equation given by Eq. (4.24) . Then the fluid lying just inside this surface exerts a stress $-\eta(\partial u/\partial r)$ on the fluid just outside, where $\partial u/\partial r$ is to be evaluated at $r = a$; the sense of this stress, and of the velocity u , are regarded as positive when along the direction of rotation. The torque about the axis associated with this stress, per unit length of cylinder, is seen to be (Prob. 5.9)

$$L = -2\pi r^2 \eta \frac{\partial u}{\partial r} , \qquad (5.74)$$

where $\partial u/\partial r$ is evaluated at $r = a$.

Now consider a body of fluid contained between two imaginary cylindrical surfaces, the inner one at $r = a$ and the outer one at $r = b$. This body will be acted on by stresses and associated torques, on both surfaces. The torque on the body exerted by stress at the inner surface is given by L in Eq. (5.74) , evaluated at $r = a$. But the torque exerted on this same body by the liquid just outside the surface $r = b$, will be given by the negative of L in Eq. (5.74), evaluated at $r = b$. Now in steady state flow the body must be in equilibrium and hence the net torque on it must be zero. Zero net torque obtains only if L, expressed in Eq. (5.74), yields the same value at $r = a$ and at $r = b$. Considering $\underline{a}$ and b to have arbitrary values we reach the conclusion: A condition for flow in the Couette arrangement is that L should be constant, that is, independent of r . We therefore obtain from Eq. (5.74)

$$r^2 \frac{\partial u}{\partial r} = C , \tag{5.75}$$

where C is independent of r . We now seek a solution of Eq. (5.75) which satisfies the boundary conditions

$$\begin{aligned} u &= 0 \quad \text{at} \quad r = R_2 \\ u &= U \quad \text{at} \quad r = R_1 , \end{aligned} \tag{5.76}$$

where U is the velocity at the surface of the (real) inner cylinder. It can be verified (Prob. 5.9) that the required solution is

$$u = \frac{U}{A} \left(\frac{1}{r} - \frac{1}{R_2}\right) ; \quad A = \left(\frac{1}{R_1} - \frac{1}{R_2}\right) . \tag{5.77}$$

In the Couette viscosimeter means are provided for determining the applied torque L and resulting velocity U for a test liquid. When these are known the viscosity coefficient for the liquid can be calculated from the formula (Prob. 5.9)

$$\eta = \frac{AL}{2\pi U} . \tag{5.78}$$

If (R_2-R_1) is much less than either R_1 or R_2 the expression for u in Eq. (5.77) can be written (Prob. 5.9) as

$$u = G(R_2 - r) , \tag{5.79}$$

where G is the velocity gradient given by $U/(R_2-R_1)$. This expression for u is the same as that for simple shearing flow in Eq. (5.41) if we associate (R_2-r) with y .

PROBLEMS FOR CHAPTER 5

5.1 Show that when S_{12} depends only on y, while S_{11} and S_{13} are zero, the x component of the volume force is as given in Eq. (5.5). Refer to the volume element of Fig. 5.1 and follow arguments analogous to those used in deriving Eq. (5.4).

5.2 (a) Use Eq. (5.16) to calculate typical pressures to which skin divers are subjected. Introduce reasonable values for data; handle units with care.

(b) Use Eq. (5.16) to calculate the typical hydrostatic pressure difference (in the blood system ignoring any pressure changes associated with blood flow) between the heart and foot of an adult human while standing. If the heart is at atmospheric pressure how much tension is developed in the wall of a cylindrical blood vessel of radius R in the foot?

(c) Calculate the difference in hydrostatic pressure (in the fluid system) between the top and bottom of a tall tree. What significance do you see in this result?

5.3 Using the relation $\rho = kp$ assumed in obtaining Eq. (5.17) one can evaluate k for an isothermal atmosphere from the perfect gas law

$$pv = R_g T.$$

In this latter equation p is the pressure, v is the volume of one mol of gas, T is the Kelvin temperature and R is a universal constant with the value 8.3×10^7 ergs mol^{-1} per Kelvin degree.

(a) Letting one mol have the mass M put the perfect gas law in the form $\rho = kp$ and obtain the expression for k given in Eq. (5.19).

(b) With this result for k find an expression for the distance ℓ of vertical ascent required to cause the pressure to decrease by a factor of e^{-1}.

The ideas involved here will be referred to later in discussing the distribution of biological particles in suspension.

5.4 Consider horizontal channel flow described by Eqs. (5.52). Suppose that at x_1 the pressure is 2.5 P_o and that at x_2 the pressure is P_o, where P_o is atmospheric pressure and $(x_2 - x_1)$ is 100 cm. Suppose also that h is 0.4 mm and η is 3.5 poise.

(a) Calculate the velocity u at y = 0, that is, at the center of the channel.

(b) Calculate numerical values of all components of the stress tensor at $x = x_1$, y = h (i.e., at a point on a wall).

5.5 Show from Eq. (5.52) that $\eta \partial u/\partial y$ evaluated at either wall is equal in magnitude to hG'.

5.6 (a) Show from Eq. (5.61) that $\eta \partial u/\partial r$ evaluated at r = R is equal in magnitude to ½ RG'.

(b) Show that this result might have been expected by considering the balance of forces on the "representative body" pictured in Fig. 5.4, when the latter extends across the entire tube (so that $r_e = R$).

5.7 Show that when the velocity u is given by Eq. (5.61) the expression for the volume V in Eq. (5.62)yields Poiseuille's law in the form of Eq. (5.63).

5.8 In a high speed centrifuge the rotor turns at angular speeds up to 60,000 rev/min. If the liquid is water and is contained in a cylindrical vessel of radius 8 cm what pressure is developed at the outer rim? (Be careful of units!)

5.9 (a) Verify the expression for the torque L in Eq. (5.74).

(b) Show that the expression for the velocity u in Eq. (5.77) satisfies the differential equation in Eq. (5.75) and the boundary conditions of Eq. (5.76).

(c) Carry out the steps required to obtain the Couette-viscosimeter equation, Eq. (5.78).

(d) Starting from Eq. (5.77) derive the approximate equation for the velocity u in Eq. (5.79).

CHAPTER 6

MECHANICS OF SUSPENDED PARTICLES AND VISCOELASTIC MEDIA

Biological tissues tend to contain large amounts of water. They often can be usefully thought of as aqueous media in which are suspended ions, molecules of moderate size (such as amino acids), macromolecules, fibers, membranes and structures built from these. When the suspended particles are small and of low concentration the suspension often behaves like a Newtonian liquid with shear viscosity coefficient greater than that for the suspending liquid. The motion of each individual particle can then be analyzed as if it were isolated from the others.

As the particle size increases, or the concentration increases, interactions become more important. The suspension then often acquires elastic properties in addition to viscous characteristics and is said to be viscoelastic. The effective viscosity may depend on the rate of shear and may vary according to the history of the medium. Of course, when the solid-to-liquid ratio is high it is appropriate to think of the medium as solid-like rather than liquid-like. Solid-like characteristics appear even for a fairly small solid-liquid ratio if, as in a gel, the solids are in the form of long molecules or fibers which entangle and form a network which responds elastically to stresses.

In this chapter we take up subjects relating to suspensions and other media which are neither pure liquid nor pure solid. First we consider forces which act on particles in suspensions, especially forces associated with compressional and viscous stresses. Simple differential equations and solutions to them are then arrived at for the motion of particles (i) in a gravitational field, (ii) in a rotating fluid and (iii) in an electric field.

We also consider macroscopic properties of media which are partly liquid, partly solid. For liquids containing relatively small concentrations of suspended particles we derive or cite expressions for the increase in macroscopic viscosity caused by the particles. For viscoelastic media, in which shear elasticity cannot be neglected, we analyze the properties of several models which give useful insight .

6.1 FORCE FROM A PRESSURE GRADIENT

Let us now consider a body such as a cell, a macromolecule, or other particle, suspended in a fluid medium. In general the body may be acted on by external forces as well as by stresses or contact forces. We treat here the effects of contact forces only; in particular, we consider compressional stresses which exist when the stress tensor is isotropic as in Eqs. (5.8) and (5.9), that is, when S_{ij} is given by $-p\delta_{ij}$, where p is the pressure.

The total force on a suspended body is determined by first analyzing the situation at an arbitrary small portion or <u>element</u> of its surface. We let the area of this arbitrary surface element be dS and also (following the rather loose language commonly used) refer to the element itself as dS. When S_{ij} is given by $-p\delta_{ij}$ with p positive, the force $\underline{dF}$ (a vector) exerted by the outer medium on the object across dS will be perpendicular to the surface element and directed inward; see Fig. 6.1 . We let θ be the angle between this inward-directed force and an arbitrary x direction. The magnitude of the vector $\underline{dF}$ is p dS where p is the pressure in the vicinity of dS, and the component of $\underline{dF}$ along the x direction is ($p \cos \theta \, dS$). To obtain the x-component F_x of the total force exerted by the outer medium on the body we add contributions from all elements dS by integrating over the entire surface S:

$$F_x = \int_S p \cos \theta \, dS \ . \tag{6.1}$$

Equation (6.1) is valid for a body of any kind and for any pressure distribution p(x,y,z). Since the x direction is arbitrary it is clear that one can use Eq. (6.1) with appropriate definition of θ to obtain the force component in any direction. A result of particular importance applies when the pressure distribution in the vicinity of the body has the form

$$p = P_o + G_p x \ , \tag{6.2}$$

where P_o and G_p are constants, the former being the value of p at x = 0. Differentiating with respect to x we obtain from Eq. (6.2) simply

$$\frac{dp}{dx} = G_p \; . \tag{6.3}$$

Thus for the distribution given by Eq. (6.2) the pressure gradient dp/dx is constant and given by G_p . Recalling Eq. (5.16) we see that the pressure distribution in a static fluid of constant density has the form of Eq. (6.2) and thus has constant gradient at all points. In other situations the gradient varies in space but slowly enough that it is nearly constant in the vicinity of a small suspended particle.

We now assume that Eq. (6.2) holds (with P_o and G_p constant) in the vicinity of a small particle or body of volume V and show that the force F_x is then given simply by

$$F_x = -G_p V \; . \tag{6.4}$$

We obtain Eq. (6.4) rather simply by carrying out the integration in Eq. (6.1) in a special way. Imagine the body to be divided into a large number of long thin pencil-like volume elements, each with long axis in the x direction, such as the element dV shown shaded in Fig. 6.2. These elements are of rectangular cross-section, cut (like french-fried potatoes) by slicing the body first along the planes perpendicular to the y axis, then along planes perpendicular to z . The volume element is bounded at one end by a portion dS_1 of the surface S , and at the other by a portion dS_2 . We define angles θ_1 and θ_2 in the figure; thus θ_1 is the (smallest) angle between the direction of x and the direction of the inward normal to S_1 , while θ_2 is defined similarly for S_2 . We require that θ_1 and θ_2 lie in different ranges:

$$0 \leqslant \theta_1 \leqslant 90^\circ \; ; \quad 90^\circ \leqslant \theta_2 \leqslant 180^\circ \; . \tag{6.5}$$

Let the pressure over dS_1 be essentially constant and equal to p_1, and similarly let that at dS_2 be essentially p_2 .

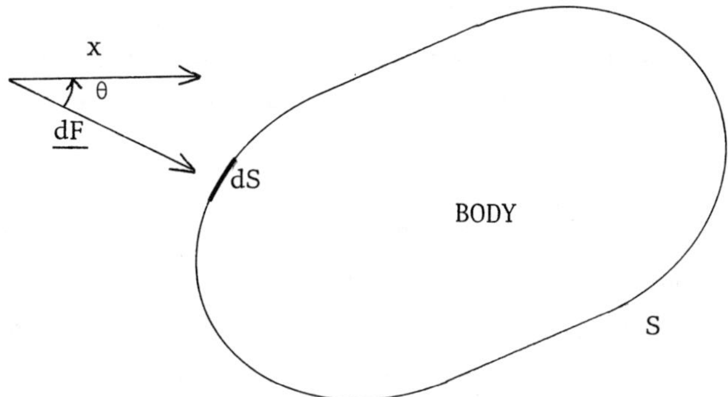

Fig. 6.1 The vector dF shows the force contribution on a suspended body when pressure acts on a surface element dS .

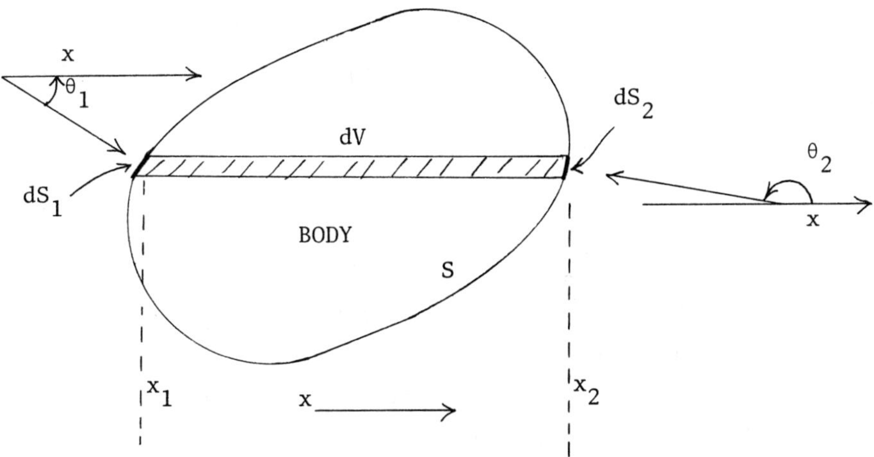

Fig. 6.2 To determine the force on a body we first determine the force on a typical volume element dV .

Returning to the task of calculating F_x from the integral in Eq. (6.1) we now consider the contribution dF_x to this integral, associated with the volume element dV and hence with the pair of surface elements dS_1 and dS_2 defined above. Evidently

$$dF_x = p_1 \cos\theta_1 \, dS_1 + p_2 \cos\theta_2 \, dS_2 \; . \qquad (6.6)$$

From Eqs. (6.5) it follows (except for zero values) that $\cos\theta_1$ is always positive while $\cos\theta_2$ is always negative. From simple geometry we see that the magnitude of either $dS_1 \cos\theta_1$ or $dS_2 \cos\theta_2$ is just the cross-sectional area of the volume element, which we shall call $d\sigma$. Hence from Eq. (6.6) the force on the element dV along x becomes

$$dF_x = (p_1 - p_2) \, d\sigma \; . \qquad (6.7)$$

Now from Eq. (6.2) it is easily shown that $(p_1 - p_2)$ is equal to $G_p(x_1 - x_2)$, where x_1 and x_2 are x-values for dS_1 and dS_2 (Fig. 6.2). Letting dV be the volume of the element (and not just its name) we note that

$$dV = d\sigma \, (x_2 - x_1) \; . \qquad (6.8)$$

Hence we obtain

$$dF_x = -G_p \, dV \; . \qquad (6.9)$$

Equation (6.9) gives the contribution to F_x from the pair of surface elements associated with an arbitrary volume element. We may obtain F_x itself by summing over all such contributions. Since Eq. (6.9) is expressed in terms of a volume differential dV we see that the summation may be accomplished by carrying out a volume integration over the volume V of the body. Our procedure has thus had the effect of replacing the surface integral given in Eq. (6.1) by a volume integral . We obtain

$$F_x = \int_V dF_x = \int_V -G_p \, dV = -G_p \, V \; ,$$

the result anticipated in Eq. (6.4).

6.2 VISCOUS DRAG

When an object immersed in a viscous fluid moves relative to the fluid, the latter exerts a force on the object which tends to oppose the relative motion, and heat is generated. Such motion occurs, for example, when biological particles are sedimented in a gravitational field or in a centrifuge. The opposing force, frequently called a <u>drag force</u>, is also important in connection with Brownian movement and diffusion. Suppose the velocity of the object relative to the fluid is along the x direction and has the magnitude U_r. Then at slow speeds it is found for a variety of objects that the rate of heat generation dQ/dt or $\dot{Q}$ is

$$\dot{Q} = fU_r^2 \; ; \tag{6.10}$$

also the drag force F is along x and given by

$$F = -fU_r \; . \tag{6.11}$$

Here f is the <u>frictional coefficient</u> and is a positive quantity; it depends on the shape and size of the object as well as the viscosity coefficient η of the liquid; in some conditions f also is affected by boundaries of the fluid-filled space in which the particle moves. By U_r is meant the difference between the velocity of the object and the velocity which the fluid (at the site of the object) would have if the object were not there. The minus sign in Eq. (6.11) means that the force is in the opposite direction to the relative velocity.

As an illustration for which f is easily calculated (but probably not of much practical importance) suppose a flat sheet is pulled lengthwise between parallel walls such that on each side of the sheet its distance from the adjacent wall is h. A fluid of viscosity η fills the space between the sheet and the walls. If the sheet is pulled steadily with velocity U and the walls are fixed the magnitude of the velocity gradient will be constant everywhere in the fluid, and equal to U/h. (The reader may satisfy himself that this is consistent with the idea that every volume element of fluid is subject to zero net force from viscous stress. Prob. 6.2) A stress ηG then exists on each side of the sheet; hence we have for the total retarding force or drag F on unit area of the sheet, and the frictional coefficient f ,

$$F = 2\eta G = 2\eta \frac{U}{h} = fU \ ; \qquad f = \frac{2\eta}{h} \ . \tag{6.12}$$

The work done per unit time will be FU or fU^2 per unit area. This work adds no potential energy to the system; nor is the kinetic energy increased, since the flow is steady. Hence the energy supplied in pulling the sheet against the viscous drag force F must be converted entirely into heat. The rate of heat production per unit area of the sheet is then

$$\dot{Q} = FU = fU^2 = \frac{2\eta U^2}{h} \ . \tag{6.13}$$

It is worth noting that, since the volume of fluid associated with unit cross-sectional area of the sheet is just 2h, the rate of heat production per unit volume of fluid is

$$\dot{Q}_v = \frac{FU}{2h} = \frac{\eta U^2}{h^2} = \eta G^2 \ . \tag{6.14}$$

The result in Eq. (6.14) may be generalized: In any parallel flow situation the rate of heat generation per unit volume in a given region of fluid is equal to the product of the viscosity coefficient and the square of the velocity gradient in that region.

In the example just discussed the drag force and frictional coefficient are dependent on the distance h from the object (the sheet) to an adjoining boundary. For objects such as molecules or cells in a blood vessel or a test tube f is independent of distance from boundaries if the distance to the nearest boundary is large compared to the object's dimensions. Theory for the drag force on spheres and ellipsoids has been developed by Stokes and others, assuming boundary effects are negligible. This theory is much too lengthy and detailed for us to take up here. Instead we shall content ourselves with listing some results of special interest.

For a solid sphere of radius R the frictional coefficient is

$$f = 6\pi\eta R \,. \tag{6.15}$$

This is the result given by Stokes. It has been shown valid when the Reynold's number N_R is much less than one, where

$$N_R = \frac{\rho R U_r}{\eta} \,; \tag{6.16}$$

here ρ is the fluid's density.

For a liquid sphere the expression for f in Eq. (6.15) should be multiplied by a factor, yielding

$$f = 6\pi\eta R \cdot \frac{2\eta + 3\eta'}{3\eta + 3\eta'} \,, \tag{6.17}$$

where η' is the viscosity of the fluid in the sphere. When η' is much greater than η the expression for f reduces to the Stokes result for a solid sphere, Eq. (6.15).

Results for an ellipsoid have been given by Lamb (1945). We shall not quote the general procedure for determining f, but instead cite expressions obtained when the ellipsoid reduces to a disk of radius a. When the relative velocity is broadside-on to the disk Eq. (6.15) may be used for f if R is replaced by 0.85 a; when the velocity is edgewise to the disk R is replaced by 0.566 a .

Burgers (1938) derived results for a long rod of length L and radius a . The frictional coefficient is

$$f = \frac{2\pi\eta L}{\ell n \frac{L}{a} - 0.72} \tag{6.18}$$

when the relative velocity is parallel to the rod's axis and

$$f = \frac{4\pi\eta L}{\ell n \frac{L}{a} + 0.5} \tag{6.19}$$

when the relative velocity is normal to the axis.

Long flexible "chain" macromolecules are discussed in Section 6.4.5. For the "impermeable" coils described there the frictional constant is given by an expression similar to that for a rigid sphere, Eq. (6.15), except that the radius is replaced by an "effective radius" R_e . It is shown from physical statistics that for a given kind of chain molecule the <u>square</u> of R_e is proportional to the length of the molecule and hence to its mass. A more detailed treatment of frictional constants for chain molecules is given by Morawetz (1965).

6.3 FORCED MIGRATION OF A PARTICLE

Here we consider a small body suspended in a viscous fluid and acted on by gravity or a force field of external origin such as an electric field. We suppose this external field exerts a force F_e on the body or particle in the x direction. Other forces may also act on the small body as a result of stresses exerted by the surrounding fluid at the body surface. In a fluid where the stress-strain relation of Eq. (4.24) applies the stress will be partly compressional and partly viscous in nature. Let F_c represent the total force on the particle in the x direction as a result of compressional stresses and let F_v be the corresponding force from viscous stresses. From Newton's second law we have that the sum of forces is equal to the product of the mass of the body and its acceleration. For a body of density ρ_o and volume V the mass is ρ_o V. Let the instantaneous velocity along x be v and let the instantaneous acceleration along x be $\partial v/\partial t$. Then the equation of motion for the particle becomes

$$F_e + F_c + F_v = \rho_o V \frac{\partial v}{\partial t} . \tag{6.20}$$

If the external force field is gravitational, then at points near the earth we have, if the x direction is downward,

$$F_e = \rho_o g\, V. \tag{6.21}$$

Of course, if the x direction is horizontal F_e is zero. We showed in Sect. 6.1, Eq. (6.4), that in a fluid where the pressure gradient along x is $\partial p/\partial x$ the force from compressional stresses is

$$F_c = -V \frac{\partial p}{\partial x} . \qquad (6.22)$$

In a gravitational field with the x axis downward the pressure gradient $\partial p/\partial x$ is given by ρg where ρ is the fluid density; see Eq. (5.12). The force F_c is then directed upward. In a rotating system as in a centrifuge with x directed radially outward we learn from Eq. (5.71) that $\partial p/\partial x$ (equal to $\partial p/\partial r$) has the magnitude $\rho\omega^2 x$, where ω is the angular velocity of the centrifuge and x is distance from the axis of rotation; the force F_c is then directed inward (toward the axis of rotation).

For F_v we have from Section 6.2, Eq. (6.11), that when the relative velocity U_r is along the x direction and is not too great

$$F_v = -fU_r , \qquad (6.23)$$

where f is the frictional constant. We can now write special forms of Eq. (6.20) for various situations.

6.3.1 Sedimentation under Gravity

First consider a particle acted on by gravity alone in a motionless liquid and take the x direction to be downward. Then from the last four equations, since $\partial p/\partial x$ is ρg and U_r is just v, we have

$$\rho_o gV - \rho gV - fv = \rho_o V \frac{\partial v}{\partial t} . \qquad (6.24)$$

Rearranging, this can be put in the form

$$\frac{\partial v}{\partial t} + \frac{v}{\tau} = A , \qquad (6.25a)$$

where

$$A = \frac{g(\rho_o - \rho)}{\rho_o} ; \qquad \tau = \frac{\rho_o V}{f} . \qquad (6.25b)$$

One readily finds by substitution that Eqs. (6.25) are satisfied by

$$v = A\tau \, [1 - \exp(-t/\tau] . \qquad (6.26)$$

The solution given in Eq. (6.26) satisfies the condition that $v = 0$ at $t = 0$. In Fig. 6.3 we see a plot of the velocity v *vs* time, according to Eq. (6.26). We see that v increases rapidly at first, then more slowly; it gradually approaches a limiting value U_∞ given by

$$U_\infty = A\tau \; . \tag{6.27}$$

At a time t equal to τ the velocity reaches the value

$$(1 - e^{-1})U_\infty \quad \text{or} \quad 0.63\ U_\infty \; ;$$

the time τ is therefore a measure of the time required for the velocity to reach its limiting value. For most situations of interest τ is very short. (See Prob. 6.4). An important quantity in the biophysical literature is the sedimentation coefficient s, defined by

$$s = \frac{U_\infty}{g} \tag{6.28}$$

This coefficient is thus a measure of the speed with which a particle will sediment at its limiting velocity. Since s has the units of time, its cgs unit is simply the second. A convenient reference unit for biological particles is the Svedberg defined as 10^{-13} sec. If S gives the sedimentation coefficient in Svedbergs and s gives the same coefficient in cgs units we see that

$$S = 10^{13}\ s \; . \tag{6.29}$$

Combining expressions from Eqs. (6.28), (6.27) and (6.25b) we find an expression for s in terms of properties of the particle and the medium in which it moves:

$$s = \frac{(\rho_o - \rho)\ V}{f} \tag{6.30}$$

Here ρ is the density of the fluid and ρ_o that of the particle. For a sphere of radius R the coefficient f is given by Stokes' expression, Eq. (6.15), and we have

$$s = \frac{2R^2}{9\eta}\,(\rho_o - \rho) \tag{6.31}$$

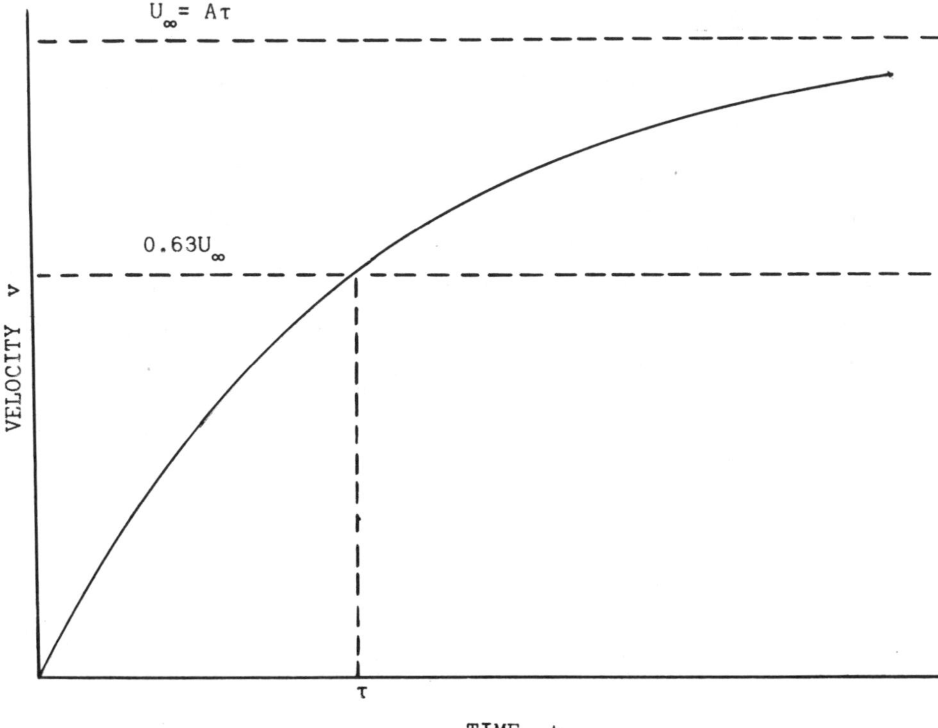

Fig. 6.3 Plot of velocity v *versus* time t from Eq. (6.26) .

Examining Eq. (6.31) we see that if ρ, ρ_o and η are known one can determine the radius R of a particle by measuring its sedimentation coefficient s. This method is indeed much used in study of biological particles. For a particle as large as a red cell one might measure the limiting velocity U_∞ of fall through a liquid under the influence of gravity, then solve for s from Eq. (6.28). But for macromolecules U_∞ would be much too small to be measured in a gravitational field.

6.3.2 Centrifugation

To obtain sufficiently rapid sedimentation of protein molecules and similar particles one uses a high speed centrifuge or ultracentrifuge. To understand the principle involved here suppose liquid of density ρ is caused to rotate with angular velocity ω about a vertical axis. Choosing x as the (horizontal) radial distance outward from the axis of rotation we see that the pressure gradient in the liquid is given from Eq. (5.71) by $\rho\omega^2 x$; hence, since the x direction is horizontal F_e and F_c are given by

$$F_e = 0 \; ; \quad F_c = -\rho\omega^2 xV \; . \tag{6.32}$$

The outward acceleration a_x is given by

$$a_x = \dot{v} - \omega^2 x \; , \tag{6.33}$$

where $\dot{v}$ is $\partial v/\partial t$ and the term $-\omega^2 x$ gives the centripetal acceleration, that is, the acceleration which results from the rotation. Equation (6.20) requires modification by substituting a_x from Eq. (6.33) for $\partial v/\partial t$; when this is done and use is made of Eqs. (6.23) and (6.32) we have (since $u = U_r$)

$$-\rho\omega^2 xV - fv = \rho_o V(\dot{v} - \omega^2 x) \; . \tag{6.34}$$

Rearranging (Cf Eqs. (6.25)) one finds that the equation can be written

$$\dot{v} + \frac{v}{\tau} = A' \; , \tag{6.35a}$$

where τ is given, as before, by Eq. (6.25b) but A' has the value

$$A' = \frac{\omega^2 x(\rho_o - \rho)}{\rho_o} \; . \tag{6.35b}$$

We see that Eq. (6.35) is identical to the first of Eqs. (6.25) except that A' replaces A and hence $\omega^2 x$ replaces g. For given x the solution will therefore be approximately as in Eq. (6.26) except that A' replaces A. The ratio A'/A is significant in centrifugation theory, and is known as the "g-factor"; hence

$$\text{g-factor} = \frac{\omega^2 x}{g} \; . \tag{6.36}$$

The limiting velocity in the centrifuge is $A'\tau$; its ratio to the limiting velocity U_∞ in a gravitational field is given by the g-factor, which is usually much larger than unity. Centrifugation therefore has the effect of increasing the limiting velocity, as if the value of g were increased. It is common to describe a centrifugal field as equivalent to a gravitational field of Xg, where X is the g-factor of Eq. (6.36). Commercially available ultracentrifuges yield values of $\omega^2 x$ up to 300,000 g, that is, they yield g-factors up to 300,000.

6.3.3 Electrophoresis

Migration of particles in electric fields or electrophoresis is important biologically, not only because electric fields are ubiquitous in living matter, but also because of laboratory techniques which are used for separation and characterization of particles. If a particle has a positive net charge q and is in an electric field of intensity E it is acted on by a force qE in the direction of the field. Suppose the particle is suspended in a motionless viscous liquid in which there are no free ions. It is easily seen that Eq. (6.20) then becomes, for motion along a horizontal direction,

$$Eq - fv = \rho_o V \frac{\partial v}{\partial t} , \qquad (6.37)$$

which may be compared with Eq. (6.24). The expression for v in Eq. (6.26) is a solution of Eq. (6.37) when τ has (as before) the value $\rho_o V/f$, while A has the value $Eq/\rho_o V$. For a biological cell or macromolecule in aqueous solution the charge q varies with the hydrogen ion concentration $[H^+]$ in the solution, becoming increasingly positive as $[H^+]$ increases and becoming negative at small values of $[H^+]$. The value of $[H^+]$ at which q is zero is the isoelectric point. In an electrolyte the effective value of q is reduced considerably by oppositely charged ions which are attracted to the particle and form a layer around it.

6.4 VISCOSITY OF A SUSPENSION

When particles are suspended in a liquid they tend to create constraints to the motion. The effective viscosity η of the suspension is therefore higher than that of the viscosity η_o of the pure liquid. When the ratio ϕ of the volume of particles to that of the total suspension is much less than unity it is frequently found that η increases linearly with ϕ according to

$$\eta = \eta_o(1 + \kappa\phi) , \tag{6.38}$$

where κ is a constant whose value depends on the nature of the particle. When ϕ is not small Eq. (6.38) usually does not give the viscosity accurately, and sometimes is not even approximately correct. Whether Eq. (6.38) holds or not the <u>specific viscosity</u> η_{sp} of a suspension is defined as

$$\eta_{sp} = \frac{\eta - \eta_o}{\eta_o} . \tag{6.39}$$

Of course if we do not require κ to be constant Eq. (6.38) can be forced to apply to any situation by making an <u>ad hoc</u> choice of κ . From Eqs. (6.38) and (6.39) we can write κ as η_{sp}/ϕ . Hence if η and η_o (which determine η_{sp}) and ϕ are known the quantity κ can be calculated. It is useful to define a limiting value for κ which applies as the concentration approaches zero. Calling this value κ_o we have

$$\kappa = \frac{\eta_{sp}}{\phi} ; \quad \kappa_o = \lim_{\phi \to 0} \frac{\eta_{sp}}{\phi} . \tag{6.40}$$

Another parameter is much in use, the intrinsic viscosity $[\eta]$ defined by

$$[\eta] = \kappa_o /\rho , \tag{6.41}$$

where ρ is the density of the material comprising the particle. For a macromolecule surrounded by a layer of bound water the density ρ is an average over a volume including the bound water.

Theoretical expressions for κ , η_{sp} and $[\eta]$ have been given for particles of various shapes and kinds. The theory is too lengthy for us to treat here, except for two kinds of hypothetical particles for which the analysis is relatively easy. The first kind is a plane sheet, assumed oriented along the flow; the second is a dumbbell, that is, a pair of spheres connected with a thin rigid shaft. Neither of these shapes is a very good approximation to a real particle, although membranes of endoplasmic reticulum might be thought of as sheets, and rod-like viruses or macromolecules might be crudely approximated by dumbbells. Our main interest here in sheets and dumbbells is that they provide interesting and instructive illustrations for the calculation of κ .

For both we refer to the arrangement pictured in Fig. 5.2 and discussed in Section 5.4. A fluid fills the space between a fixed surface at $y = 0$ and a parallel one at $y = h$, the latter one being caused to move along x with velocity U. This arrangement provides a means for defining the viscosity coefficient η of the fluid, whether the latter be a homogeneous pure liquid or a suspension: η is just the ratio of the stress S applied to the upper plate to the quantity U/h. If the fluid is a pure liquid L with viscosity η_o the velocity gradient will be constant throughout the space and equal to U/h. But if the fluid is a suspension of particles in the liquid L the liquid volume will be decreased and the average value of the velocity gradient will be somewhat greater than U/h .

6.4.1 Sheets

Suppose specifically that the particles are solid plane sheets all oriented with planes perpendicular to the y direction, and that these are allowed to flow freely with the liquid. If there are a number of sheets of varying thickness ε_i the total thickness of the space occupied by these sheets is just the sum of the ε_i , which we call d. The thickness remaining for the liquid L is now reduced to (h - d). Since (in steady flow) the stress must be the same across any plane of constant y between y = 0 and y = h it follows that the velocity gradient G must be constant everywhere in the liquid L and given by

$$G = \frac{U}{h - d} \tag{6.42}$$

The stress S on the plate at y = h , assumed to be in contact with the liquid L, must then be

$$S = \eta_o G = \frac{\eta_o U}{h - d} . \tag{6.43}$$

If, as anticipated earlier, we <u>define</u> the effective viscosity η of the suspension as the ratio of S to the quantity U/h we obtain

$$\eta = \frac{Sh}{U} = \frac{\eta_o}{1 - (d/h)} . \tag{6.44}$$

Note that (d/h) is just the volume ratio called ϕ in Eq. (6.38). When ϕ is small we know from a series expansion (see Appendix B) that $(1 - \phi)^{-1}$ is approximately $(1 + \phi)$ and hence obtain

$$\eta = \eta_o(1 + \phi) . \tag{6.45}$$

We see that for a suspension of these hypothetical sheet-particles the relation $\eta(\phi)$ is of the type indicated in Eq. (6.38) ; for the constants κ and $[\eta]$ we have

$$\kappa = 1 ; \quad [\eta] = \frac{1}{\rho} \tag{6.46}$$

6.4.2 Dumbbells

Here our particle, a dumbbell, consists of a pair of spherical bodies A and B, each of radius R and volume V, connected by a thin rigid rod of length 2b. See Fig. 6.4. This particle is suspended in a liquid where the velocity u is along x and given by Gy , as in Eq. (5.41) and in the arrangement of Fig. 5.2. For simplicity we assume the dumbbell lies in the xy plane. If the midpoint

O of the rod is at y_o the velocity in the fluid at O is Gy_o; we assume the entire particle (rod plus bodies A and B) travels without rotation at the same velocity Gy_o. If the rod is inclined at an angle θ to the x direction the body A has the y coordinate $y_o + \varepsilon$, where ε is $b \sin\theta$. At this value of y the velocity is greater than at O by the amount $G\varepsilon$. Similarly at the body B the velocity is less than at O by $G\varepsilon$.

Continuing the assumption that A and B are both travelling with the same velocity Gy_o it follows that the fluid at $y_o + \varepsilon$ is moving relative to the body A with speed $G\varepsilon$ in the positive x direction. On the other hand the fluid at $y_o - \varepsilon$ is moving relative to the body B with speed $G\varepsilon$ in the negative x direction. See Fig. 6.4. Suppose the frictional constant is f for each of the bodies A and B. Then by Eq. (6.10) heat will be generated at each body in the amount fU_r^2. Since the relative velocity U_r is $G\varepsilon$ for each body the total heat generated per second for the two bodies A and B is

$$\dot{Q} = 2fG^2\varepsilon^2 = 2fG^2b^2 \sin^2\theta . \tag{6.47}$$

Heat is also generated along the rod connecting the spheres A and B, but this is negligible if the rod is sufficiently thin. Another condition for validity of Eq. (6.47) is that

$$\sin\theta \gg \frac{R}{b} . \tag{6.48}$$

When the condition in Eq. (6.48) is not satisfied the contribution $\dot{Q}$ given by Eq. (6.47) is small, and another small contribution to the heating must be taken into account. This is the heat generation which would occur if the connecting rod were removed from each dumbbell, leaving only the spherical bodies; these bodies would then be free to move independently. See Prob. 6.7

If there are n identical particles per unit volume (all with the same orientation) the rate of heat generation per unit volume $\dot{Q}_v$ because of the particles is

$$\dot{Q}_v = 2nfG^2b^2 \sin^2\theta . \tag{6.49}$$

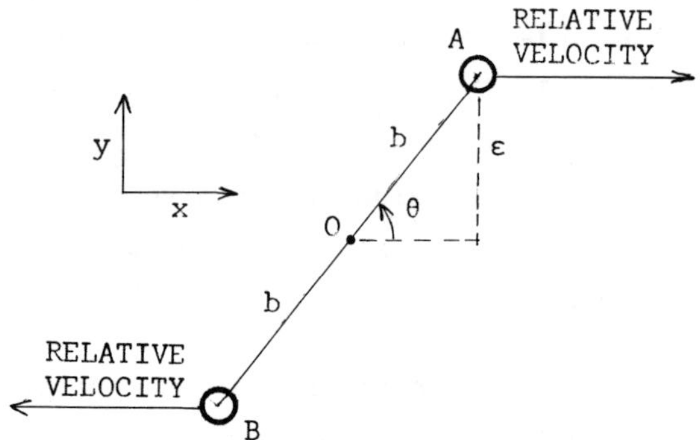

Fig. 6.4 A dumbbell of length 2b is suspended in a flowing liquid whose velocity is along x and increases with y. Arrow at body A or B shows the velocity of fluid relative to that body, if the dumbbell does not rotate.

If the dumbbells have random orientations (but lie in the xy plane) we can obtain the average value $< \dot{Q}_v >$ by noting that the average of $\sin^2 \theta$ is ½ :

$$< \dot{Q}_v > = nfG^2 b^2 . \qquad (6.50)$$

We see that the dumbbells contribute to the heat production at a rate which is proportional to G^2 . But according to Eq. (6.14) the heat generation in the absence of particles is also proportional to G^2 . Let the total heat production per unit volume be $\dot{Q}_t$ for a suspension of dumbbells in a liquid whose viscosity (in the absence of dumbbells) is η_o . Then by adding the contributions we obtain (on the average)

$$\dot{Q}_t = \eta_o G^2 + nfG^2 b^2 = (\eta_o + nfb^2)G^2 . \qquad (6.51)$$

Here $\eta_o G^2$ is the contribution, according to Eq. (6.14) for a pure liquid of viscosity η_o in the absence of particles. Clearly the total heating $\dot{Q}_t$ in the suspension, given by Eq. (6.51) is the same (for given G) as it would be in a pure liquid of increased viscosity. Thus $\dot{Q}_t$ is just ηG^2 if for η we choose

$$\eta = \eta_o + nfb^2 . \qquad (6.52)$$

This may be put in the form of Eq. (6.38) if we note that here the volume fraction ϕ occupied by particles, each of volume 2V, is

$$\phi = 2nV \ . \tag{6.53}$$

We then obtain Eq. (6.38) with κ given by

$$\kappa = \frac{fb^2}{2\eta_o V} \tag{6.54}$$

Since each of the bodies A and B is a sphere of radius R we use the expression $6\pi\eta_o R$ from Eq. (6.15) for f and the appropriate expression for V to obtain κ and $[\eta]$:

$$\kappa = \frac{9b^2}{4R^2} \ ; \quad [\eta] = \frac{9b^2}{4\rho R^2} \ . \tag{6.55}$$

A generalization of this result is taken up in Prob. 6.8

As noted earlier, in the derivation leading to Eq. (6.55) a contribution to the heating has been neglected. This is just the heating which occurs with independent spheres (i.e., without the connecting rod of the dumbbell). In the next subsection we see, Eq. (6.56), that this contribution by itself leads to a value of 2.5 for κ . Hence a reasonable criterion for the validity of Eq. (6.55) is that the value of κ given there should be large compared to 2.5; this will be true if $b \gg R$. Also if $b \gg R$ the condition stated in Eq. (6.48) will be met, except for small values of θ . The result obtained in Eq. (6.50) by averaging the expression in Eq. (6.49) is then fairly accurate since Eq. (6.49) is adequate over most of the θ-range.

In the foregoing treatments of sheets and dumbbells we have taken up "model" situations which are instructive in that they permit calculation of κ and $[\eta]$ with relative ease. Much more realistic models for biological particles are afforded by spheres and ellipsoids. It is found for these as with dumbbells (see, e.g., Eq. (6.47)) that each particle causes a contribution to the heating proportional to G^2. But for spheres and ellipsoids the mathematics is much too lengthy for inclusion here. In spite of their importance we therefore content ourselves with relatively brief presentations of results for these basic situations.

6.4.3 Spheres

For a suspension of rigid spheres Einstein (1906) carried through a lengthy analysis and derived a result in the form of Eq. (6.38) where

$$\kappa = 2.5 \; . \tag{6.56}$$

The value 2.5 probably applies fairly well even to a non-spherical particle if its contour is smooth and its deviation from sphericity is not too great. For spherical droplets containing fluid of viscosity η' Taylor (1932) obtained the expression

$$\kappa = 2.5 \frac{(\eta' + 0.4\,\eta_0)}{(\eta' + \eta_0)} \; ; \tag{6.57}$$

when η'/η_0 is large this reduces to the Einstein result. Equation (6.57) is valid when the velocity gradients are small; when these are large the droplets become deformed and offer less constraint to the flow. Hence κ is a decreasing function of the flow velocity. The deformations will be discussed further in a later Chapter.

6.4.4 Ellipsoids

For rigid asymmetrical particles the constraint they offer to the flow will depend on their orientation. For small particles this orientation will not usually be fixed, but will change randomly with time because of Brownian movement (a kind of thermal agitation which is discussed in a later Chapter. In a velocity gradient as, for example, in the situation of Section 5.4, the particles tend to orient themselves (often in a rather complicated rotating motion) so as to reduce the constraint they offer to the flow. The simplest situation exists when the velocity gradient G is small and the orientation is random. For this condition Simha (1940) has derived expressions for κ, applicable to suspensions of ellipsoidal particles with "axial ratio" p; here p is the ratio of length along the axis of symmetry to the equatorial diameter. Thus we have $p = 1$ for a sphere, $p > 1$ for a prolate ("cigar-shaped") ellipsoid and $p < 1$ for an oblate ("disc-shaped") ellipsoid. We shall not quote the very lengthy expression Simha obtained for κ. Instead we show in Fig. 6.5 plots from Morawetz (1965) based on the Simha results. Ellipsoids clearly contribute more to the viscosity of a suspension than do spheres (for a given volume fraction), the contributions being more from prolate than from oblate asymmetry. For very large p (i.e., for long rods) Simha obtained

$$\kappa = 0.933 + \frac{p^2}{5} \left[\frac{1}{3 \ln 2p - 4.5} + \frac{1}{\ln 2p - 0.5} \right] \quad (6.58)$$

In Fig. 6.6 a plot is shown of κ versus p, based on Eq. (6.58). For very small p (i.e., for discs) Simha's expression reduces to

$$\kappa = \frac{0.68}{p} . \quad (6.59)$$

These expressions are valid only for relatively small values of the velocity gradient G; as G increases κ tends to decrease.

6.4.5 Flexible Macromolecules

Results given above for spheres and ellipsoids have proved useful in analysis of viscosity measurements on suspensions of biological cells, viruses and macromolecules. But of course many biological particles are not even approximately

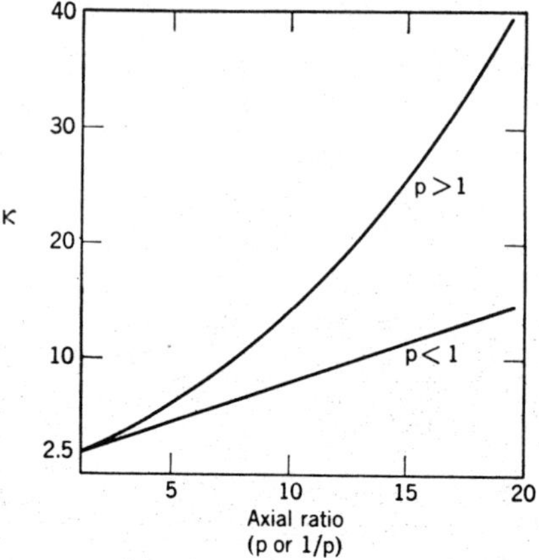

Fig. 6.5 Dependence of κ on the axial ratio p for ellipsoidal particles. From Morawetz (1965).

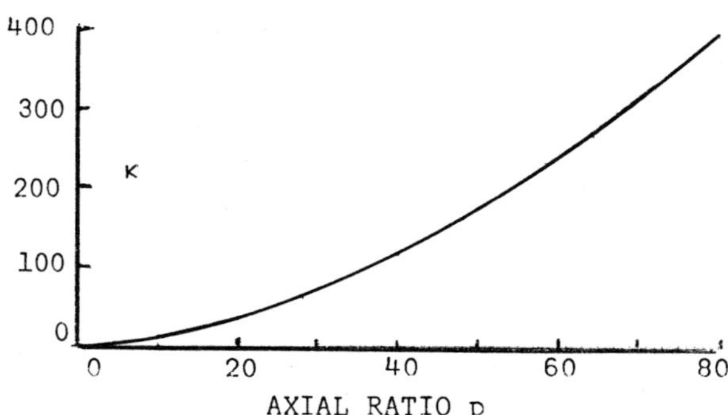

Fig. 6.6 Dependence of κ on the axial ratio of ellipsoidal solute particles, for large p. From Eq. (6.58).

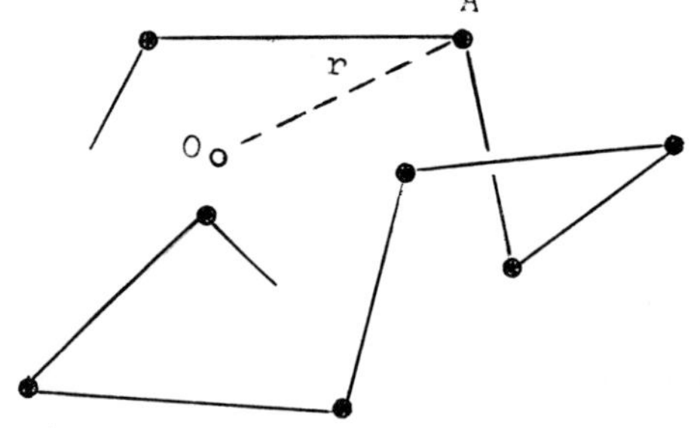

Fig. 6.7 Flexible-chain model of macromolecule. Spheres, represented by filled circles, are not necessarily in plane of figure. Center of mass at O. Projection of OA on plane of figure is r.

ellipsoidal. In particular, some large molecules, especially polysaccharides, are better regarded as flexible threads or filaments. Much theory has been developed for such macromolecules on the basis of various models. A particularly interesting one is the <u>flexible chain</u> model: here the macromolecule is represented by a number N of identical spheres of radius R and volume V connected by thin rigid links each of length <u>a</u>, consecutive links being free to rotate about each other. See Fig. 6.7 . The assemblage of spheres and links is immersed in a liquid whose viscosity η_o is typically not greatly different from that of water. If the chain is very loosely wound the liquid can move freely past the spheres and the molecule is said to be <u>free draining</u>. On the other hand if the chain is very tightly coiled liquid can hardly move in or out of it and the coil can often be regarded as <u>impermeable</u>.

6.4.5.1 Impermeable Coils

A tightly coiled chain is typically treated as equivalent to a rigid sphere of radius R_e and volume V_e ; this <u>effective</u> volume V_e may include much water and thus may be much greater than the actual volume V_a of the molecule alone. Taking this picture literally and defining κ as in Eq. (6.40) we let this parameter be 2.5 , the Einstein value for a rigid sphere, but replace ϕ by an <u>effective</u> volume fraction ϕ_e , based on the effective volume V_e . Letting ϕ_a be the actual volume fraction, based on the volume V_a , it is clear that ϕ_e/ϕ_a is equal to V_e/V_a . Hence we have

$$2.5 = \frac{\eta_{sp}}{\phi_e} = \frac{V_a}{V_e}\frac{\eta_{sp}}{\phi_a} \quad . \tag{6.60}$$

There is no direct way to determine ϕ_e or the ratio η_{sp}/ϕ_e . Instead experiments lead fairly directly to the quantity η_{sp}/ϕ_a which we shall now call κ. We have then

$$\kappa = \frac{\eta_{sp}}{\phi_a} = 2.5\,\frac{\phi_e}{\phi_a} \quad . \tag{6.61}$$

Measurement of κ then leads directly to a determination of the ratio ϕ_e/ϕ_a which is often of much interest (see, e.g., Section 7.5,). The intrinsic viscosity is given, as in Eq. (6.41), by κ_o/ρ , where κ_o is the limiting value of κ as ϕ approaches zero.

The value for R_e or V_e depends on the solvent, that is, on the liquid in which the macromolecule is immersed. In a poor solvent the atoms of the macromolecule attract each other more than they are attracted to atoms of the solvent and the macromolecule tends to contract to a relatively small value of R_e . In a good solvent the reverse is true and R_e has a relatively large value. Of special interest are intermediate solvents, called <u>Θ-solvents</u>, in which there is effectively no net force tending to cause contraction or expansion. The orientations of the various links in the chain are then random. Probabilities for various configurations of the chain can be determined by physical statistics, in analogy with topics taken up in Chapters 1-3. It is found that R_e is proportional to $\sqrt{N}$ where N is the number of links in the chain. It follows (Prob. 6.10) that for the impermeable chain molecule

$$[\eta] = C\sqrt{m}\ , \tag{6.62}$$

where m is the mass of the molecule and C is a constant. The conclusion which follows from Eq. (6.62), that the intrinsic viscosity should be proportional to the square root of the molecular weight, has been found approximately true for a wide variety of polymers in Θ-solvents (Morawetz, 1965).

6.4.5.2 Free Draining Coils

Opposite to the situation of an impermeable coil is that of a chain of links and spheres so loosely coiled that the fluid flows by each sphere as if the others were not there. This model is valid if the distance <u>a</u> between spheres is large compared to the radius R of each sphere. Debye (1946) has considered the behavior of this model in a constant-gradient velocity field. Fundamental to the theory is the frictional drag on each sphere, treated in a manner somewhat analogous to the theory for dumbbells in Sect. 6.4.2. It is assumed that the frictional drag on the thin links is negligible.

Suppose the fluid motion is along x, with velocity u given by Gy as in Eq. (5.41). Then, according to a conclusion reached by Debye, the molecule will rotate with angular velocity $d\theta/dt$ or $\dot{\theta}$ (see Fig. 6.8) given by $-G/2$. This is perhaps not very surprising since $-G/2$ is just the angular velocity of an element of the fluid itself in the flow field under consideration. (See Prob. 4.6).

Assume the center of mass of the molecule is at O and moves with the same velocity as the fluid at O. Then as in Fig. 6.8 consider a sphere A whose center has the coordinates x, y, z relative to the center of mass of the molecule. Let r be the distance to A from the z axis, so that r^2 is just $x^2 + y^2$. The velocity of A because of rotation will then be $r\dot{\theta}$ in magnitude; its component along x will be $-r\dot{\theta}\sin\theta$ or $-y\dot{\theta}$ and its component along y will be $x\dot{\theta}$. Letting $\dot{\theta}$ be $-G/2$ we find (Prob. 6.11) that

$$U_x = \tfrac{1}{2}Gy \; ; \quad U_y = \tfrac{1}{2}Gx \; , \qquad (6.63)$$

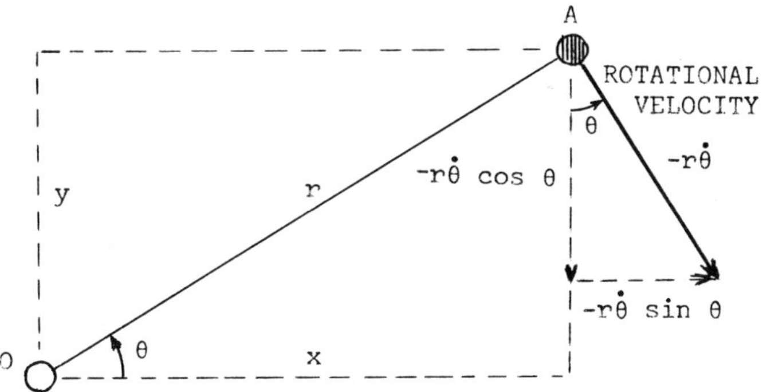

Fig. 6.8 Velocity of shpere A (in flexible-chain model) as a result of rotation.

where U_x and U_y are, respectively, the x and y components of the velocity of the fluid relative to the sphere A. The magnitude of the relative velocity is the square root of the sum of the squares of U_x and U_y and hence is just $\frac{1}{2}Gr$. Letting the frictional constant be f the heat generated near the sphere A is $f(\frac{1}{2}Gr)^2$. If there are n molecules per unit volume, each including N identical spheres, the heat generated per unit volume is

$$\dot{Q}_v = \tfrac{1}{4}\, nNfG^2 < r^2 > , \tag{6.64}$$

where $< r^2 >$ is the average value of r^2 for the N spheres in a given molecule. We see that the expression for $\dot{Q}_v$ in Eq. (6.64) resembles a related expression for dumbbells in Eq. (6.50) but differs by numerical constants, and also differs in that $< r^2 >$ appears instead of b^2. By proceeding in the same way as in Sect. 6.4.2 we find (Prob. 6.11) that if ϕ is defined as nNV (V being the volume per sphere) the constant for the flexible chain is given by

$$\kappa = \frac{f <r^2>}{4\eta_o V} , \tag{6.65}$$

which may be compared with the corresponding result for dumbbells, Eq. (6.54). Using expressions for f and V suitable for a sphere of radius R in a liquid with η_o as shear viscosity coefficient we obtain (Prob. 6.11)

$$\kappa = \frac{9 <r^2>}{8R^2} . \tag{6.66}$$

For a chain of N links, each of length a, it is characteristically found that $< r^2 >$ is proportional to Na^2, that is,

$$<r^2> = cNa^2 , \tag{6.67}$$

where the dimensionless constant c is of the order of unity but depends on the nature of the "junctions" or "hinges" between consecutive links. We then have for κ and the intrinsic viscosity $[\eta]$

$$\kappa = \frac{9cNa^2}{8R^2} ; \quad [\eta] = \frac{9cNa^2}{8\rho R^2} , \tag{6.68}$$

where ρ is the density of the material in each sphere. The mass m of the entire molecule is equal to Nm_o where $m_o = \rho V$ is the mass of each sphere; the "molecular weight" is just m multiplied by Avogadro's number. We see that κ and $[\eta]$ are each proportional to N and hence to m, for given values of $\underline{a}$, R and ρ . Hence as Debye pointed out, results for the free-draining flexible-chain model of a macromolecule lead to the conclusion that κ and $[\eta]$ are each proportional to the molecular mass m . This proportionality is in contrast to the prediction for impermeable coils (Eq. 6.62) that $[\eta]$ is proportional to the square root of m . The dependence of viscosity on m for a given kind of molecule and solvent evidently depends on the nature of the molecule, and on its interaction with the solvent. Results for a wide variety of molecule-solvent combinations have been fitted approximately by the empirical equation

$$[\eta] = c'm^{\alpha} , \tag{6.69}$$

sometimes called the modified Staudinger equation. For many situations involving macromolecules Eq. (6.69) applies with values of α between 0.5 and 1.0 ; thus for these situations α ranges between the values for impermeable and free-draining coils, respectively. Presumeably the more nearly impermeable a long flexible molecule is, the more nearly its α value will approach 0.5. Conversely, for a more open molecule α will have a greater value.

Evidently some information about the configuration of a given kind of macromolecule in a given solvent can be gained from a determination of α . Such a determination requires that the molecule be available in a range of lengths,that is, in a homologous series, so that $[\eta]$ can be measured as a function of m . Among molecules of biological interest this is true, for example, of some of the polysaccarides. Experiments show that for many flexible chain molecules α is in the neighborhood of 0.5 , showing that these approximate the behavior expected for impermeable coils. (Morawetz , 1965).

In deriving these results for the free-draining coil the "Einstein contribution" has been neglected. This contribution comes from the heating which would occur if the links of the chain were removed so that the spheres were free to move independently. In the latter situation we know from the Einstein expression in Eq. (6.56) that κ would be simply 2.5 . A reasonable criterion for the validity of Eqs. (6.68) is then simply that κ should be large compared to 2.5 . (Compare with similar considerations for the dumbbell, Sect. 6.4.2)

6.5 CONCENTRATED SUSPENSIONS; VISCOELASTIC MEDIA

Results in previous sections are applicable primarily to suspensions in which the suspended particles are relatively small and, on the average, widely separated from each other. When this is true the typical distance between a particle A and its nearest neighbor is large compared to the particle dimension. The behavior of A in response to a force can then be treated as if A were alone in the suspending liquid, as was done in the treatment of particle migration in Section 6.3. Also the heat generation from viscous drag on A can be calculated as if A were alone, as was done in Sections 6.2 and 6.4. Under these conditions it follows that for a suspension containing many particles a "superposition principle" holds, that is, the total heat generated is just the sum of single-particle contributions. For a suspension of identical particles the specific viscosity η_{sp} is then proportional to the number n of particles per unit volume or to the volume fraction ϕ .

As the concentration of particles increases the fluid motion near a given particle is affected more and more by the presence of its neighbors, and the superposition principle becomes less and less valid. This can be seen in an expression derived by Simha (1952) for the specific viscosity of a suspension of rigid spheres, valid for an intermediate range of ϕ :

$$\eta_{sp} = 2.5\phi + 12.6\phi^2 \; . \tag{6.69}$$

When $\phi \ll 1$ the ϕ^2 term can be neglected and Eq. (6.69) reduces to the result expected from the linear equation, Eq. (6.38), with the Einstein value for κ , Eq. (6.56). With increasing ϕ the ϕ^2 term becomes significant; of course, η_{sp} is then no longer proportional to ϕ and the superposition principle fails.

A new characteristic of the suspension appears when the particles exert force on each other so that certain configurations or structures are preferred. If such a suspension is disturbed it will tend to return to its preferred or equilibrium configuration after a certain characteristic time. This tendency to return to an equilibrium state after a disturbance can often be accounted for by assuming that the constitutive relation in Eq. (4.22) holds, with a nonzero value for the shear elastic coefficient μ . Such a medium is characterized by Scott-Blair (1969, Ref. 7.17) as _visco-elastic_ .

Other media respond to an application of stress by an initial deformation followed by flow which continues as long as the stress is applied. Recovery from the initial deformation occurs as soon as the stress is removed, but the flow deformation is not reversible. Scott-Blair has characterized this type of medium as _elastico-viscous_.

We shall use the adjective _viscoelastic_ (without a hyphen) as a rather general term to designate any medium which combines characteristics of viscosity and shear elasticity. Thus a viscoelastic medium might be either visco-elastic or elastico-viscous; on the other hand, it might have properties that do not fit strictly into either category. Most biological tissues are viscoelastic in this general sense.

Insight into the behavior of viscoelastic media is gained by considering imaginary mechanical models made from springs and other elements. In Section 6.6 we treat models for the response of a medium to tangential stress such as S_{12} . In Section 6.7 we take up models which have been much used to represent response to a tensile stress such as S_{11} .

6.6 MODELS FOR RESPONSE TO TANGENTIAL STRESS

6.6.1 Visco-elastic Model

First consider the arrangement of Fig. 6.9 . Here is pictured a parallel-plate arrangement similar to that of Figs. 4.1 and 5.2 . The space between the plates is filled with a Newtonian liquid whose coefficient of shear viscosity is η . A spring is attached at one end to the upper plate <u>via</u> a rigid member extending downward from it, and is similarly attached at the other end to the lower plate. This spring represents elastic elements in the suspension, which may consist of membranes, fibers, etc., which form a tangled network. We should suppose that the springs are fairly uniformly distributed, say, one spring per unit area of either plate. A stress S_{12} applied to the upper plate in the direction shown in the figure will (i) stretch the springs and (ii) produce flow in the fluid. Letting the displacement of the upper plate (in the direction of the arrow in Fig. 6.9) be ξ we let G_K be ξ/h, a quantity which is equal to the average displacement gradient in the space between the plates. Letting the "effective" shear elastic modulus be μ the portion of the stress required to stretch the spring may then be set equal to μG_K . Similarly the portion of stress required to produce the velocity gradient $\partial G_K/\partial t$ or $\dot{G}_K$ is $\eta\dot{G}_K$. Hence we have

$$S_{12} = \mu G_K + \eta\dot{G}_K \; ; \qquad (6.70)$$

this equation is consistent with the constitutive equation in Eq. (4.22) if we equate ε_{12} with $\frac{1}{2}G_K$. If S_{12} is constant and equal to S_o a solution of this equation is

$$G_K = G_\infty[1 - \exp(\frac{-t}{\tau})] \qquad (6.71a)$$

where

$$G_\infty = \frac{S_o}{\mu}; \quad \tau = \frac{\eta}{\mu} . \qquad (6.71b)$$

This solution is chosen to satisfy the condition that $G_K = 0$ when $t = 0$. This condition is appropriate for a situation where the stress S_{12} is zero until $t = 0$, when it is given the constant value S_o . Since G_K in Eq. (6.71a) has the same time dependence as v in Eq. (6.26) a plot of G_K vs t will have the same appearance as the plot in Fig. 6.3 . Just after $t = 0$ the strain G_K rises rapidly; with increasing time the curve levels off as G_K approaches the limiting value G_∞ , that is, S_o/μ . In this limiting condition the medium behaves as if it were a homogeneous elastic solid with rigidity modulus μ . The characteristic time τ is called the retardation time of the system. When $t = \tau$ the value of G_K has increased to nearly two-thirds of its limiting value; specifically, at $t = \tau$ we have

$$G_K = 0.632\ G_\infty \ . \qquad (6.72)$$

Equation (6.70) and its solution in Eqs. (6.71) are of the kind associated with a model of viscoelastic media credited both to Lord Kelvin, the Scottish mathematician-physicist (1824-1907), and the German physicist W. Voigt (1950-1919).

6.6.2 Elastico-viscous Model

Another kind of medium is represented by Fig. 6.10 . As in Fig. 6.9 we again consider an arrangement where the "model medium" is contained in the space between parallel plates, one of which can move in its own plane relative to the other. Here the medium is layered: one layer, of thickness ε , is a Newtonian liquid with viscosity

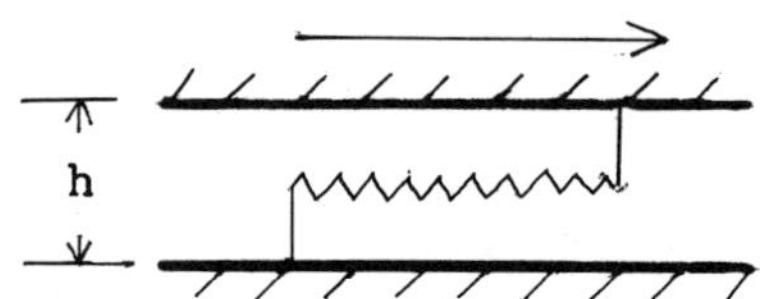

Fig. 6.9 Visco-elastic model; see Section 6.6.1 . A stress in the direction of the arrow is exerted by the upper plate on the medium between the plates.

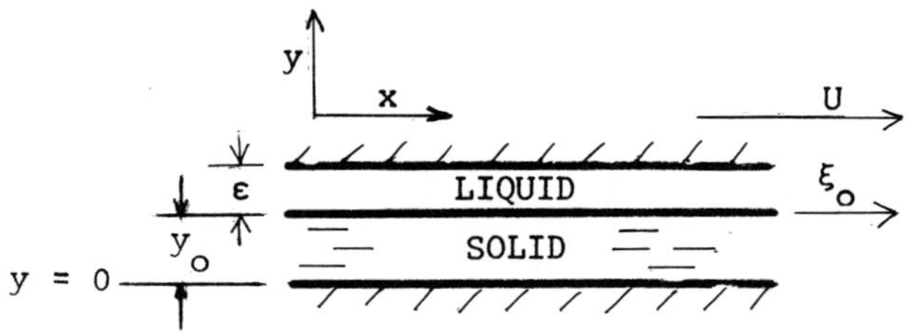

Fig. 6.10 Elastico-viscous model; see Section 6.6.2 .

coefficient η ; the other, of thickness y_o , is an elastic solid with rigidity modulus μ . Suppose that at time $t = 0$ a velocity U is imparted to the upper plate in the x direction, and that this velocity is maintained indefinitely thereafter. If the thickness ε is small a flow field $u(y)$ will be established rather quickly in the liquid layer such that the velocity gradient $\partial u/\partial y$ or $G(t)$ is uniform, although it varies (relatively slowly) with time. If the displacement of any point on the solid-liquid interface is $\xi_o(t)$ the velocity gradient in the liquid at any time t is

$$G(t) = \frac{U - \dot{\xi}_o}{\varepsilon} \quad , \tag{6.73}$$

where $\dot{\xi}_o$ is $\partial\xi_o/\partial t$. The stress exerted across any plane of constant y in the liquid layer is just ηG ; this same stress must be exerted on the solid across the interface at $y = y_o$. The entire solid will respond quickly to this stress; a displacement field $\xi(y)$ is then set up in the solid in which the gradient $\partial\xi/\partial y$ or $G_o(t)$ is uniform in space although it varies (again, relatively slowly) with time. Specifically, the displacement gradient in the solid becomes

$$G_o = \frac{\xi_o}{y_o} \quad . \tag{6.74}$$

At any time t the stress S_{12} across any plane of constant y in the solid must be equal to S_{12} in the liquid. Hence at all points in the space between the plates the stress is

$$S_{12} = \eta G = \mu G_o \quad . \tag{6.75}$$

We now proceed to obtain a differential equation in S_{12} . First we write the obvious equation

$$U = (U - \dot{\xi}_o) + \dot{\xi}_o \quad , \tag{6.76}$$

where the "dot" again represents the operation $\partial/\partial t$. We then notice from Eqs. (6.73) and (6.74) that

$$U - \dot{\xi}_o = \varepsilon G \ ; \quad \dot{\xi}_o = y_o \dot{G}_o \ . \tag{6.77}$$

Substituting expressions from Eqs. (6.77) into Eq. (6.76) and using the relationships between G, G_o and S_{12} from Eq. (6.75) we have

$$U = \frac{\varepsilon}{\eta} S_{12} + \frac{y_o}{\mu} \dot{S}_{12} \ . \tag{6.78}$$

The average velocity gradient G_M over the entire region between the plates is just U/h ; dividing through by h in Eq. (6.78) we obtain an equation which can be written

$$G_M = \frac{S_{12}}{\eta^M} + \frac{\dot{S}_{12}}{\mu^M} \tag{6.79a}$$

where

$$\eta^M = \frac{\eta h}{\varepsilon} \ ; \quad \mu^M = \frac{\mu h}{y_o} \ . \tag{6.79b}$$

The constant η^M has the units of a viscosity coefficient and μ^M has the units of a rigidity modulus. Equations (6.79 are consistent with a model proposed for viscoelastic media by the British physicist James Clerk Maxwell (1831-1879). A solution for S_{12} is

$$S_{12} = \eta^M G_M \left[1 - e^{-t/\tau}\right] \ , \ \tau = \eta^M/\mu^M \ . \tag{6.80}$$

This solution satisfies the condition that S_{12} is zero at t = 0 . As with the expression for G_K in Eq. (6.71a) we see that for a time after t = 0 the stress S_{12} increases at a constant rate. As time goes on S_{12} levels off to a constant value, namely, $\eta^M G_M$; in this limiting condition the situation is as if the medium were a Newtonian liquid with viscosity η^M . Note that $\eta^M G_M$ is equal to $\eta U/\varepsilon$ and hence is equal to the final value for S_{12} in the liquid layer. (For the velocity gradient in this layer is U/ε after ξ_o reaches a constant value). The time τ, is clled the <u>relaxation time</u>, is given by the same expressions as is the retardation time τ for the Kelvin-Voigt model except that η and μ are replaced by the larger values η^M and μ^M , respectively.

6.7 MODELS FOR RESPONSE TO TENSILE STRESS

6.7.1 Springs and Dashpots

A variety of mechanical models for viscoelastic media have been arrived at by different justapositions of elementary units, much as electrical circuits are constructed from units of resistance, capacitance and inductance. One such unit is a simple spring which is assumed to obey Hooke's law; when a tension F is applied to this unit it promptly extends an amount ξ such that

$$F = k\xi , \tag{6.81}$$

where k is the spring constant. As would be expected these units are meant to represent elastic elements in the medium.

A second important unit is the "dashpot". When a tension F is applied to this unit it extends with a velocity $\partial\xi/\partial t$ or $\dot{\xi}$ such that

$$F = b\dot{\xi} , \tag{6.82}$$

where b is a frictional constant. The dashpot represents viscous processes; it behaves as if the force F caused an object to be pulled with velocity $\dot{\xi}$ through a viscous liquid. Figure 6.11 shows the representations for the spring and dashpot; here they are shown "in series", an arrangement known as a "Maxwell body". In a series arrangement the same force F is applied across each unit, and the extension ξ_M of the entire body is the sum of the separate extensions of the spring and dashpot. From a solution of Eq. (6.82) we see that the extension of the dashpot is Ft/b where t is the time during which the tensile force has been applied; using this result together with Eq. (6.81) we obtain

$$\xi_M = \frac{F}{k} + \frac{Ft}{b} . \tag{6.83}$$

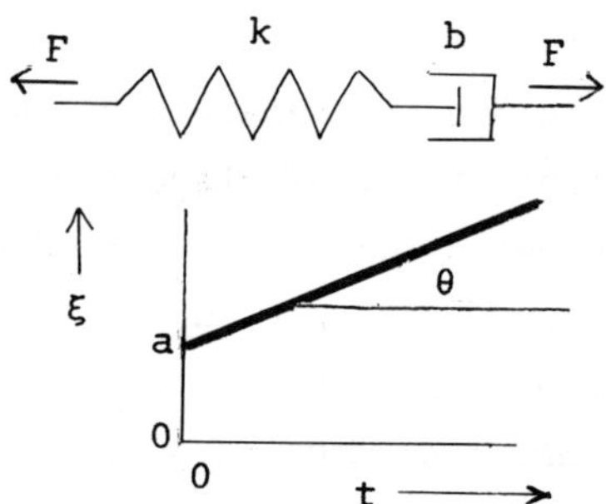

Fig. 6.11 Maxwell body; spring and dashpot in series. Plot based on Eq. (6.83).

$$a = F/k \;; \quad \theta = \tan^{-1}(F/b)$$

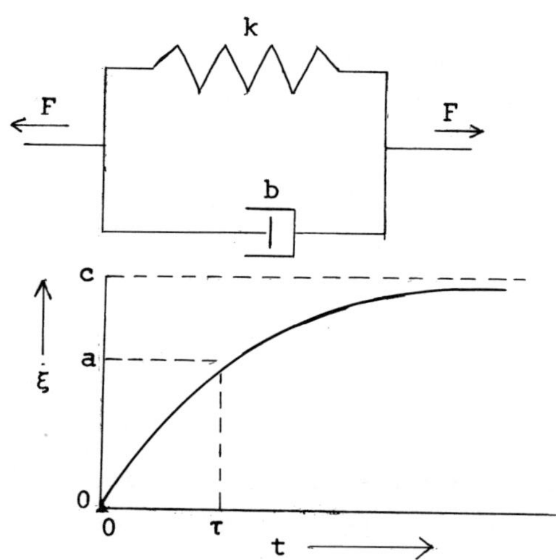

Fig. 6.12 Kelvin-Voigt body: spring and dashpot in parallel. Plot based on Eq. (6.85).

$$a = 0.63\ F/k; \quad c = F/k \;; \quad \tau = b/k \;.$$

A plot of ξ_M against time is shown in Fig. 6.11 . It is implicit in Eq. (6.83) that the spring extension F/k occurs instantly when the force F is applied. For t > 0 the extension ξ_M increases linearly with time because of the action of the dashpot.

In Fig. 6.12 is shown a "parallel" arrangement of spring and dashpot, which we shall call a "Kelvin-Voigt body". Here it is assumed that the spring and dashpot have the same extension $\xi(t)$ and that the total tensile force F across the unit is the sum of forces across the separate units. We have then that

$$F = k\xi + b\dot{\xi} \ , \tag{6.84}$$

of which a solution is

$$\xi = \xi_K = \frac{F}{k}\ [1 - e^{-t/\tau}], \tag{6.85}$$

where $\tau = b/k$. A plot appears in Fig. 6.12 (which may be compared with the similar plot in Fig. 6.3). We see that ξ is zero at t = 0, then increases steadily at first, but levels off to the value F/k (as if only the spring were present) at large values of the time. The constant τ, the retardation time, is the time required for the displacement to reach 0.63 F/k .

A useful model for some purposes consists of a series connection of a Maxwell body and a Kelvin-Voigt body. This four-element model is sometimes called a "Burgers system" after the Dutch physicist J. M. Burgers (1895-). See Fig. 6.13. By superposing ξ_M from Eq. (6.83) and ξ_K from Eq. (6.85) we obtain as the total extension of this combination, when subjected to an applied tensile force F:

$$\xi = \frac{F}{k_1} + \frac{Ft}{b_1} + \frac{F}{k_2}\,[1 - e^{-t/\tau}] \ , \tag{6.86}$$

where $\tau = b_2/k_2$. A plot of the expression in Eq. (6.86) appears in Fig. 6.13. When the force F is applied ξ jumps immediately to the value F/k_1 expected for the k_1 spring. It then increases

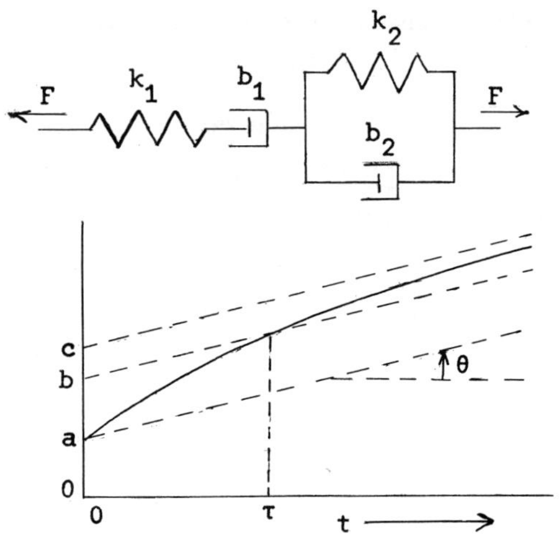

Fig. 6.13 Burgers system. Plot based on Eq. (6.86)

$a = F/k_1$; $b = a + 0.63\ F/k_2$

$c = a + F/k_2$; $\tau = b_2/k_2$

$\theta = \tan^{-1}(F/b_1)$

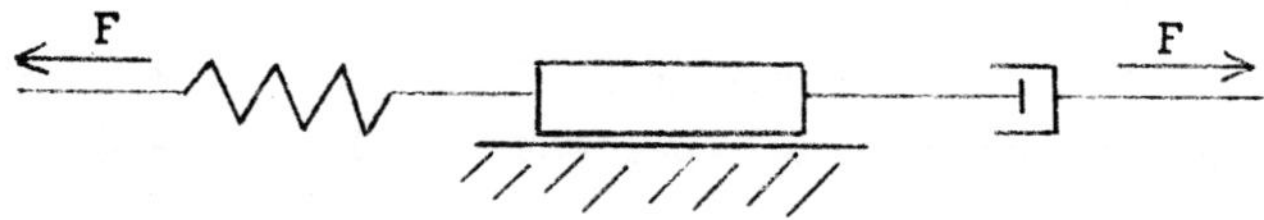

Fig. 6.14 Bingham system. See Section 6.7.2 .

steadily after $t = 0$, but with velocity decreasing until a limiting value F/b_1 is reached, as expected for the b_1 dashpot. Extrapolating the limiting straight line to $t = 0$ gives an intercept greater than F/k_1 by the amount F/k_2 , as expected for the k_2 spring. The b_2 dashpot plays its role in determining the relaxation time τ ; the latter is an indicator of the time required for the extension of the Kelvin-Voigt body to reach its limiting value.

Much use has been made of springs and dashpots to represent the viscoelastic behaviour of biomaterials; see Chapters 7 and 13.

6.7.2 Sliders

Another useful model includes an element, a "slider", which is immovable at small stresses but which moves freely, without friction or any other restraint, for stresses above a certain "yield" value. Figure 6.14 shows a model containing such a slider; this is often called a "Bingham system", after the rheologist E. C. Bingham.

PROBLEMS

6.1 Given Eq. (6.4), derive Archimedes' principle.

6.2 Suppose that in a Newtonian liquid the velocity u is along x and is given by Gy, where G is constant. Define a volume element (of convenient shape) in the liquid, consider the forces exerted by viscous stress on its faces, and show that the sum of these forces is zero.

6.3 Refer to Eq. 6.14, and consider it applied to a sheet being pulled along a channel containing water. If the sheet is pulled with velocity 10^3 cm/sec in a channel where h = 0.1 mm:

(i) Determine the volume rate of heating $\dot{Q}_v$.

(ii) Calculate the rate at which the temperature of the water would rise if there was no transport of heat from the region in which it is generated.

6.4 a) Calculate the sedimentation coefficient in water for (i) a spherical cell of radius 3 μm and (ii) a spherical or <u>globular</u> macromolecule of radius 30 A. Assume for each that its density is 1.05 gm/cm^3 . Give units both in seconds and in Svedbergs.

b) Calculate the g-factor required in a centrifuge if the cells are to be sedimented through 2 cm in 5 minutes.

c) Similarly calculate the g-factor required if the macromolecules are to be sedimented at the rate of 1 mm/hr.

d) Caclulate the characteristic time τ for both the cell and the macromolecule

6.5 (a) Using Eq. (6.16) and results of Section 6.3 show that for a spherical particle falling with limiting velocity in a gravitational field the Reynold's number N_R is given by

$$N_R = \frac{2}{9}\,\frac{R^3}{\eta^2}\,\rho g\,\Delta\ ,$$

where Δ is the magnitude of $(\rho_o - \rho)$

(b) If the particle substance has density 1.2 gm/cm^3 find the largest radius R for which the Stokes result, Eq. 6.15, holds in water; that is, determine R for which $N_R = 1$.

(c) Determine as in (b) the value of R for which $N_r = 1$, but for a particle sedimenting in a centrifuge where the g-factor is 300,000 .

6.6 The differential equation for sedimentation in a centrifuge, Eqs. (6.35), is identical to that for sedimentation in a gravitational field, Eqs. (6.25), except that A' replaces A. One might suppose, then, that the velocity v of migration in a centrifuge would be obtained from the expression Eq. (6.26) by simply replacing A by A'. The expression would then be

$$v = \frac{dx}{dt} = A'\tau[1 - \exp(-t/\tau)]\ ,$$

where the limiting velocity is

$$A'\tau = \frac{\omega^2 x\,(\rho_o - p)\,V}{f} = s\omega^2 x\ .$$

But x varies with time, and the above expression for dx/dt is not an exact solution of Eqs. (6.35).

a) Show, if you can, that the expression is nevertheless sufficiently accurate for most purposes.

b) When $t \gg \tau$ the velocity dx/dt in this expression reaches its limiting value $s\omega^2 x$. Show that the logarithm of x then increases linearly with the time. (This result is used in analyzing data on sedimentation in a centrifuge).

6.7 According to Eq. (6.14) the heat generation quantity $\dot{Q}_v$ in a fluid of viscosity η_o flowing with velocity gradient G is $\eta_o G^2$.
(a) Show that the increase $\Delta\dot{Q}_v$ in this quantity when particles are suspended in the fluid is

$$\Delta\dot{Q}_v = \eta_o \kappa\phi G^2 ,$$

where ϕ is the volume fraction occupied by the particles and κ is the constant defined in Eq. (6.38).
(b) Arrive at an expression for $\Delta\dot{Q}_v$ applicable to n_s spheres of radius R per unit volume.
(c) If the connecting rods are removed from the n dumbbells per unit volume considered in Eq. (6.49), the result will be a suspension of n_s spheres per unit volume, where $n_s = 2n$. Let $\Delta\dot{Q}_{vs}$ be the result obtained for $\Delta\dot{Q}_v$ in (b) when $n_s = 2n$. Also let $\dot{Q}_v$ be the rate of heat generation for a suspension of oriented dumbbells as given in Eq. (6.49). Then show that the condition $\Delta\dot{Q}_{vs} << \dot{Q}_v$ is equivalent to the condition in Eq. (6.48) that sin θ is much greater than R/b .

6.8 In arriving at Eqs. (6.50) and (6.55) it was assumed that the dumbbells always lie in the xy plane. Suppose this restriction is removed. Then to make Eq. (6.49) apply to this more general situation it is only necessary to replace $\sin^2\theta$ by $\cos^2\phi$, where ϕ is the angle between the dumbbell axis and the y direction. Show that then (since $\cos^2\phi$ averaged over a hemisphere is 1/3) the expressions for $< \dot{Q}_v >$, κ and $[\eta]$ in Eqs. (6.50) and (6.55) should all be multiplied by 2/3 .

6.9 Following arguments similar to those which lead to Eq. (6.54) obtain the expression for κ applicable to dumbbells which all have the same orientation θ :

$$\kappa = \frac{9b^2 \sin^2\theta}{2R^2} .$$

6.10 Justify the "square root" law of Eq. (6.62), given that R_e is proportional to $\sqrt{N}$. As intermediate steps in a proof you can show that while the molecular mass m and (actual) molecular volume V_a are each proportional to N, the "effective" molecular volume V_e is proportional to $N^{1.5}$.

6.11 Complete the arguments leading to Debye's results for free draining coils in Eqs. (6.63), (6.65), and (6.66) .

6.12 Comparing Eqs. (6.71) and (6.85) we note the similarity of these solutions for the visco-elastic model of Fig. 6.9 and the Kelvin-Voigt model of Fig. 6.12 . In this problem you are to find an expression applicable to the Maxwell body, Fig. 6.11, which is similar to that given in Eq. (6.80) for the elastico-viscous model, Fig. 6.10 . This can be done by noticing that the total rate of extension $\dot{\xi}$ of the model is the sum of extension rates for the spring and dashpot.

(a) Show that

$$\dot{\xi} = \frac{\dot{F}}{k} + \frac{F}{b}$$

where F, b and k have the meanings indicated in Fig. 6.11.

(b) From the analogy between this equation and Eq. (6.79) find an expression for F analogous to that for S_{12} in Eq. (6.80). Assume $\dot{\xi}$ is constant.

(c) Show that if $\dot{\xi}$ is zero the solution for F is proportional to $\exp(-t/\tau)$, and interpret this result.

CHAPTER 7

MECHANICAL CHARACTERISTICS OF BIOLOGICAL MEDIA

7.1 INTRODUCTION

We have now reached a position where, with background provided by previous chapters, we can explore a number of interesting areas in biological mechanics. In this chapter we consider mechanical properties of body fluids and basic structural elements; where possible we seek out the physical basis for their properties and also their significance. The word "explore" was chosen advisedly to suggest that we shall be investigating territory which is as yet only partly charted. Biophysical mechanics is an area of active research in which new facts and ideas are being discovered even as this book is being written. Some data and interpretations will probably be outdated soon, and others as years go by. Much of the material in this chapter should therefore be thought of as introductory and by no means as the "final word".

A rather lengthy list of references to books and journals has been provided for this chapter. These give original sources for data and theories and also are recommended for readers who wish to pursue topics in greater depth. In addition, recent issues of journals and annuals such as those cited should be consulted for contributions and reviews which have appeared since the present writing. In developing this textbook one of the author's goals has been to prepare the student to read original journal publications and other technical literature related to the areas emphasized. The articles cited for this chapter are examples of contributions relative to biophysical mechanics. However in reading this and other material the student will be confronted with a problem in scientific communication which has become especially aggravated in modern times: many of our journals and books are intelligible only to limited groups of scientists since they are written for readers with advanced background in specialized topics. Most of the publications cited contain material which is readable by students who have mastered previous subject matter in this book; but they also contain varying amounts of material which requires more background. Some of the latter will be more accessible to the student after taking up later chapters of this book. Other material is beyond the scope of this course. In the important and interesting activity of reading "the literature" it is clear that one must be selective.

In characterizing substances mechanically it is especially important to know the applicable constitutive relations, that is, the relationships between S_{ij} and ε_{ij}, the components of stress and strain (or quantities obtained by operating on these, such as time derivatives of S_{ij} and ε_{ij}). In this chapter we examine information available on constitutive relations and, when these are expressible as linear equations, take note of numerical values for the constants. Experimental values for elastic constants come partly from rather direct determinations of stress and strain, but mostly from measurements of the velocity of elastic waves. The theory for propagation of such waves is taken up in a later chapter, where relations between elastic constants and wave velocities are derived.

For some biomaterials theory is available whereby numerical values for elastic constants are calculated from basic principles. An example is keratin for which good information exists on its chemical composition and macromolecular structure. For keratin functions relating stress to strain have been derived, and absolute values for the Young's Modulus calculated. It is interesting to compare such predicted functions and constants with experimental results on biomaterials, such as wool fibers, which contain keratin. In so doing valuable information is gained about internal structure of the biomaterial.

For fluids which are Newtonian or approximately so experimental determinations of the shear viscosity coefficient η are usually made by use of viscometers which operate with steady flow. Examples are capillary viscometers and Couette viscometers for which relevant theory is given in Chapter 5. For pure liquids insight on the origin and physical significance of shear viscosity comes from reaction-rate theory, although it is probably not yet possible to calculate absolute values of η with accuracy. For suspensions experimental results can be compared with theory such as that discussed in Chapter 6.

In selecting topics for this chapter emphasis is placed on those for which the background provided in previous chapters is sufficient, or nearly so. We shall reserve for later chapters certain material on elastic constants, nonNewtonian behavior of fluids and viscoelasticity, for whose analysis principles of thermal physics are required.

Our first topic is simply water, since this unique liquid is so important to biology. In spite of its familiarity, we shall see that water is not a simple substance and, in fact, is still an object of active research. Powerful experimental and theoretical techniques of physics and chemistry have been directed toward elucidation of water structure.

We then take up modifications of water properties which occur when solutions or suspensions are formed with water. While changes in compressibility are usually small in dilute solutions or suspensions, large increases in viscosity may occur in suspensions of very elongated macromolecules, even at low concentrations.

As examples of real biological fluids we consider blood, about which considerable information exists, and synovial fluids. The latter substance is important to biological lubrication and has interesting mechanical properties.

As an example of elastic biomaterials we give special attention to keratin and bone. We have already referred to keratin, a protein which is an important constituent of tough materials such as hair and nails. Its macromolecule has been well characterized by an array of physical and chemical studies, and has also been the object of basic theoretical work. Also many studies have been made on mechanical properties of bone, as would be expected. Its structure is an interesting one, and poses challenging questions.

Membranes have great significance in biology, for several reasons: They form envelopes with unique characteristics for separating intracellular environment from extracellular space. In folded configurations they also provide large areas of surface, thus facilitating vital biochemical processes. In addition they form tough envelopes and walls for body organs and vessels. Of special scientific interest are the basic units which are sometimes called "elementary" membranes, and also synthetic membranes which are used as models. In concluding this chapter we consider some aspects of the mechanical properties of biomembranes, and then treat briefly the properties of a complex tissue.

7.2 WATER

It is common knowledge that life as we know it depends on the existence of water. Speculation about life at a given location in outer space usually focuses on the question of whether water exists at that location. There is no known living organism, plant or animal, primitive or advanced, that does not contain water. In the adult human body the ratio of total water mass to the total mass of fat-free tissue remains surprisingly constant at about 73%. We consider here some physical properties of water, some of which mark it as a most unusual substance. Presumably all of these properties can ultimately be accounted for in terms of known physical principles, but this is far from true at present. Some aspects of the subject are well developed, and interesting theories exist for others. Future research will undoubtedly bring increased understanding of this important substance.

All of the properties of water must arise ultimately from the nature of its most common molecule H_2O, where O is the isotope of oxygen with molecular weight 16. (Naturally occurring water contains about 0.2% of other isotopes.) The shape and size of the H_2O molecule are known with high accuracy from spectroscopic data. Its nuclei form an isosceles triangle with OH distances 0.957 A and with HOH angle equal to 104.5^o, somewhat greater than a right angle. This triangle of positively charged nuclei is surrounded by a cloud of negative electronic charge. From wave mechanics it has been possible to define the cloud by calculating the average charge density at various points. Much of the cloud determined in this way follows the contours of the nuclear triangle HOH. A more surprising feature is a pair of lobes whose principal directions OA and OA' extend from the O nucleus in a plane which is perpendicular to the HOH plane and bisects the angle HOH. See Fig. 7.1; here the HOH triangle is labelled HOH' . As suggested in the figure, OA' is the mirror image of OA in the HOH' plane. It is significant that the directions OH, OH', OA and OA' are such that any three correspond roughly to the edges of a tetrahedron with corner at O (between which edges the angle would be 109.5^o). Much importance is attached to the lobes OA and OA'; together these are often referred to as the lone hybrid pair or lone electron pair. These lobes carry a net negative charge while positive charge predominates near H and H' .

There is special significance in the fact that an H_2O molecule often forms attachments to other molecules along the directions OA and OA' . Typically such an attachment occurs by means of an H atom in the second molecule. Suppose, for example, that the second molecule is also H_2O . Suppose further that the O nucleus and one of the H nuclei of molecule 2 are both along OA of molecule 1, oriented such that this H lies between the two O nuclei. The linear O-H-O configuration so formed results in an attractive force between the two molecules. This force is known as the hydrogen bond and is extremely important in biology. The energy required to break a hydrogen bond varies, but for water is about 5 kcal per mole of bonds; this corresponds to an energy per bond of 3.5×10^{-13} ergs or 0.22 electron volts. For comparison the energy in a quantum of red light is about 1.8 electron volts.

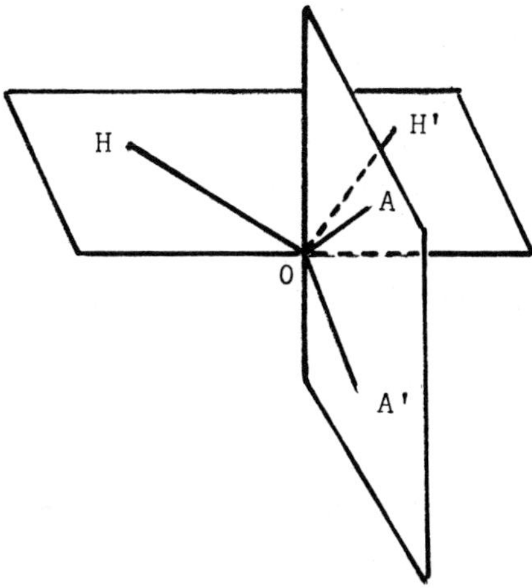

Fig. 7 .1 Geometrical relations for H_2O molecule; OA and OA' are principal directions for lone electron pair . The planes HOH' and AOA' are orthogonal; the first plane bisects the angle AOA' and the second, the angle HOH'. The molecule has net positive charge near the H nuclei and net negative charge in the directions of A and A'.

We turn now to aggregated forms of H_2O . There are a number of possible solid forms known as ice I, ice II, etc., each with its own characteristics. Ordinary ice is ice I whose crystal structure is classified as <u>hexagonal</u>. Here every O nucleus is at the center of a tetrahedron formed by four more O nuclei each 2.76 A away. This structure is remarkable in that every water molecule, say, W is hydrogen-bonded to its four nearest neighbors. We could see how this comes about if we were to examine models in three dimensions. We should then see that the molecules can be oriented so that the O-H directions from molecule W are (approximately) in line with lone pair lobes on two of its neighbors; this permits the forming of two O-H-O hydrogen bonds. Furthermore the lone-pair lobes of W can be directed toward O-H directions of other neighbors, forming two more O-H-O hydrogen bonds. Because of this effective bonding the molecular cohesive energy of ice I is high. However, the lattice structure is a rather open one; H_2O is unlike most substances in that its solid form is less dense than its liquid form. This latter fact is fortunate since it leads to the vitally important result that ordinary ice floats on water.

While the main structural features of ice I are well established the same is not true of liquid water. A variety of models have been hypothesized for the structure of water, in order to account for its properties. Most models are based on the idea that water has a structure related to ice I, but differing from it in ways which make the individual H_2O molecules more mobile. In subsequent discussion we shall be referring to ideas involved in these models.

We now take up some properties of water, emphasizing mechanical properties and others which seem especially relevant to this book. Electrical, magnetic, optical and other characteristics are described at length in special volumes on water (Dorsey, 1940; Eisenberg and

Kauzmann, 1969). Data are listed in Table 7.1 . Factors for converting units are given in Table A.2 . At 0°C the density is 0.917 gm/cm^3 . As remarked earlier it is not surprising that water should be more dense than ice I since this ice has a hexagonal lattice structure, a rather open one. It has been suggested that when melting occurs the lattice expands a little and breaks up or bends to some extent; many of the H_2O molecules then take on orientations and positions such that they fill open spaces in the lattice. After melting the space is then packed more fully with H_2O molecules and the substance is, of course, more dense than ice I. When melting has occurred and the temperature is raised further, spaces in the lattice are filled more and more but also the average distance between lattice molecules increases because of intermolecular vibration. The first of these effects causes the density ρ to increase and the other causes it to decrease. For temperatures below 4°C the first tendency dominates somewhat, while the second dominates above this temperature. Thus ρ increases slowly in the temperature range 0° to 4°C , is maximum at 4°C, then decreases with further increase of temperature. In Chap. 13 the velocity of sound c_s for a liquid is expressed in terms of its density ρ and its bulk modulus of elasticity B. A good method for determining B is to calculate it from this equation, since the velocity of sound and density have both been measured accurately. Solving for B we have

$$B = \rho c_s^2 \; . \tag{7.1}$$

From Section 4.3 it is clear that the modulus B can be defined as the magnitude of the ratio $\Delta p/\theta$, where Δp is the increment in pressure p associated with a small fractional change in volume, the dilatation θ . We see that B has the units of stress; it is proportional to the (negative) pressure required to produce a given amount of dilatation. Hence B measures the extent to which a medium resists compression or expansion. The <u>compressibility</u> β is defined as the reciprocal of B ;

TABLE 7.1

PROPERTIES OF WATER

T, temperature in ^{o}C ; ρ , density in g/ml ; c_s , velocity of sound in m/sec ; B_{ad} , adiabatic bulk modulus of elasticity, in dyn/cm^2 when tabulated values are multiplied by 10^{10}; β_{ad}, adiabatic compressibility in cm^2/dyn when tabulated value is multiplied by 10^{-11}; β_{is} , isothermal compressibility in cm^2/dyn when tabulated value is multiplied by 10^{-11} ; η , coefficient of shear viscosity in cP ; σ , surface tension in dyn/cm^2 .

T	ρ	c_s	B_{ad}	β_{ad}	β_{is}	η	σ
0	0.9999	1402.7	1.967	5.083	5.02	1.793	75.6
5	1.0000	1426.5	2.035	4.915		1.518	74.9
10	0.9970	1447.6	2.089	4.787	4.78	1.309	74.2
15	0.9991	1466.3	2.148	4.656		1.144	73.5
20	0.9982	1482.7	2.194	4.557	4.58	1.008	72.8
25	0.9970	1497.0	2.234	4.476	4.57	0.894	72.0
30	0.9956	1509.4	2.268	4.409	4.46	0.800	71.2
40	0.9922	1529.2	2.320	4.310	4.41	0.653	69.6
50	0.9880	1542.9	2.352	4.252	4.40	0.549	67.9
60	0.9832	1551.3	2.366	4.226	4.43	0.469	66.2
70	0.9778	1555.1	2.365	4.229	4.49	0.407	64.4
80	0.9718	1554.8	2.349	4.257	4.57	0.357	62.6
90	0.9653	1550.8	2.322	4.308	4.68	0.316	60.7
100	0.9584	1543.4	2.283	4.380	4.80	0.283	58.8

obviously it has the units of reciprocal stress and is a measure of the extent to which a medium will decrease in volume when subjected to pressure.

In defining B and β a consideration arises which we have ignored up to now. In a specific definition it is necessary to state the conditions under which a dilatation θ occurs. In a static method the liquid whose modulus is to be measured would be contained in a stiff-walled cylindrical vessel and subjected to pressure by means of a piston, while the fractional change in volume is measured. If this change occurs at a constant temperature the magnitude of the ratio $\Delta p/\theta$ yields the <u>isothermal bulk modulus</u> which we identify as B_{is} . The reciprocal of B_{is} is β_{is} , the <u>isothermal compressibility</u>. In a sound wave it is typically true that the change in volume of each fluid element occurs nearly adiabatically, that is, without any appreciable exchange of heat between the fluid element and the fluid which surrounds it. The temperature then changes somewhat during a change in volume and the ratio $\Delta p/\theta$, the <u>adiabatic bulk modulus</u> B_{ad} , is different from the isothermal modulus. Correspondingly the <u>adiabatic compressibility</u> is β_{ad} given by the reciprocal of B_{ad} . For gases β_{is} may exceed β_{ad} by a ratio as large as 1.67, but in liquids the difference is much smaller.

In Table 7.1 values of c_s , B_{ad} , β_{ad} and β_{is} are given for distilled water as a function of temperature. Values of the isothermal compressibility β_{is} are less accurate than those for the adiabatic compressibility β_{ad} ; at most temperatures β_{is} exceeds β_{ad} by a few percent. Both compressibilities exhibit minima with respect to temperature in the range above $30^{\circ}C$; explanation of these minima requires discussion somewhat too extended for our purposes here. The relatively high compressibility near $0^{\circ}C$, is attributed to the fact that at near-freezing temperatures much of the water retains the relatively open hexagonal lattice structure of ice I . This lattice is more compressible than the "close-packed" structure assumed by the other, more disoriented H_2O molecules; the latter increase in number as the temperature rises.

A glance at Table 7 .1 shows that the viscosity of water is far more dependent on temperature than is the density or compressibility. Thus in a change of temperature from 0° to $100^{\circ}C$ the density ρ decreases by 4% , the compressibility β_{ad} decreases by about 15% and the shear viscosity coefficient η decreases by more than a factor of 6 . This reflects a profound difference in the nature of viscosity from the static properties of density and compressibility. In the constant-gradient flow field treated in Section 5.4 the shear viscosity coefficient η gives the stress required to produce unit velocity gradient. We see that the units of η are those for the product of stress and time. In general viscosity is a measure of the stress required to produce a given rate-of-strain, while an elastic modulus measures the stress required to produce unit static stress.

Understanding the concept of viscosity in a liquid requires consideration of the process involved when the liquid undergoes shearing flow of the type discussed in Section 5.4 . According to theories advanced by Glasstone, Laidler and Eyring (1941, Ref. 7.7), Frenkel (1955, Ref. 7.8) and others, shearing flow of a liquid occurs as molecules execute a series of jumps between sites at which they are momentarily at equilibrium. The rate at which the jumps occur depends critically on the height E_v of an "energy barrier" which the molecule must leap over in passing from one site to the next. When E_v , also called the activation energy, is independent of temperature (as is true for "normal" liquids) the computed viscosity coefficient η has the form

$$\eta = B_v \, e^{E_v/R_g T} \; ; \tag{7.2}$$

here R_g is the gas constant (see Appendix D), T is the Kelvin temperature, and B_v is a quantity whose temperature variation can often be neglected. Equation (7.2) is found to apply to many ordinary liquids; for these a plot of ln η vs 1/T yields a straight line from whose slope E_v can be calculated. However, as also appears in other connections, water is not an ordinary liquid. Because of the hydrogen bonding between H_2O molecules water is classed as an associated liquid. For such liquids the viscosity is high relative to nonassociated liquids formed from molecules of comparable size; also the activation energy E_v is strongly temperature dependent. A plot of ln η vs 1/T based on the data in Table 7.1 shows that for water E_v has values 5.5, 4.2, and 2.8 at the Centigrade temperatures 0°, 30° and 100° , respectively, where the units of E_v are kcal mol^{-1} (Prob. 7.1). The energy barrier E_v is then roughly comparable to the energy of a hydrogen bond, suggesting that a molecule must overcome restraint offered by such bonds as it jumps from one site to another. As the temperature rises the lattice apparently becomes more and more disorganized and the hydrogen bonding less effective.

Also listed in Table 7.1 is the surface tension σ for water relative to an air interface (which is little different from that for water relative to vacuum). This quantity has the units of force/length as if it applied to the tension in a stretched membrane. The concept of a liquid-gas interface behaving as if a stretched membrane existed at the interface is useful, for example, in considering the tendency of a surface to return to its original shape after a disturbance. However the analogy is not always a reliable one. It is perhaps more basic to consider the fact that σ also has the units of energy/area and think of σ as representing surface energy. In fact σ represents the work required to form unit area of new surface. To form new surface, molecules must be brought from the environment which exists in the interior of a liquid to the altered environment which exists at the surface.

For water σ is unusually large. This can be attributed mostly to the effective hydrogen-bonding which exists in the interior, and the fact that a molecule at the surface is hydrogen-bonded to fewer neighbors. As the temperature rises σ decreases significantly, probably because the hydrogen bonding is reduced at higher temperatures.

7.3 DILUTE SOLUTIONS AND SUSPENSIONS IN WATER

Body fluids are mostly water in which are varying percentages of ions, minerals and other suspended matter.

7.3.1 Ions in Solution

When an electrolyte such as NaCl is added to water the density of the solution increases and the compressibility decreases. This is explained by the tendency of H_2O molecules to form a hydration sheath (one or more molecules thick) of relatively high density and low compressibility around each ion. Data on density ρ , sound velocity c_s and adiabatic compressibility β_{ad} for NaCl solutions are given in Table 7.2 as a function of molarity F, the latter being defined as concentration in moles per liter. The compressibilities are computed from measurements of density and sound velocity (Weissler and Del Grosso, 1951, Ref. 7.9). When it is desired to suspend red cells or other animal cells in a relatively simple liquid, a "physiological saline" solution consisting of 0.15 molar NaCl is often chosen to meet osmotic requirements (see Sections 2.7 and 11.2). From Table 7.2 we see that for this solution the density is about 0.6% greater and the compressibility about 2% less than for distilled water.

Also in Table 7.2 are data on the shear viscosity coefficient η in NaCl solutions (relative to that of water). It can be seen (Prob. 7.3) that η increases essentially linearly with concentration, so that Eq. (6.38) applies, as if the Na^+ and Cl^- ions were macroscopic particles.

TABLE 7.2

PROPERTIES OF Na Cl SOLUTIONS

F, molarity; ρ, density in g/ml ; c_s , sound velocity in m/sec ; β_{ad}, adiabatic compressibility, in cm^2/dyn when number in table is multiplied by 10^{-12} ; η , ratio of shear viscosity coefficient to that for water at the same temperature. Viscosity data for 18°C, other data for 30°C .

F	ρ	c_s	β_{ad}	η
0.0	0.9956	1510.0	44.05	
0.1	0.9996	1516.4	43.50	1.008
0.2	1.0039	1522.8	42.96	
0.25				1.020
0.4	1.0117	1534.4	41.98	
0.5				1.040
1.0				1.080

7.3.2 Molecules in Solution

Sucrose, with chemical formula $C_{12}H_{22}O_{11}$, and molecular weight 342.17 is an example of a molecule of modest size which is important biologically; it is widespread in plants and is an important nutrient for man. Solutions of sucrose are important in centrifugation techniques. When sucrose molecules are added to water the density of the solution increases and, especially at the higher concentrations, the coefficient of shear viscosity increases greatly. Table 7.3 gives data on the density and viscosity for sucrose solutions at concentrations up to 60%.

TABLE 7.3

PROPERTIES OF SUCROSE SOLUTIONS

c, concentration in % by weight; ρ , density at 20°C in g/ml ; η^{10} , η^{20} , η^{30} , coefficient of shear viscosity in cP at 10°C , 20°C etc. From Handbook of Biochemistry , 2nd edition, The Chemical Rubber Co. (1970). Density of sucrose is 1.588 g/ml .

c	ρ	η^{10}	η^{20}	η^{30}
0	0.999	1.0308	1.004	0.798
1	1.003	1.343	1.030	0.818
2	1.006	1.380	1.057	0.834
3	1.010	1.420	1.086	0.860
4	1.014	1.462	1.116	0.883
5	1.018	1.506	1.148	0.907
10	1.038	1.771	1.337	1.048
20	1.081	2.642	1.946	1.493
40	1.177	9.167	6.163	4.372
60	1.287	110.8	58.5	33.84

In the lower ranges of concentration the viscosity follows approximately the linear relation of Eq. (6.38); see Prob. 7 .4 . At higher concentrations the viscosity rises rapidly and does not at all follow the linear relation. This is not surprising in that Eq. (6.38) applies only to non-interacting particles, that is, particles so far apart that each one moves as if it were alone. When the concentration is 20% or more every sucrose molecule is rather near its neighbors and, as it moves, must make its way through a crowd. Also η is very strongly dependent on temperature, especially at the higher concentrations; such temperature dependence is expected in a general way on the basis of concepts discussed in connection with Eq. (7 .2).

7.3.3 Macromolecules in Solution

Much is learned about the large molecules of biology by studying them in dilute solutions with physical and physicochemical techniques. When the solution is dilute the macromolecules are far apart so that each behaves as if it were alone. Among the properties measured we consider here primarily those that are essentially mechanical, including the density, compressibility, sedimentation coefficient and viscosity. From these and other measured properties calculations can be made or inferences drawn on mass, volume, shape, dimensions and internal structures of the macromolecules.

In studying dilute suspensions the scientist is proceeding in a typical way to obtain basic information. He simplifies and idealizes his situation in order to obtain results which are reproducible and amenable to interpretation. It is true that his situations are then not "real" ones. Thus in actual biological tissues macromolecules are usually in highly concentrated media and often in aggregated forms. The macromolecule is then obviously not "alone" and indeed some of its properties differ from those measured in dilute solutions. Nevertheless the information gained from the basic studies proves to be extremely helpful in understanding the more complex problems of "real life". We consider here some representative information on several important biological macromolecules. Additional information on this subject is given in later sections.

First we consider <u>globular</u> molecules; these are roughly spherical and are fairly compact. A very important example is hemoglobin, often symbolized by Hb. It is this protein which gives blood its red color and which has the vital function of transporting oxygen. Hemoglobin is one of the first molecules to have its structure revealed in detail by Xray analysis. See Fig. 7.2. About 55 A in mean dimension, it has a molecular weight of 68,000 and density of 1.3 gm/cm^3 . Its sedimentation coefficient S in water is about 4.3 Svedbergs. For a solution of hemoglobin in water the parameter κ is reported to be 5.3 . The excess of 5.3 over the Einstein value (2.5) for rigid spheres (Eq. 6.56) is attributed partly to asymmetry of the Hb molecule and partly to water which is bound to the Hb molecule and effectively increases its radius. (Prob. 7.6) · Measurements on sound velocity for Hb solutions have been made by Carstensen and Schwan (1959). From their data we estimate (Prob. 7.7) that the adiabatic compressibility β_{ad} decreases by about 0.7% for each addition of 1 gm Hb to 100 ml of water.

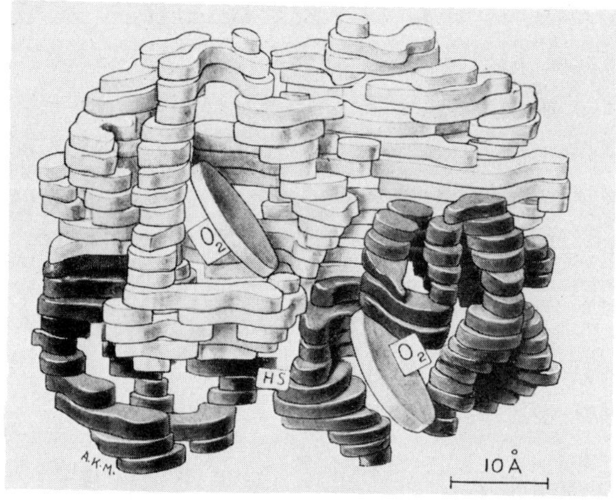

Fig. 7.2 Model of the oxygenated hemoglobin molecule arrived at from X-ray studies. It consists of four subunits, two of which are shown in white and two in black. Oxygen is carried by four hemes shown as discs (only two visible). From M. Perutz, New Scientist and Science Journal 50, 676-679 (1972). A print of this figure was kindly provided by Dr. Perutz.

An example of a fibrous molecule is collagen, a protein which forms the substance of bone, cartilage, tendons, ligaments, skin and other "connective tissues". This molecule is rod-like with length about 3000 A, much greater than its diameter, about 13 A. See Fig. 7.3. Its molecular weight is about 300,000 to 350,000, its density is 1.4 g/cm^3 and its sedimentation coefficient S is about 3.5 Svedbergs. For a solution of collagen in water $[\eta]$ is 1400 cm^3/gm. (Ananthanarayanan and Veis, 1972). Such a large value of $[\eta]$ is not surprising in view of the large length-width ratio for collagen. From Eq. (6.58) and Fig. 6.6 for ellipsoids and Eq. (6.55) for dumbbells it is clear that asymmetry leads to large value of κ and $[\eta]$, regardless of the total mass or volume of the molecule.

As a flexible macromolecule we consider dextran, a polysaccharide which is synthesized by microorganisms, and is used as a plasma substitute for human beings suffering from loss of blood. This material is available commercially in various ranges of molecular weight from 10^4 to 2×10^6. These flexible polymers tend to form random coils in suspension so that the analysis of Section 6.4.5 is pertinent. For solutions of dextran in water the intrinsic viscosity is given for fractions with varying molecular weight in Table 7.4. These data have been fitted by the equation

$$[\eta] = 0.243\ \overline{M}^{0.42} . \tag{7.3}$$

This is of the type of Eq. (6.69), where α has the value 0.42, somewhat lower than the value, 0.5, expected for impermeable random coils. The discrepancy between 0.42 and 0.5 may be partly because the fractions are not homogeneous, but include molecules with a range of molecular weights.

Changes in the basic structure of a macromolecule in solution can sometimes be followed very effectively by measurements on the viscosity of the solution. An example has been given by Doty (1959, Ref. 7.13) for a molecule of DNA, the polynucleotide that carries the genetic code.

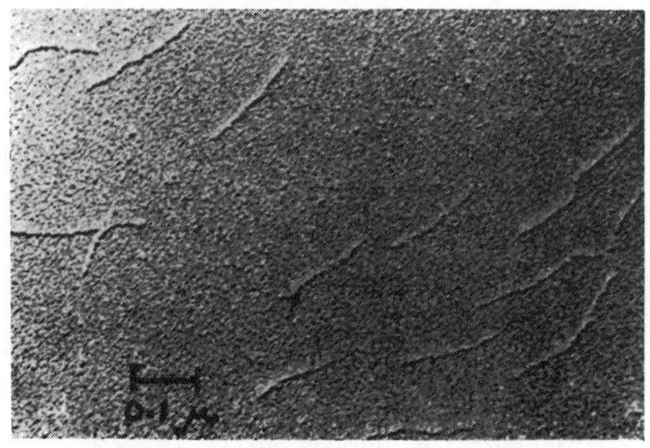

Fig. 7.3 Electron micrograph of collagen macromolecules. From C. E. Hall, cited by F. O. Schmitt, Ref. 7.36.

TABLE 7.4
VISCOSITY OF DEXTRAN SOLUTIONS

$\bar{M}$, weight-average molecular weight; $[\eta]$, intrinsic viscosity in ml/g at 20°C. Data from Pharmacia, Uppsala, Sweden.

$\bar{M}$	$[\eta]$
10,000	10
20,000	16
40,000	19
70,000	26
110,000	32
150,000	35
250,000	42
500,000	48
2,000,000	70

This molecule has been shown to consist of a double stranded helix; it is somewhat stiff, and is sometimes many microns in length. When the temperature is raised to about 80° or more the molecule "melts" and becomes much more flexible, evidently because of weakening in the hydrogen bonding between the two strands. This structural change is shown dramatically by a twenty-fold decrease in $[\eta]$; see Fig. 7.4 .

7.4 BLOOD

We now take up properties of real biological fluids. Some of the media which exist in living systems behave as Newtonian liquids under some circumstances; that is, they are governed by Eq. (4.24) with η constant. But very commonly biological fluids deviate from this behavior in one way or another and thus are non-Newtonian. The non-Newtonian characteristics result from large molecules, biological cells or other small structures suspended in the liquid.

Human blood is a most complex substance containing water, inorganic ions, dissolved gases, organic molecules of varying sizes, protein macromolecules and cells. If a tube is filled with blood and spun in a centrifuge the cells travel to the bottom part of the tube. The packed cell mass constitutes about 45 percent of the volume, a percentage which is called the hematocrit; an almost clear, faintly yellow fluid called the plasma makes up the remaining 55 percent. For whole blood the density ρ is about 1.055 gm/ml and for plasma ρ is about 1.027 gm/ml . Most populous among the cells are the erythrocytes or red cells, numbering roughly 5×10^9 per cubic centimeter. Some characteristics of these cells are described in Section 2.7 . The plasma contains 91-92 % water, about 7% proteins and 1-2% miscellaneous smaller molecules and ions.

Presence of the solids makes the blood a little less compressible than water, but the difference is small. From measurements on the velocity of sound in horse blood at 37°C, Urick (1947) found the adiabatic compressibility to be 40.9 for plasma and 34.1 for erythrocytes, compared to 43.4 for distilled water, all in units of 10^{-12} cm^2/d . For suspensions of cells in plasma with hematocrit ϕ a linear law of additivity is assumed, both for the density and the compressibility, so that

$$\rho = \phi\rho' + (1 - \phi)\rho'' \tag{7.4a}$$

and

$$\beta_{ad} = \phi\beta'_{ad} + (1 - \phi)\beta''_{ad} \ . \tag{7.4b}$$

Here the hematocrit ϕ gives the volume fraction occupied by cells and hence $(1 - \phi)$ gives that occupied by plasma; ρ and β_{ad} give the density and adiabatic compressibility for the suspension; ρ' and β'_{ad} give the corresponding values for the cells; ρ'' and β''_{ad} are for plasma. Equations (7.4) are as expected for a suspension of any phase, say A, in a liquid phase L if the density and compressibility of A and L in the suspension are the same as when A and L exist separately. (Prob. 7.8) . Support for use of Eqs. (7.4) has been given by Ahuja and Bugliarello (1971). Using Eq. (7.4b) and the Urick data cited above, the plot of Fig. 7.5 is obtained; as the hematocrit increases from 0 to 100% the compressibility decreases from the value for plasma to that for erythrocytes. At 45% hematocrit the compressibility of blood is 7% less than that for plasma and 13% less than that for water.

While blood is therefore only a little less compressible than water, its response is very different to stresses that cause, or tend to cause, flow. It is true rather generally for biological fluids that solids in suspension affect flow properties much more than they do elastic properties. For such fluids under, at least, some circumstances the Newtonian relationship of Eq. (4.24) between stress and rate-of-strain no longer holds. If the concept of viscosity is retained at all, it is with the understanding that the coefficient η is not constant, but varies with conditions. Reported values of the apparent viscosity η for blood vary from 2 to more than 100 times that of water, depending on the condition of the measurement. Nevertheless it has

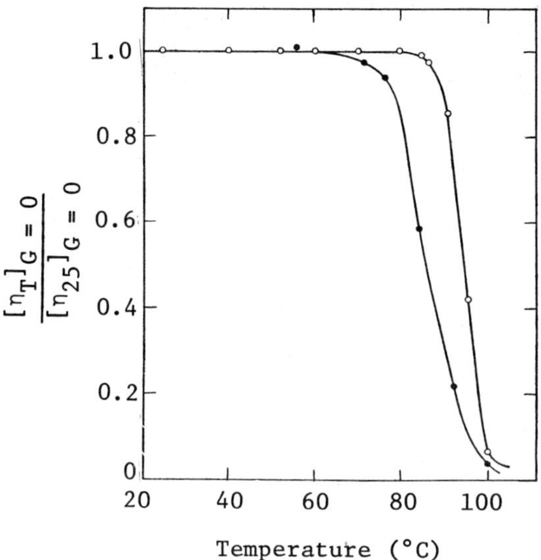

Fig. 7.4 Intrinsic viscosity of two samples of DNA at various temperatures relative to that at 25°C . The striking decrease in viscosity in the range 80° - 100°C is attributed to conversion of the molecule from a relatively stiff one to a flexible coil. From Doty, Ref. 7.13 .

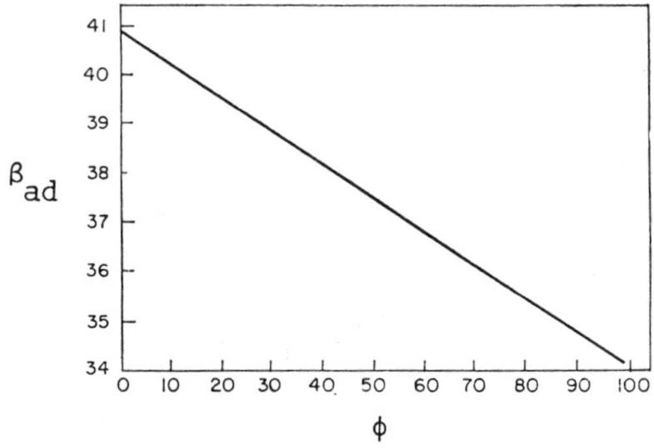

Fig. 7.5 Adiabatic compressibility β_{ad} versus hemacrit ϕ for horse blood. Units of β_{ad} are $10^{-12} cm^2/dyn$. From Ahuja and Bugliarello, Ref. 7.20 .

been shown that in the range of parameters which normally apply in most of the circulatory system of the human body the blood does in fact behave as a Newtonian fluid, characterized by Eq. (4.24). See Burton (1965, Ref. 7.16).

When the cells are removed by centrifugation, the remaining plasma behaves as a Newtonian liquid under ordinary conditions (when the rate of strain is not too low or too high), its coefficient η then being less than twice that of water. Some aspects of the behaviour of plasma are puzzling, however, Evidence has been given that the apparent viscosity of plasma depends on the nature of the inner surface of the tube through which the flow occurs. (Copley, 1960, Ref. 7.21).

The excess of the viscosity for plasma over that for water comes about primarily because of the plasma proteins. These include the (serum) albumin (3.2%), (serum) globulin (3.1%) and fibrinogen (0.45%). Of these the albumin has molecular weight 68,000 and is important in maintaining the osmotic pressure of the blood. (See Section 11.2). The globulins are so-called because of their approximate spherical shape. By electrophoretic means they separate into three fractions called, α-, β-, and γ- globulins, each of which can be fractionated further. See Section 6.3.3. Ranging in molecular weight from less than 100,000 to more than a million, some of the globulins are important as antibodies in protection against disease. For albumin the density is 1.36 g/ml, the sedimentation coefficient S is 4.6 Svedbergs and the intrinsic viscosity $[\eta]$ is 4.2 ml/g . For γ globulin the molecular weight is 160,000 , the density is 1.35 g/ml , S is 7 Svedbergs and $[\eta]$ is 6 ml/g . Measurements of velocity of sound in plasma show that the velocity increases with increasing protein concentration, at the rate of 4 m/sec for each additional gram of protein per 100 ml of solution (Carstensen and Schwan, 1953). This means (Cf Prob. 7 .7) that β_{ad} decreases by about .8% for each additional gram per 100 ml, that is, at a slightly faster rate than for hemoglobin.

The fibrinogen macromolecule is especially interesting. It has a threadlike shape, being about 700 A long and 38 A in diameter with molecular weight 400,000; its density is 1.38 gm/ml and S is 7.6 Svedbergs. The intrinsic viscosity [η] is 23 ; this relatively high value of [η] is reasonable in view of the large length/width ratio. (See Section 6.4.4.) Fibrinogen plays a vital function in blood clotting. When the clotting action is initiated the molecules aggregate to form fibers of <u>fibrin</u>; these tangle into a mesh which traps blood cells thus constituting a clot. Onlv 0.4 gm of fibrin are required to convert 100 ml of blood plasma into a clot. When fibrinogen is removed from the plasma the remaining fluid is called <u>serum</u> .

We return now to consideration of whole blood; this substance is not, in general, a Newtonian fluid. In fact Merrill and 5 coworkers (Ref. 7.23, (1963), hereafter referred to as MGCSBW , have concluded that blood is an example of a <u>plastic fluid</u> which under some conditions behaves like a solid. Confining samples of blood in the narrow space between cylindrical surfaces of radius R_1 and R_2 , as in Fig. 5.6 , they subjected the sample to shearing stress of varying magnitude and observed the response. Making the approximation that the spacing $R_2 - R_1$ is much less than either R_1 or R_2 the situation "seen" by a volume element of blood is similar to that discussed for a solid in Section 3.6 and for a liquid in Section 5.4 . Taking the "1" axis to be along the direction of motion near the volume element of interest and taking the "2" axis to be along the radial direction we see that the stress which arises is the diagonal component S_{12} . MGCSBW found that when the shearing stress S_{12} is very low, less than 1 d/cm^2 , blood behaves like a solid. Under these conditions application of the stress does not cause flow but only an elastic deformation, as if the rigidity modulus μ were nonzero and the stress-strain relation in Eq. (4.7) were applicable. Flow occurs when S_{12} exceeds a yield stress S_y , a behavior characteristic of the class of media called plastic fluids.

This yield-behavior is evidently to be attributed to the red cells in the blood. As part of their experiments MGCSBW separated the cells from the plasma by centrifugation, then added them to the plasma again in various concentrations. Their yield stress S_y was then found to increase with cell concentration according to the approximate empirical expression

$$S_y = 5 \times 10^{-7} (\phi - \phi_c)^3 \ . \qquad (7.5)$$

Here ϕ is the hematocrit, giving the cell concentration as the ratio of total cell volume to total volume of suspension; ϕ_c is a (small) critical value of ϕ which is constant for blood from a given donor but varies somewhat from donor to donor. Equation (7.5) applies only in the range of ϕ extending from ϕ_c to about 45%, the normal value for human blood. When ϕ is less than ϕ_c the plasma-cell combination flows at all values of the stress, that is, it no longer exhibits plastic-fluid behavior. As might be expected, the flow behavior approaches that of the plasma itself as ϕ approaches zero.

When the stress exceeds the yield value S_y the fluid takes part in Couette flow discussed in Section 5.4. However at the lower values of S_{12} the velocity gradient $\partial u/\partial y$ (which by Eq. (3.34) is just twice the rate-of-strain $\dot{\varepsilon}_{12}$) is not proportional to the stress. Instead for a plot of stress versus the velocity gradient the slope decreases with increasing stress. See Fig. 7.6. We shall call this slope the differential viscosity following Haynes (1962); it is equal to the increase in stress required to produce unit increase in velocity gradient. We see similar results for suspensions of macromolecules in Section 7.5 . Also analogous results are obtained in experiments on flow of red-cell suspensions with different hematocrit values through cylindrical tubes, where the flow is of the Poiseuille-type discussed in Section 5.6. See Fig. 7.7; note that in this figure the axes are reversed in that the quantity (pressure drop) analogous to stress is plotted horizontally and the flow (analogous to strain) is plotted vertically. There are several aspects of blood flow which lead to such nonlinear curves as appear in Figs. 7.6 and 7.7.

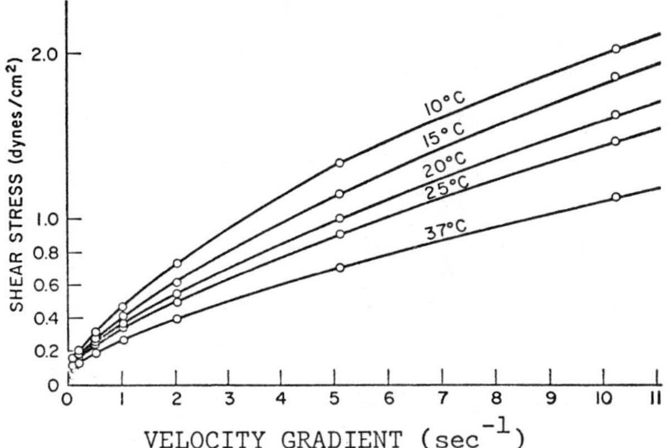

Fig. 7 .6 Measurements of stress S_{12} versus velocity gradient G for human blood in Couette flow for various temperatures. Less increment of S_{12} is required to produce unit increment of G as G increases. From Merrill, et al, Ref. 7 .23 .

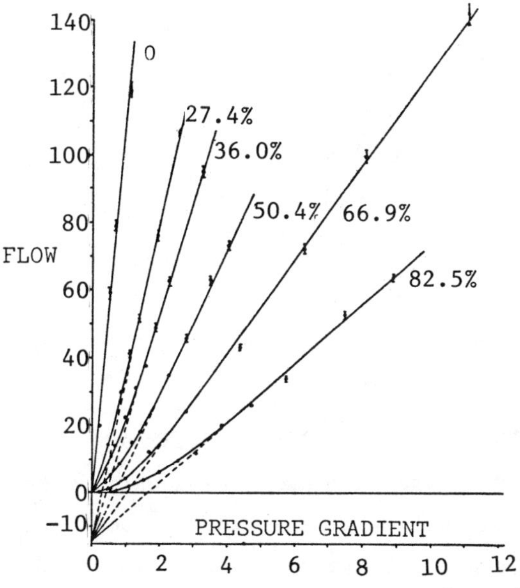

Fig. 7 .7 Measured flow of red-cell suspensions (in uits of 10^{-3} ml/sec) versus pressure gradient (in mm H_2O per cm) in glass tube of radius 185 μm . Hemacrit values indicated on curves. Less increment of pressure gradient is required to cause unit increase of flow as the flow increases, an effect analogous to that seen in Fig. 7 .6 . From Burton, Ref. 7 .16 .

Here we mention just one. In normal blood at rest or flowing slowly the cells are closely packed and the disc-like red cells interact with each other to form loose structures which, while weak, tend to inhibit the flow. The tendency to form such structures competes with a counter effect of the flow which tends to randomize the orientation. It is perhaps reasonable that as the ordering decreases, as is true with increasing velocity gradient, the differential viscosity should also decrease.

Quantitative theory is lacking for data such as that in Figs. 7.5 and 7.6 ; in the absence of such theory, curves have been fitted to the data on an empirical or semi-empirical basis. Thus Scott Blair (1969, Ref. 7.17) has shown that data on blood are fitted very well by the Casson equation, a "square-root" law; letting G be velocity gradient and S_c the stress this is

$$G^{\frac{1}{2}} = a_1 \left(S_c^{\frac{1}{2}} - b_1\right) , \qquad (7.6)$$

where a_1 and b_1 are positive constants, and also S_c and G are understood to be positive; the equation applies when $S_c^{\frac{1}{2}}$ is equal to or greater than b_1 . On the other hand Haynes (1962) has discussed another equation which fits well the data on flow in cylindrical tubes; an analog to this equation is

$$G = a_1 S_c - a_2\left(1 - e^{-kS_c}\right) , \qquad (7.7)$$

where a_1, a_2 and k are constants. A "plausibility argument" for Eq. (7.7) is taken up in Prob. 7.9 .

According to Burton (1965, Ref. 7.16) the conditions which apply in the normal human body are such that the blood behaves as a Newtonian fluid. The criterion for Newtonian flow of the blood through a tube is found to be that RG' should exceed 40 dyn/cm^2 when G' is the magnitude of the pressure gradient. It can be shown (Prob. 5.6) that this criterion corresponds to a statement that the tangential viscous stress at the tube wall should exceed 20 dyn/cm^2 . Under these conditions the velocity profile for blood passing through the tube follows approximately the parabolic law we saw earlier, for instance, in Eq. (5.61); this prediction was tested by Haynes (see Fig. 7.8).

In the Newtonian range the viscosity η of blood at any given temperature for a normal hematocrit of about 50% is three to four times that of water at the same temperature. The ratio of viscosity for a given blood sample to that for water has been found to be independent of temperature to within a few percent over the range 10 to 40° C (Merrill, et al., 1963). Hence the viscosity of blood at a given temperature in this range (under Newtonian conditions) can be obtained directly from the values for water, Table 7.1). The above situation, namely that blood is a Newtonian fluid under normal physiological conditions, greatly reduces difficulties in understanding blood flow in the circulatory system.

Nevertheless many interesting complications remain. One, pointed out by Copley (1960), and referred to earlier in connection with plasma, is that the flow rate for a given pressure gradient and tube radius is greater in a tube coated with fibrin than in a glass tube. Copley suggests that blood vessels in the body may very well have an inner lining which somehow facilitates the flow. Another complication is that tubes in the circulatory system are flexible; the radius R of a tube varies according to the pressure difference across its wall and also is probably to some extent controlled by contractile elements in the wall. Since the flow rate is proportional to R^4 for given driving conditions (see Eq. 5.63) adjustment of R gives sensitive control of the flow. Another is that the capillaries are very small, with dimensions comparable to the dimension of a red cell; hence cells pass through these single-file and the blood then hardly seems like a "homogeneous suspension". Still another complication is the <u>sigma effect</u>, whereby the apparent viscosity in a small tube is smaller than that in a large one. Related to a number of these phenomena occurring in flow down a tube is the tendency of particles in suspension under some circumstances to migrate toward the axis of the tube. Analysis of blood flow in the circulatory system is a many-faceted problem, for which we can take no more space here.

7.5 SYNOVIAL FLUIDS

A novel class of biological fluids are the substances which facilitate working of the joints in animal skeletons. Our knees and elbows, for example, are lubricated with remarkable effectiveness which it has so far not been possible to duplicate outside the body. Partly responsible for the effectiveness of this biological lubrication is apparently the bone itself which has unique characteristics of elasticity, sponginess and porosity. But also of critical importance is the synovial fluid which exists in the joints.

In a typical adult knee joint there is about 1 ml of this fluid; its physical and chemical properties vary considerably from one individual to another. The density ρ varies from 1.008 to 1.015 gm/ml. Using a viscometer in which the velocity gradient G is variable Caygill and West (1969) obtained data for a number of adults, some of which are shown in Fig. 7.9. For normal adults the viscosity is high (sometimes up to 60 cP) at low G-values, but drops considerably as G increases. In patients suffering from rheumatoid arthritis the viscosity is much less dependent on the velocity gradient G, and tends to be low. Although proteins (mostly albumin and globulin) are present to the extent of 2 to 3 grams per 100 ml of fluid, the unusual viscous characteristics come almost entirely from hyaluranic acid macromolecules, here called HA for convenience, present in an amount varying from 0.1 to 0.3 gm per 100 ml. Measurements show that the intrinsic viscosity $[\eta]$ of HA is over 1000 times as great as for the proteins. In addition to protein and HA macromolecules, synovial fluids contain inorganic ions, for example, Na^+, in concentrations comparable to those in other body fluids such as blood.

It has been shown also that the unusual lubricating properties of synovial fluids come primarily from the HA macromolecules. These long molecules with diameter about 6 A are polysaccharides (possibly bonded to smaller molecules of protein) formed by linking together many

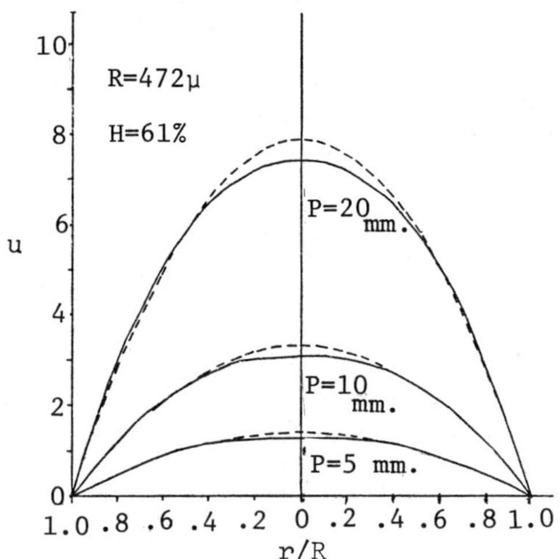

Fig. 7.8 Solid line: velocity u (cm/sec) of blood at various distances r from the axis of a tube with radius R = 472 μm . Dashed line: theoretical curve based on parabolic law such as Eq. 5.61 . Slight flattening of curve near r = 0 is attributed to tendency for erythrocytes to concentrate near axis. From Burton, Ref. 7.16 .

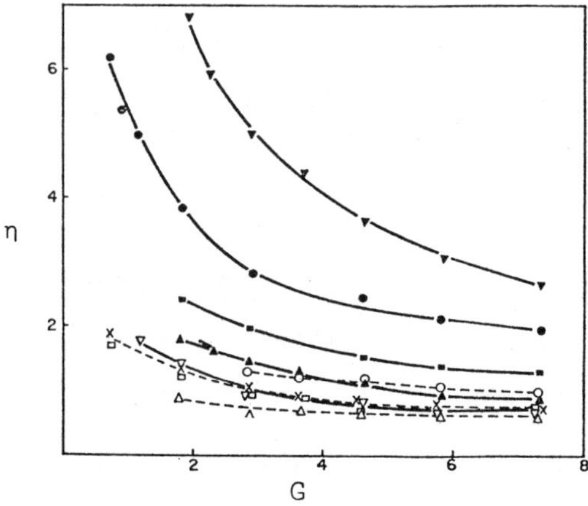

Fig. 7.9 Viscosity η (in units of 10 cP) versus velocity gradient G (in units of 10^2 sec^{-1}) for human synovial fluid. Solid lines for normal individuals and dashed lines for patients with rheumatoid arthritis. From Caygill and West, Ref. 7.25 .

disaccharide units to reach lengths as great as 25 μm. In synovial fluids from various sources the HA molecules are characterized by molecular weights up to more than 10^7 daltons; the density of HA has been given as 1.56 gm/ml.

A trio of investigators, Preston, Davies and Ogston (1965) made very extensive measurements on HA molecules separated from synovial fluid; their fluid was obtained from the knee and ankle joints of cattle or oxen. Suspending the HA in NaCl solutions they determined its properties by a variety of physical and chemical techniques. From light-scattering measurements they found the molecular weight to be about 14×10^6 daltons. (For the theory of light scattering techniques see, e.g., Morawetz, 1965. Ref. 7.27). Considering the macromolecule as a flexible coil they were able to determine the root-mean-square radius R_{rms}, that is, its radius of gyration; this quantity is often used as a measure of coil dimension. They found R_{rms} to be about 0.24 μm at pH 6.8, decreasing considerably when the hydrogen ion concentration $[H^+]$ is increased above this value. From the large magnitude of R_{rms} we can see that the HA molecule must be a coil that extends over a large volume which includes much water. (Prob. 7.10). The fact that R_{rms} decreases, that is, the coil contracts, as $[H^+]$ increases is attributed to the fact that the HA molecules have fixed negative charges distributed along their length. When H^+ ions are present the repulsive forces between negative charges are to some extent reduced; the molecule then coils itself into a smaller "effective" volume V_e such that equilibrium is established between tendencies for expansion and counter-tendencies for contraction.

Using Couette viscometry Preston and his coworkers found that HA solutions are far from being Newtonian liquids. The intrinsic viscosity $[\eta]$ was found to be a decreasing function of velocity gradient G and also a decreasing function of NaCl concentration. At 0.2 M NaCl and zero velocity gradient (results for the latter condition obtained by extrapolation) $[\eta]$ proved to be about 14×10^3 ml/gm.

The very large observed value for $[\eta]$ gives further evidence that each HA molecular chain extends itself through a very large volume of water, more than a thousand times its own volume. We refer to theory for the viscosity of flexible macromolecules, Section 6.4.5. Probably the HA molecule conforms to neither of the extreme models called (1) an impermeable coil or (2) a free-draining coil. But if as an example we consider the first model we refer to Eq. (6.61). Letting κ (specifically its limiting value κ_o) be 10^4 we see that ϕ_e/ϕ_a is 2500. Since this ratio is also equal to V_e/V_a we have reached the conclusion that the effective volume V_e (which includes water "trapped" by the coil) is 2500 times greater than the actual molecular volume V_a.

In respect to the observation that $[\eta]$ is a decreasing function of velocity gradient G for HA, this is expected behaviour for asymmetric particles and long flexible coils. Orienting tendencies on asymmetrical particles and deforming tendencies on flexible coils are believed responsible for this G-effect. Theory exists for some aspects of the problem, but we defer discussion of it to Chap. 14.

The tendency for $[\eta]$ to decrease with increasing NaCl concentrations is another example of behaviour resulting from negative charge fixed along the HA coil. Here the Na^+ ions play a role similar to the H^+ ions considered earlier, tending to reduce electrostatic repulsion between different parts of the HA molecule. Presence of these ions thus allows the coil to assume a smaller volume, and a reduced viscosity results.

The sedimentation coefficient S for HA was found by Ogston and Stanier (1950) to be 13 Svedbergs in 0.2 M NaCl at low values of the HA concentration. This is to be compared with S values in the range 4-7 Svedbergs for albumin and globulin, the main protein components of the synovial fluid. However the sedimentation constant for HA is very dependent on the HA concentration and Preston, et al (1965) express uncertainties about the S value for this molecule.

In other studies by Laurent and others (1960) HA was obtained from the bovine eye and its properties determined by a variety of techniques. Here the HA macromolecules were found to be nonuniform in molecular weight M. By separating the HA into fractions, each with its own fairly uniform value of M, it was possible to study HA characteristics as a function of M. The S values varied from 2.5 to 7 Svedbergs, the molecular weight from 10^5 to 1.5×10^6 daltons and the intrinsic viscosity $[\eta]$ from 250 to 2450 ml/gm. They found that data on $[\eta]$ and M fit fairly well to the equation

$$[\eta] = 0.036\ M^{0.78} , \qquad (7.8)$$

an example of a modified Staudinger equation, Eq. (6.69) with $\alpha = 0.78$. It is worth noting that α lies between the value, 0.5 , for an impermeable coil and the value, 1.0, for a free-draining coil. The authors determined R_{rms} from light scattering; values varied from 0.05 to 0.14 μm and they found R_{rms} approximately proportional to $M^{0.6}$. For comparison, theory for a perfectly random coil has R_{rms} proportional to $M^{0.5}$.

Bland (1969) has shown that the viscosity of whole synovial fluid depends on the electrolyte which bathes it. Starting with normal concentrations of Na^+ and other ions in synovial fluid, all contained in a permeable dialysis bag, he allowed the fluid to come gradually to equilibrium with electrolyte-free water in which the bag was immersed. As this occurred the ion concentration slowly fell, approaching zero; at the same time water entered the bag, increasing the fluid volume. Taking viscosity measurements every hour during this procedure he found that η rose by a factor of two or more, both for human and bovine fluid. This result is attributed to fixed negative charges on the HA coils, whose repulsive forces are reduced by the Na^+ ions; this electrostatic interaction is of the kind we discussed earlier in connection with solutions of HA alone.

Returning to the topic of biological lubrication brought up at the beginning of this section, a number of theories have been advanced. These appeal in part to the peculiar non-Newtonian behaviour of HA, and in part to the spongy and porous nature of the bones which form the ball and socket of an animal joint. Basic questions remain and the problem of understanding why our joints are so remarkably effective is a challenging one. (See, e.g., McCutchen, 1965).

7.6 ELASTICITY FROM BOND STRETCHING

Elasticity of a medium refers to its tendency to return to an original configuration after a deformation. Fluids, while elastic with respect to volume changes, are commonly defined as media which do not "spring back" after shear deformation, the kind of deformation exemplified in Sect. 3.6 and Fig. 3.7 . Hence when the constitutive relation of Eq. (4.22) applies to a fluid it is usually with the understanding that the rigidity modulus μ is zero. There are many media for which the assumption of zero-μ is a useful one but there may be none for which it is precisely true. Substances which we usually call "biological fluids", such as blood and synovial fluid, exhibit nonzero rigidity modulus under some circumstances, and hence are not always "true"fluids.

We now turn our attention from fluids (true or not) to substances which are elastic in a more general sense, and first consider how elasticity originates. Consider a solid like rubber in which long flexible molecules are linked together in some way. Each such molecule may be fixed at certain points or junctions and still be relatively flexible along its length between junctions. An interesting model for the molecule is the random chain referred to in Section 6.4.5 . Thermal fluctuations cause the configuration of the chain to be constantly changing, but in such a way that the average distance between prescribed points A and B on the chain has a most probable value r_m . If the distance between A and B is altered so that it differs from r_m , thermal motion will tend to reduce the difference to zero; in other words, the chain tends to return to its original length after a deformation and is thus elastic. Elasticity which arises in this way is said to be of thermal or entropic origin. We shall defer more detailed discussion of this kind of elasticity to Chapter 12.

In a crystalline solid like ice elasticity arises in a different way. Here the atoms assume equilibrium positions relative to one another (subject to small fluctuations) such that energy is required to change these relative positions. Most of the energy can be accounted for in terms of chemical bonds between each atom or ion and its near neighbors. If ice is deformed, the distances between atomic nuclei are altered and the bonds are said to be stretched or compressed, as if they were tiny springs. External forces are required to produce such a deformation; the work done by these forces increases the potential energy of the solid. If the external forces are removed the solid will tend to minimize its potential energy as all interatomic "springs" return to their equilibrium lengths. In this situation, then, the elastic behavior of the solid arises from stretching or compressing interatomic bonds.

While these bonds can to some extent be understood in terms of classical electrostatic interactions between electrons and nuclei, the theory of wave mechanics is required for satisfactory treatment of interatomic forces. Characteristic of wave mechanics is the picture of atomic electrons, not as particles whose orbits can be computed with precision, but rather as continuous clouds whose density varies from point to point. Another characteristic is the prediction of energy levels for systems under consideration. The isolated hydrogen atom H, consisting only of a proton and electron, is a relatively simple system and has been treated thoroughly by wave mechanics. It is found that the H atoms may have any of the negative energy levels U_n given by

$$U_n = -\frac{|U_1|}{n^2}, \tag{7.9}$$

where n is any positive integer 1, 2, etc. The constant $|U_1|$ is the magnitude of the lowest energy level, i.e., the value of $|U_n|$ for n = 1; it has the value 13.56 electron volts. When n = 1 the H atom is in its ground state and the negative charge density at a distance r from the proton is given by ρ where

$$\rho = C\,e^{-x} \; ; \quad x = \frac{2r}{a_o} \; ; \qquad (7.10)$$

here a_o has the value 0.529 A while C is a constant such that the integral of ρ over all space (outside the proton) is equal to the electronic charge e. Various interpretations can be given to the fact, illustrated by Eq. (7 .10) that wave mechanics predicts continuous density of charge. One may think of the electron itself as a cloud spread over all space so that its charge is distributed continuously. Alternatively one may choose to think of the electron as an indefinitely small object with charge e moving rapidly about so that the probability of finding it in a given small region of volume dv is $\rho\, dv/e$.

When wave mechanics is applied to molecules the result is again a prediction of discrete energy levels and continuous charge distributions. However, the applicable formulae are more complex than those in Eqs. (7 .9) and (7 .10) since vibrational and rotational energy are involved. The problem of a hydrogen molecule H_2 , consisting of two protons and two electrons is considerably more difficult than that of the isolated hydrogen atom, but it has nevertheless been investigated extensively by wave mechanics. In order to understand the behavior of such a molecule when it is deformed the energy levels of the system of two protons and two electrons has been determined as a function of the distance r between the protons. As is true also for other diatomic molecules in which the atoms are alike the lowest energy level U_1 , i.e., the ground state energy level, is found to follow rather closely a function suggested by P. M. Morse. The Morse function is

$$U_1 = D\,e^{-2a(r-r_o)} - 2De^{-a(r-r_o)} , \qquad (7.11)$$

where D, r_o and <u>a</u> are constants which vary from molecule to molecule. For H_2 the values which have been determined are 4.454 electron volts for D , 0.75 A for r_o and 1.94 A^{-1} for <u>a</u> . A plot of Eq. (7 .11) is in Fig. 7 .10 . The function U_1 has a positive value at r = 0, is minimum with value -D at $r=r_o$ and approaches zero when r becomes large compared to both r_o and a^{-1} .

When the molecule is part of a larger body it will, in general, be subject to tension or compression exerted by adjoining atoms. Consider a force F applied to each proton, directed outward from the center of the molecule, that is, in a direction tending to increase r . Suppose that in increasing the distance from r' to r" the average force required is $\bar{F}$ and that the energy U_1 increases from U' to U" . Then U" - U' must be equal to the work done, that is, to the product of $\bar{F}$ and (r"-r') . Hence we can write

$$\bar{F} = \frac{U'' - U'}{r'' - r'} \quad . \tag{7.12}$$

Proceeding as we have done many times before in previous chapters we now consider the limiting form of Eq. (7.12) as r" approaches r' . The left hand side becomes the stretching force F required to maintain the separation r equal to r' , while the right hand side becomes $\partial U_1/\partial r$, evaluated at r' . Since r' is an arbitrary point we can write simply

$$F = \frac{\partial U}{\partial r} \, , \tag{7.13}$$

it being understood that both sides of this equation are to be evaluated at the same value of r . Thus the force F required to maintain a given separation r is given by the slope of the U vs r curve at that value of r . From Fig. 7.10 we see that when r is greater than r_o the slope dU/dr is positive and the force F is a stretching force or a tension. In the opposite situation, when r is less than r_o, the slope is negative and F is compressional. When r is just equal to r_o the force is zero and the molecule is at equilibrium. At a critical value of r, indicated on the graph as r_c , the slope is positive and maximum. If the molecule is stretched with a force equal to or greater than this maximum value it will stretch indefinitely and the molecule will separate into two hydrogen atoms. The energy required to bring about this separation is called the dissociation energy and is just the constant D .

Near the minimum it is convenient to represent the function $U_1(r)$ as a power series as in Eq. (B.7), Appendix B . Near $r = r_o$ we have

$$U_1(r) = U_o + \left(\frac{\partial U}{\partial r}\right)_o (r - r_o) + \tfrac{1}{2}\left(\frac{\partial^2 U}{\partial r^2}\right)_o (r - r_o)^2 , \qquad (7.14)$$

where U_o and $(\partial U/\partial r)_o$ are, respectively, the values of U_1 and $\partial U_1/\partial r$ at the minimum, that is, at $r = r_o$. Omitted on the right side of Eq. (7 .14) are terms proportional to $(r-r_o)^n$ where n is 3 or greater. For the sake of a systematic treatment we have included a term involving the constant $(\partial U/\partial r)_o$ as coefficient; however, this term is really zero, since $\partial U/\partial r$ is zero at the minimum point $r = r_o$. Near this minimum, when terms for n greater than 2 are negligible, the force F from Eqs. (7 .13) and (7 .14) is simply

$$F = \frac{\partial U}{\partial r} = C_o(r - r_o) , \qquad C_o = \left(\frac{\partial^2 U}{\partial r^2}\right)_o = 2a^2 D . \qquad (7.15)$$

Thus the force required to produce the extension $(r-r_o)$ is proportional to the extension; this linear relation is the molecular basis for Hooke's law. Using values already given for a and D we find that for H_2 the force constant C_o is 5.36×10^5 dyn/cm.

From information on interatomic forces it is possible in principle to compute elastic constants for media in which the restoring forces arise from bond stretching. Consider such a medium subject to tension along the "1" dimension and imagine a plane, PP in Fig. 7 .11 , passing through the medium with its normal along this direction. By definition the stress component S_{11} gives the force per unit area across this plane. This force is exerted between atoms across plane PP ; for simplicity suppose the force comes from bonds normal to PP connecting atoms as suggested in Fig. 7 .11 . If there are n_b bonds per unit area, each representing a force of magnitude f_b then S_{11} is just $n_b f_b$. Let r be the spacing between paired atoms as shown in Fig. 7 .11, and let r_o be the equilibrium value of r. Then assuming f_b has the same form as F in Eq. (7 .15) we can write

$$S_{11} = n_b\, C_o(r - r_o) . \qquad (7.16)$$

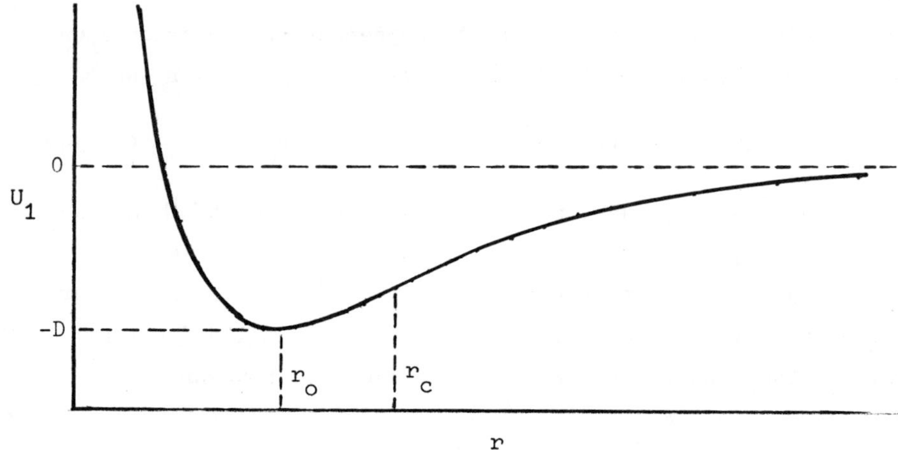

Fig. 7.10 Morse curve, based on Eq. 7.11, for H_2 molecule. U_1, ground state energy; r, distance between protons; r_o, r_c and D defined in text.

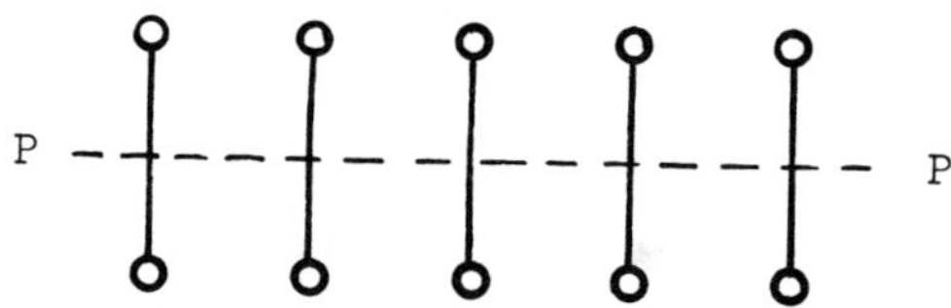

Fig. 7.11 Model of medium in which elasticity arises from bond stretching. Atoms are pulled upward by neighbors above and downward by neighbors below.

The fractional increase in spacing is $(r-r_o)/r_o$ and we know from Section 3.3 that this is equal to the strain ε_{11} . Hence we obtain a simple relation between stress and strain:

$$S_{11} = C_{11}\ \varepsilon_{11}\ , \qquad (7.17a)$$

where

$$C_{11} = n_b\ C_o r_o\ . \qquad (7.17b)$$

Equation (7 .14) is a form of Hooke's Law which applies to many circumstances; C_{11} has the dimensions of an elastic modulus. For purposes of an order-of-magnitude estimate we take the molecular force constant C_o to be 10^5 dyn/cm and the equilibrium distance r_o to be 1 A or 10^{-8} cm. Assuming one bond across the plane PP per atom pair, and about 10^{15} pairs per square centimeter (10 A^2 projected area per atom) we obtain for C_{11} the value 10^{12} dyn/cm^2. This is indeed the order-of-magnitude of commonly defined elastic constants for solids such as glass and steel. However, the model of Fig. 7 .11 is, of course, very naive and quite inadequate for most materials.

In subsequent sections we consider elastic moduli for specific biological media. Before proceeding we note here some facts about commonly used elastic constants, recalling material from Chapter 4. Suppose the medium is such that the constitutive relation of Eq. (4.22) is valid. The strain in a body in the 1 direction is accompanied by strain in the 2 and 3 directions, to an extent depending on the dimensions of the body, the conditions at its boundary and the nature of the medium. If the body is a thin solid filament or wire surrounded by fluid and stretched along its length the lateral strain quantities ε_{22} and ε_{33} are each equal to $-\nu\ \varepsilon_{11}$ (see Section 4.4) and the constant C_{11} is just the Young's Modulus E. If the body is indefinitely wide ε_{22} and ε_{33} may be negligib le; then C_{11} is equal to $(\lambda + 2\mu)$. If the body is a static fluid (as in Section 4.3) both the stress and strain tensors are isotropic, so that ε_{22} and ε_{33} are each equal to ε_{11} ; then C_{11} is equal to 3λ , and λ reduces to B , the bulk modulus of elasticity.

7.7 KERATIN

Among the tougher substances in animal bodies are hair, wool, horn, feathers, claws and nails. In all of these a basic constituent is keratin, a protein macromolecule which apparently occurs in either of two configurations, called the α or β forms, or as a random coil. From Xray diffraction data it has been concluded that the α form is a helix (probably a "coiled coil") of 5 A radius with 5.4 A spacing between turns. Calculations have been made (Enomoto and Krimm, 1962) of the effective Young's modulus E for the keratin molecule as it would be if isolated. In these calculations details of the chemical structure are considered and use is made of independent information on interatomic spacings and force constants; the value of E is then found to be 4.46×10^{10} dyn/cm^2 for the α form.

It is thought that the α helices aggregate to form 3-stranded ropes. In hair and wool these ropes of keratin are packed into microfibrils about 70 A in diameter and, in turn, the microfibrils are arranged side by side in the interior of a fibril. Of course some space exists between the keratin units and hence the E-value calculated for an isolated helix would not be expected to apply accurately to the entire fiber. Assuming tentatively that whatever fills the non-keratinous space contributes nothing to the restoring force Enomoto and Krimm calculated E for the microfibril to be 4.1×10^{10} dyn/cm^2 and E for the total fibril to be 1.82×10^{10} dyn/cm^2. But the orderly crystalline packing which Enomoto and Krimm considered in their calculation does not extend along the entire length of a hair or wool fiber. Instead such order exists only in small regions, and these alternate with noncrystalline regions which are more extensible. For a fair comparison with theory the measurement of strain should not be an average over the whole fiber but should be specifically for the crystalline regions.

Such measurements have been made by Xray techniques. When the wool fibers are stretched by applying tension it is found for the crystalline regions that the stress is proportional to the strain, that is, Hooke's law is obeyed, provided that the longitudinal strain does not exceed 0.02 . The crystalline keratin is shown by Xrays to be in the α form under these low-strain conditions. For these same conditions experimental values of E have been obtained for the crystalline regions of wool, ranging from 5×10^{10} dyn/cm^2 at 20°C to 10^{11}dyn/cm^2 at temperatures approaching absolute zero.

Since these observed values are 2 or 3 times higher than that predicted by ignoring the material in the spaces between keratin "ropes" and microfibrils, Enomoto and Krimm conclude that this material, far from being negligible, actually has a very high elastic modulus, and makes a major contribution to the overall value of E for the crystalline portion of the fiber. If this conclusion proves to be correct it will provide another example, of which there are many in physics, where careful comparison of experiment and theory has led to new information.

At high values of strain the keratin tends to assume its β form. It has been suggested that this is a "pleated sheet" arrangement in which parallel macromolecules aggregate side by side in a sheet with folds in and out like the expansible sides of a camera bellows or an accordion. For this condition Enomoto and Krimm calculate E to have the relatively high value of 36.6×10^{10} dyn/cm^2.

The complete stress-strain curve for keratin, ranging from low to high values of strain, is an interesting nonlinear one which we shall take up in Chap. 12.

7.8 BONE AND TEETH

One role that our skeleton plays is similar to that of the frame of an automobile; it holds components of our body in place and protects them from external blows and pressures. Also our legs endow us with high mobility and our arms and fingers provide extremely effective means for manipulating our environment. Our teeth allow us to choose from a wide range of fare for our nourishment and, sometimes, provide means of self-defense.

The hard tissue that makes up bones and teeth derives its strength from submicroscopic crystals which are enmeshed in fibers of the protein collagen. These crystals have been shown by electron microscopy (Glimcher, 1959) to be from 200 to 400 A long and 15 to 30 A wide. They evidently consist of apatite, probably hydroxyapatite $Ca_{10}(PO_4)_6(OH)_2$.

Collagen fibers are visible with ordinary optical microscopy, having diameters of the order of microns. Each fiber is a parallel array of fibrils, each from several hundred to a few thousand Angstroms in width. Finally, the fibril is formed as an aggregate of fundamental units, the collagen macromolecules. These latter are of molecular weight about 360,000 daltons; in solution the molecules behave as fairly rigid rods of length about 2800 A and width 14 A. Each rod is composed of 3 helical polypeptide chains wound together as in a rope. In aggregating to form native collagen the macromolecules form an interesting staggered array as shown in Fig. 7.12. A consequence of this staggering is a spatial periodicity with repeat distance of about 640 A. In electron microscopy of stained preparations the periodicity results in a series of lateral bands spaced 640 A apart along the fibril axis. Under suitable conditions the structures can be dissolved and the collagen molecules separated from each other in solution; they can then be reaggregated in a variety of interesting forms depending on the chemical environment and other conditions. (Schmitt, 1959).

In the cortex of long bones the fibers of collagen, encrusted with crystals of hydroxyapatite, are arranged in spiral patterns about small canals. The cylindrical structures so formed are the osteones, basic units of bone. These osteones are oriented with their axes approximately parallel to the bone axis. Because of this orientation we do not expect bone to be completely isotropic in its elastic properties. Thus, choosing the "3" direction to be along the bone axis we should expect, for example, that the ratio between S_{33} and ε_{33} for given experimental conditions would differ from that between S_{11} and ε_{11} for similar conditions. Referring to the general linear constitutive equation, Eq. (4.1), this suggests that a_{3333} differs from a_{1111} , unlike the situation for a completely isotropic medium. But it is reasonable to assume that, on the average, bone is isotropic in a plane perpendicular to the bone axis; by this assumption the ratios S_{11}/ε_{11} and S_{22}/ε_{22} , for example, are the same in comparable experiments. For this kind of symmetry it is found that the number of independent elastic constants required is five, in contrast to the two constants which suffice for a completely isotropic solid. (Prob. 7.13).

All of these elastic constants were determined for dried bone (bovine phalanx and femur) by Lang (1970) who measured velocities of compressional sound waves at 5 MHz and shear waves at 2.25 MHz in several directions. The velocity of compressional waves was found to be 4400 m/sec along the bone axis and 3300 m/sec in the transverse direction. The density of the bone was found to be 2.00 g/cm^3 . Using formulae analogous to those of Section 13.2 Lang computed the Young's modulus E to be $2.2 \times 10^{11} dyn/cm^2$ along the axis and about half as great 1.15×10^{11} dyn/cm^2, in the transverse direction. Values of the rigidity modulus μ were also calculated and found not to be as dependent on direction; in the axial and transverse directions μ was found to be 5.4 and 4.4 respectively, both in units of 10^{10} dyn/cm^2. Elastic constants have also been determined for hydroxyapatite alone; average values of E and μ are 11.4 and 4.45 , respectively, each in units of 10^{11} dyn/cm^2, and thus are higher than for bone (Katz, 1971). On the other hand the Young's modulus for collagen alone is much lower, being quoted by Katz as approximately 1.2×10^{10} dyn/cm^2. Similar results on elastic constants in bone have been obtained by Brash and Skoreski (1970) using a method involving low frequency vibrations; based on measurements for a number of specimens they obtained a mean value of 2.34×10^{11} dyn/cm^2 for E along the bone axis.

It is perhaps not surprising that material with especially high values for the elastic coefficients are to be found in teeth. The enamel of teeth (the outer layer) contains about 95% hydroxyapatite, the remainder being partly a protein similar to keratin. For human tooth enamel E is about 8×10^{11} dyn/cm^2; for bovine tooth enamel E and μ are 7.4 and 3.0, respectively, each in units of 10^{11} dyn/cm^2. In dentine (the layer just under the enamel) hydroxyapatite occupies about 41% of the volume, while 48% is collagen. In human dentine E is 1.83×10^{11} dyn/cm^2; for bovine dentine E and μ are 2.1 and 0.80 respectively, each in units of 10^{11} dyn/cm^2 (Katz, 1971).

7.9 MEMBRANES

Basic to any discussion of living systems is the subject of membranes. These play a necessary role in forming envelopes of cells and of intracellular bodies such as nuclei and mitochondria. These membranous envelopes have special properties for controlling internal environment by selective passage or transportation of ions. Membranes also are crucial in providing mechanical support for enzymes which catalyze the various steps of biochemical reactions; membranes thus act as floors, walls and shelves of factories for synthesizing the necessities of life.

A vast literature exists on the subject of biological membranes, as studied by means of optical birefringence, Xrays, electron microscopy and other techniques. Also means have been found for producing artificial membranes, that is, "model" membranes, which have life-like properties; many studies on these have been reported. In general research on biomembranes and model membranes is very active. Because of the importance of membranes we can expect new insight on life processes as well as new medical applications as a result of such research. Several articles on biomembranes are included in the list of references for this chapter. (Fox, 1972; Singer and Nicolson, 1972; Vanderkool and Green, 1971). These describe contemporary results and also cite older literature.

Our present concern is with mechanical properties of biomembranes. Unfortunately this is a subject on which we do not have enough information, either from theory or from experiment. A theoretical calculation of elastic constants, for example, would require a detailed knowledge of membrane structure at the molecular level; this knowledge is being sought vigorously but is still a distant goal. And, not surprisingly, there are difficulties in making direct measurements on mechanical constants. One problem is that biological membranes are usually available only in microscopic quantities; another is that for experimental results to be meaningful, they must be obtained under conditions that are life-like or which can be related to life-like conditions.

A biophysicist, like a physicist, is happy when he can explain many facts with just a few basic principles, preferably only one. It would be pleasing, therefore, if it should be found that all membranes are of the same design, and this a simple one. While present indications are somewhat against this, it does appear that many membranes have common features. Thus the primary constituents of membranes are proteins and phospholipids. The phospholipid molecules are several Angstroms in diameter and may be about 20 to 25 A long; they have the distinctive characteristic of being <u>amphipathic</u>. By the latter characteristic is meant that one end, which tends to carry a net electrical charge, is hydrophilic (water soluble) while the other is hydrophobic (insoluble in water). Under some conditions when phospholipids alone are in aqueous solution they aggregate to form bilayers, each containing two parallel layers of molecules. These molecules are oriented along the normal to the layer plane; hydrophilic ends point outward from the bilayer while the rather flexible hydrophobic ends are in the interior. See Fig. 7.13. It appears from optical and Xray studies that the phospholipids in biomembranes are also arranged in this orderly bilayer configuration; the existence of such phospholipid bilayers constitutes an important unifying principle for biomembrane construction. This has implications for mechanical properties since the bilayer is believed to have characteristics similar to those of a highly viscous liquid.

At present, however, the orientation and distribution of the proteins in biomembranes is not as well known. Some of the protein macromolecules are external to the phospholipid bilayers; of these some are probably long macromolecules such as, for example, collagen and are oriented parallel to the bilayer plane. These long units on the surface probably contribute to the tension developed in the membrane when it is stretched. It is known that other proteins are more globular in shape and extend into the interior of the membrane. These play a role in transport of ions and small molecules through the membrane, and will be discussed further in a later chapter.

Turning now to mechanical properties Rand (1964) has given information for the red cell membranes; this is primarily a protein-phospholipid structure of the basic type we have been discussing. Electron microscopy suggests a thickness for the membrane of about 100 A. Rand obtained his information by

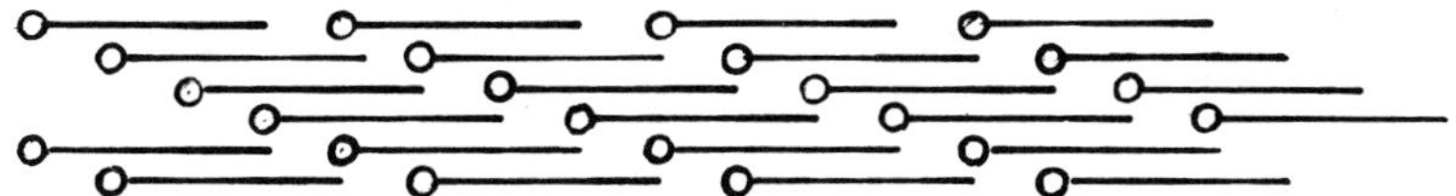

Fig. 7 .12 Staggered array of collagen macromolecules in native fibers.

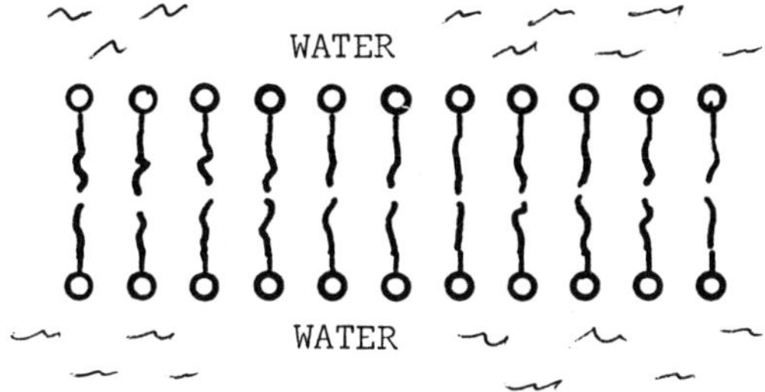

Fig. 7 .13 Phospholipid bilayer. Circles show hydrophilic ends of amphipathic molecules ; tails are hydrophobic and somewhat flexible .

analyzing data obtained with the methods described in Section 2.7 . Suction was applied to part of the cell exterior, thus causing an immediate increase in the tension σ throughout the cell membrane. To interpret his findings Rand assumed that the membrane responded to the increase in σ in a manner expected for the Burgers model referred to in Section 6.7 . According to this model the membrane stretches a certain amount governed by the spring k_1 immediately after increase of the tension. It then stretches further by extending the spring k_2 ; this process occurs more slowly, with time constant given by b_2/k_2 , where b_2 is a constant characteristic of the dashpot in parallel with the spring k_2 . After the latter spring is fully extended the membrane still continues to stretch at a rate determined by the dashpot b_1 . A plot showing the time course of the stretching process is shown in Fig. 6.13 .

Unfortunately, as Rand was careful to point out, it has not been possible to give specific interpretation of the spring and dashpot units in terms of membrane structures. Nevertheless, by assuming these units have cross-sectional areas comparable to that of the membrane cross-section (membrane thickness assumed to be in the range 100 A to 1000 A) he obtained rough values for the Young's modulus E and shear viscosity coefficient η :

For springs k_1 and k_2 : $E = 7.3$ to 300 in units of 10^6dyn/cm^2
For dashpot b_1 : $\eta = 5.8$ to 240 in units of 10^7 poise
For dashpot b_2 : $\eta = 1.6$ to 90 in units of 10^9 poise .

Stress-strain data were obtained in a somewhat more direct manner for <u>sarcolemma</u>, the membrane of muscle fibers. (Fields, 1970). This membrane is more complex than the red cell membrane and has been shown (Mauro and Adams, 1961) to have four components: (1) a layer, about 100 A thickness, interpreted as the <u>plasma membrane</u> (i.e., a phospholipid-protein membrane of the basic kind discussed earlier); (2) an amorphous layer 300 to 500 A thick; (3) a braid-like layer of collagen filaments, thickness variable but of the order of 1000 A and; (4) an outer layer of unidentified fine filaments. Stress-strain curves are nonlinear, such that the effective Young's modulus E is relatively low at small values of strain, probably because the membrane is then "slack". At the high values of strain E reaches a constant value. However E is greater for longitudinal strain than for circumferential or transverse strain. For high strain the longitudinal value for E is found to be in rough agreement with values given in the literature for collagen, about 10^{10} dyn/cm^2, suggesting that most of the tension developed in the stretched sarcolemma comes from the collagen layer. For the same high strain conditions, the circumferential value for E is about 3 to 5 times less than for longitudinal strain. This difference between longitudinal and transverse values of E is reasonable if we assume the collagen fibers are oriented primarily in the longitudinal direction. Longitudinal stretching would then give rise to a relatively high restoring force because the collagen macromolecules are themselves **being stretched. Restoring forces in the circumferential direction would be smaller since these would arise mainly from transverse bonding between macromolecules (weaker than intramolecular bonding) and whatever small circumferential component of orientation the fibers may have. See also Section 12.4.3.**

7.10 TISSUE ELASTICITY

In general biological materials are complex in structure and hence their mechanical characteristics are difficult to describe. For example skin tissue is discussed by Dinnar (1970). Skin has two principal layers: the <u>epidermis</u> (roughly 100 μm thick), which has no blood vessels and consists essentially of dead cells; and the <u>dermis</u> (about 1 to 2 mm thick) which contains blood and lymph vessels, nerves, muscle fibers, glands and hair follicles. To represent known experimental results on the response of skin to compression Dinnar proposed a four-element model, the Burgers model of Fig. 6.13. In this model k_1 is taken to represent the epidermis, k_2 and b_2 the dermis, and b_1 the (subcutaneous) tissue just below the dermis. When edema or dropsy occurs fluid accumulates and mechanical properties of the skin are altered in a way corresponding to a decrease in k_1 and b_1 . Aging has the effect of increasing k_1 and k_2 while decreasing b_1 and b_2 .

For other tissues combinations of Maxwell and Kelvin-Voigt models are often postulated to represent mechanical behavior. The use of dashpots to represent viscous behavior reflects the fact that the media are viscoelastic. Under the simplest conditions the constitutive equations are linear, as exemplified by Eq. (4.22). The elasticity (that is the tendency for a body to return to its original shape after deformation) can be accounted for to some extent in terms of bond stretching, as in the calculations of Enomota and Krimm for keratin (Section 7.7). However, as mentioned in Section 7.6, another contribution to elasticity is thermal in origin and is sometimes called entropic elasticity ; this will be discussed in a later chapter. Properties of viscoelastic media are conveniently determined by experiments utilizing vibratory methods, whose results are best discussed in the mathematical language of complex numbers. We shall take up some aspects of this subject in a later chapter.

Life being as complicated as it is, we probably would expect mechanical properties of biological materials to extend beyond those which are describable in simple terms. In such expectations we would not be disappointed; biological media exhibit a rich variety of mechanical or rheological behavior. The constitutive equations are not, in general, linear; the response of a medium to an applied force often depends on previous history of the medium. Mechanical properties depend on the chemical and other aspects of the environment as well as on "biological" factors such as the health and age of the organism. These complications are brought up here partly as a warning (not to oversimplify) and partly as an invitation (to investigate the subject further). But we shall now turn our attention to other aspects of biological physics.

PROBLEMS

7.1 Show that when Eq. (7.2) applies the activation energy E_v can be obtained from knowledge of the viscosity coefficient η at two temperatures as follows:

$$\frac{E_v}{k} = \frac{\ln \eta' - \ln \eta}{(1/T')-(1/T)} ,$$

where η' and η give the viscosity at T' and T, respectively. The data on η for water can be fitted to Eq. (7.2) if E_v is allowed to vary with temperature. An approximate value for E_v at any temperature T can be obtained by use of the above expression if T' differs from T by only a few degrees. Use this method, with data from Table 7.1, to estimate E_v at one or more temperatures for which values of E_v are quoted in the text.

7.2 A simple form of "physiological saline" is made by adding 0.9 gm of NaCl to 100 ml of water. Show that the molarity of this solution is approximately 0.15 N.

7.3 Derivations of the linear equation for viscosity of a suspension, in the form of Eq. (6.38), are usually based on the assumption that the particles are of macroscopic size, that is, much larger than atoms. Nevertheless the equation is useful for dissolved molecules and even electrolytes. For example, refer to Table 7.2 for NaCl, and plot η vs molarity; you should find that the data follow the linear law fairly well. Assuming Eq. (6 38) is valid for these data and that each ion (or the ion plus its hydration sheath) is a rigid sphere use the relations in Section 6.4 to determine the value for ϕ which corresponds to unit molarity (F = 1). Is the result a reasonable one?

7.4 Using data from Table 7.3 plot η vs concentration for sucrose solutions at 20° C. Determine the intrinsic viscosity $[\eta]$ and discuss the result in the light of material in Section 6.4.

7.5 Data are given in Section 7.3.3 on the molecular weight, sedimentation coefficient, density and mean dimension of the Hb molecule in water. It is sometimes assumed that the Hb molecule is spherical. Check the mutual consistency of these data with that assumption by calculating (1) its molecular weight and (2) its sedimentation coefficient according to theory in Section 6.3. What is your conclusion?

7.6 (a) Explain why water bound to the Hb molecule would increase the measured value of κ .

(b) Assume the Hb molecule is a rigid sphere with radius r, and that the bound water effectively increases the radius to R. What value of R/r is required to account for the quoted value (5.3) for κ ?

7.7 In this problem we show how experimental measurements on sound velocity can be used to determine the compressibility of a dissolved substance, in this instance hemoglobin. In Eq. (7.1) B is the adiabatic value of the bulk modulus; the corresponding compressibility β_{ad} is therefore given by $(1/\rho c_s^2)$. For convenience we write β_{ad} simply as β and c_s as c and thus write

$$\beta = \frac{1}{\rho c^2} \, . \qquad \text{(i)}$$

Suppose β , ρ and c represent values of the compressibility, density and velocity of sound for a solution of Hb in water, while β_o , ρ_o and c_o give the corresponding values for water and β' , ρ' , c' give those for Hb . Let ϕ be the volume fraction of Hb . We realize that β , ρ and c are functions of ϕ , varying from β_o , etc., when $\phi = 0$ (pure water) to β' , etc. when $\phi = 1$ (pure Hb).

(a) By differentiating Eq. (i) with respect to ϕ show that

$$\frac{1}{\beta}\frac{d\beta}{d\phi} + \frac{1}{\rho}\frac{d\rho}{d\phi} + \frac{2}{c}\frac{dc}{d\phi} = 0 \; ; \qquad \text{(ii)}$$

this equation applies at any value of ϕ .

(continued on next page)

(Prob. 7.7 continued)

(b) Now apply Eqs. (7.4) to the problem by a suitable translation of symbols. Differentiating with respect to ϕ show that $d\beta/d\phi$ and $d\rho/d\phi$ are constants given by

$$\frac{d\beta}{d\phi} = \beta' - \beta_o \; ; \; \frac{d\rho}{d\phi} = \rho' - \rho_o \; . \qquad \text{(iii)}$$

(c) By substituting the expressions of Eqs. (iii) into Eq. (ii) (and rearranging) obtain, specifically for $\phi = 0$,

$$\frac{\beta_o - \beta'}{\beta_o} = \frac{\rho' - \rho_o}{\rho_o} \quad \frac{2}{c_o} \frac{dc}{d\phi} \; ; \qquad \text{(iv)}$$

here $dc/d\phi$ (if it is not constant) is to be evaluated at $\phi = 0$.

(d) From the Carstensen-Schwan measurements at 25°C and low Hb concentrations it appears that the velocity c increases by about 3.25 m/sec for each addition of 1 g Hb to 100 ml of water. Determine $dc/d\phi$, assuming ρ' to be 1.33 g/ml.

(e) Using data from Table 7.1 for ρ_o and c_o calculate the ratio β'/β_o .

7.8 Justify Eqs. (7.4) for a suspension of particles in liquid. Make use of definitions for density and compressibility. Assume these properties of particle and liquid are the same in suspension as when they exist separately.

7.9 While Eq. (7.7) is essentially empirical Haynes has given a derivation based on assumptions which are not unreasonable, and which we paraphrase here. Let G be the velocity gradient which results for a given stress S; then for blood G is a function of S. For given S the apparent fluidity ψ is defined as the ratio G/S and the differential fluidity ψ' as dG/dS. Let ψ'_∞ and ψ'_o be values of ψ' when S is infinity and zero, resp. At low

(continued on next page)

(problem 7.9 continued)

values of S considerable structure exists in the blood. This structure breaks down increasingly with increasing S until at sufficiently large values of S no structure remains and the blood behaves as a Newtonian liquid; ψ'_∞ is the fluidity under these Newtonian conditions. Haynes introduces the quantity $\Delta\psi'$, defined as $\psi'_\infty - \psi'$, suggesting that $\Delta\psi'$ is a reasonable measure of the amount of structure existing in the liquid for a given value of S. From the form of the curves in Figs. 7.6 and 7.7 we can see that $\Delta\psi'$ decreases as S increases, becoming essentially zero at large values of S. He assumed further that in a small increase of stress dS the "structural" quantity $\Delta\psi'$ decreases in such a way that the fractional change $d(\Delta\psi')/\Delta\psi'$ is proportional to dS. Specifically, the latter assumption is

$$\frac{d(\Delta\psi')}{\Delta\psi'} = -k\,dS ,$$

where k is a constant. Show that a solution of this equation can be written

$$\psi' = \psi'_\infty - (\psi'_\infty - \psi'_0)\,e^{-kS} ,$$

since $\Delta\psi' = \psi'_0 - \psi'_\infty$ when S = 0. Remember that ψ' is dG/dS , integrate the equation, use the fact that G = 0 when S = 0, and obtain the form of Eq. (7.7).

7.10 Show that a HA macromolecule of molecular weight 14×10^6 daltons and R_{rms} given by 0.24 μm in solution must extend over a volume which includes much water. Do this by showing that a sphere with the large radius found for R_{rms} could hardly be a single molecule of HA with molecular weight as given.

7.11 Calculate the molecular weight in daltons of a cylindrical molecule which is 10 μm long with a diameter of 6 A and density 1.5 g/ml.

7.12 (a) Show for the Morse function, Eq. (7.11), that U_1 is minimum when $r = r_o$.

(b) Show that the stretching force F is maximum at $r = r_c$ where

$$r_c = r_o + 0.694\ a^{-1} \ .$$

(c) Show that the maximum value of F is

$$F_{max} = \frac{aD}{2} \ .$$

7.13 Justify the statement in Section 7.8 that for the kind of symmetry possessed by bone (symmetry about an axis) there are 5 independent elastic constants in the constitutive equation, Eq. 4.1.

This problem offers exercise in the use of tensor transformations such as were discussed in Chap. 4, and which are treated in more detail in such textbooks as that by Long cited in Reference 2.1. As stated in Chap. 2, for any symmetrical tensor such as the stress tensor S_{ij} or the strain tensor ε_{ij} it is possible to find a set of axes, called principal axes, in which the tensor has diagonal form. Suppose a medium of interest (representing bone) is symmetrical about some axis, as far as elastic properties are concerned. Consider a strain field whose principal axes are 1, 2, 3 where the 3 axis coincides with the elastic axis of symmetry. Then it can be shown that the stress field has the same principal axes. In these axes we can write

$$\begin{aligned} S_{11} &= c_{1111}\varepsilon_{11} + c_{1122}\varepsilon_{22} + c_{1133}\varepsilon_{33} \\ S_{22} &= c_{2211}\varepsilon_{11} + c_{2222}\varepsilon_{22} + c_{2233}\varepsilon_{33} \\ S_{33} &= c_{3311}\varepsilon_{11} + c_{3322}\varepsilon_{22} + c_{3333}\varepsilon_{33} \, , \end{aligned} \tag{A}$$

(continued on next page)

(Prob. 7.13 continued)

the off-diagonal terms S_{12}, etc., being zero. In Eqs. (A) there are 9 elastic constants c_{1111}, etc. Now consider a transformation through 90° about the 3 axis such that the direction cosines are

$$a_{ij} = \begin{bmatrix} 0 & 1 & 0 \\ -1 & 0 & 0 \\ 0 & 0 & 1 \end{bmatrix} .$$

Then, for example,

$$S_{11}' = \sum_{i=1}^{3} \sum_{j=1}^{3} a_{1i} a_{1j} S_{ij} = S_{22} = c_{2211}\varepsilon_{11} + c_{2222}\varepsilon_{22} + c_{2233}\varepsilon_{33} ,$$

where

$$\varepsilon_{11} = \sum_{m=1}^{3} \sum_{n=1}^{3} a_{m1} a_{n1} \varepsilon_{mn}' = \varepsilon_{22}' , \text{ etc.}$$

In this problem you are to complete the development from this point. Thus obtain an expression for S_{11}' as a sum of terms involving ε_{11}', etc. Because of elastic symmetry about the 3 axis we know that the relation between stress and strain components is the same in the 1', 2', 3'axes as in the 1, 2, 3 axes. Hence as another expression for S_{11}' we have

$$S_{11}' = c_{1111}\varepsilon_{11}' + c_{1122}\varepsilon_{22}' + c_{1133}\varepsilon_{33}' .$$

Comparing the two expressions for S_{11}' you should arrive at equalities of the type $c_{1111} = c_{2222}$, $c_{1122} = c_{2211}$, etc. Proceed similarly with S_{22}' and S_{33}' as needed, obtaining a sufficient number of equalities to reduce the 9 elastic constants in Eqs. (A) to 5 independent ones.

(For a completely isotropic medium we would proceed similarly with another transformation, say, about the 1 axis, and reduce the number of elastic constants to two.)

CHAPTER 8

INTRODUCTION TO THERMOPHYSICS; CONCEPTS IN PROBABILITY THEORY

8.1 INTRODUCTION

In previous chapters we considered principles of continuum mechanics and some of their consequences. These principles are deterministic in their implication that the future of a system can be predicted with certainty. For example, the flow of a fluid through a horizontal capillary is determined, according to Eq. (5.61), when the radius of the tube, the viscosity of the fluid and the pressure gradient are known. Also the downward velocity of a particle sedimenting in a liquid under gravitation can be predicted, according to Section 6.3.1, when characteristics of the particle, the liquid and the gravitational field are known. Deterministic theory of this kind is very useful as a means of predicting accurately or describing concisely the behaviour of systems of various kinds under a wide range of conditions. But for many situations the deterministic principles of classical mechanics are either not valid or cannot be applied in any practical way. An example of such a situation is offered by the fascinating and basic phenomenon of Brownian Movement.

Early in the 19th century the botanist Brown (1828) discovered by observations through a microscope that small particles in suspension execute a ceaseless irregular motion. His first observations were on particles of the pollen of living plants. However he soon found that the same kind of random activity was exhibited by particles from plants that had been dead for many years. Furthermore he showed that coal dust and even particles from inorganic rock take part in this motion, which then seemed to be of physical (not "vitalistic") origin.

The phenomenon of Brownian Movement attracted the attention of other investigators. But its nature remained in doubt until it was shown by Einstein (1906), Smoluchowski (1906) and Perrin (1920) that the motion is as expected from the kinetic theory which was developed during the 19th century. This theory, especially successful for gases, deals with the behavior of a medium by considering the molecules of which it consists. The molecules are treated as if they were tiny billiard balls subject to the laws of classical mechanics. Special aspects of kinetic theory are usually taken up in elementary physics. It is not hard to show, for example, that the average pressure on the walls of a vessel containing an ideal gas is given by

$$p = \frac{2c}{3} < KE > , \qquad (8.1)$$

where c gives the number of molecules per unit volume in the gas and $< KE >$ gives the average translational kinetic energy per molecule. Agreement with measured properties of gases was obtained by introducing the specific assumption that

$$< KE > = \frac{3}{2} kT , \qquad (8.2)$$

where T is the Kelvin temperature, and k is the Boltzmann constant with the value of 1.38×10^{-16} ergs per degree. Combining the above two equations we obtain simply

$$p = ckT , \qquad (8.3)$$

an important result for gases, which also finds applications in biology.

Kinetic theory, such as that leading to Eq. (8.3), is deterministic in some ways but not in others. A definite value is predicted for the average pressure p. However, it would be a hopeless task to determine trajectories of all the individual molecules as they collide with each other and with bounding surfaces. Furthermore, the pressure on any part of the boundary is not constant, but fluctuates with time. It is not possible to predict the pressure fluctuation at any given time, but average values of the fluctuations can be calculated by statistical means.

We return now to the Brownian movement, involving suspended particles of, say, about 1 μm in dimension, much larger than gaseous molecules. It was assumed by Einstein and Smoluchowski that the kinetic theory postulate of Eq. (8.2) applies even to the suspended particles studied by Brown. Using this postulate together with equations from continuum mechanics, theoretical results were obtained which could be compared with observation. The theoretical approach is illustrated in Section 9.2 where a relatively simple derivation is given of an expression for the average square $< x^2 >$ of the displacement x in one-dimensional Brownian movement; see Eq. (9.17). Predictions of such theory were tested in careful and extensive observations by Perrine; good agreement was found between theory and experiment. The Brownian movement studies are important historically as background for the theory of gases, and are important here because of their relevance to biological suspensions.

Derivation of Eq. (9.17) in Section 9.2 requires the assumption that Brownian motion is random: the net displacement during any given time interval is as likely to be toward the left as toward the right. It may seem remarkable that in spite of this ignorance about how the particle position x varies with time a definite prediction is arrived at for $< x^2 >$ as a function of time. But the situation is not really unusual. It is characteristic of problems treated by statistical methods that information about average quantities is obtained, starting from assumptions of randomness. Statistical principles are useful generally for calculating properties of systems containing many "bodies" or "particles." <u>Statistical mechanics</u> is a highly developed science in which such principles are applied to gases, particles in suspension, molecular chains, etc. We do not have space here to treat advanced aspects of this subject. But we shall discuss basic principles of probability and statistics, for which we subsequently find applications to biology.

We learn from elementary physics that in any medium its ions, atoms or molecules take part in random movements, and that these correspond to the macroscopic phenomenon of heat. The idea of equivalence between mechanical energy and heat evolved in the 19th century. Theoretical analyses and experimental studies led to the development of thermodynamics, a broad subject very important to biology, of which we take up some aspects in later chapters. Thermodynamics differs from kinetic theory and statistical mechanics in that, like continuum mechanics, it is a macroscopic science and makes relatively little reference to the atoms and molecules of which a medium consists. Instead thermodynamics deals with relations between such quantities as pressure, density and temperature which are often directly observable.

In this brief introduction we have referred to several important topics (random phenomena, many-body problems and heat-energy relations), pointing out that these can be approached through the methods of kinetic theory, statistical mechanics and thermodynamics. Together these methods of approach comprise the broad science of thermophysics, with which we deal in this and several succeeding chapters. In the remainder of Chap. 8 we take up selected topics in probability and statistics. These find application in later chapters, including Chap. 9 which deals with Brownian movement and diffusion. A brief introduction to thermodynamics is presented in Chap. 10. Applications of thermophysics to transport of matter across membranes, and to bio-elasticity are taken up in Chaps. 11 and 12; further applications arise in Chaps. 13 and 14.

8.2 FLUCTUATIONS AND PROBABILITIES

For a scientist the possibility of reproducible experiments is a basic one. In fact one way to contrast science with humane studies is to point out (in spite of the adage "History repeats itself") the impossibility of repeating experience that involves human beings. Thus the New Deal political "experiment" cannot really be tried again and again, each time under exactly the same conditions; it is perhaps because of this that historians do not agree on their evaluations of political experiments. On the other hand the density ρ_w of water is a quantity that we regard as a "constant." It has been measured many times and, no doubt, all investigators agree that under ordinary conditions ρ_w has the approximate value of 1.0 g/cm^3. In fact scientific authorities might agree that at 3.98°C the density ρ_w (in the same units) is greater than 0.99996 and less than 0.99998. This is a considerably better consensus than can be obtained in predicting consequences of economic policies. Nevertheless, it may be that we have here mainly a very large difference of degree, rather than absolute difference of kind.

For it is not feasible to increase indefinitely the precision with which quantities like ρ_w can be specified. In measuring ρ_w one must make determinations of (1) the mass and (2) the volume of a given quantity of water. Consider for an example the mass determination, which would typically be done with a sensitive balance. Limitations on the accuracy with which the mass can be measured arise ultimately from fluctuations in the balance position. If we denote this position (or the balance "reading") as x, a recording of x as a function of time might appear as shown in Fig. 8.1. For this very sensitive balance we see that x maintains a value whose mean lies between the narrow limits of 1.000005 g and 1.000006 g, but which exhibits fluctuations in the sixth place after the decimal. In some situations the observed fluctuations arise from mechanical disturbances such as vibrations caused by persons walking and automobile traffic in the vicinity of a laboratory. Under these conditions isolations mounts and other devices can be used to reduce fluctuations. Other disturbances are electrical or magnetic in nature and can be reduced by various kinds of shielding. But when all this is done, either with mechanical or electronic devices, there is always an irreducible limit below which fluctuations can be reduced no further.

This basic fluctuation can sometimes be explained quantitatively in terms of the kinetic energy of individual atoms, electrons or other particles; since the average energy of a particle increases with temperature, the fluctuation is then said to be _thermal_ in origin. Fluctuations of a basic nature also arise in measuring the output of a light source or of an X-ray generator. These can often be explained in terms of the particulate nature of electromagnetic radiation. If we think of light or X-rays as composed of discrete quanta or photons, it is perhaps reasonable that a series of pulses or bursts would differ from each other because of variations in numbers of photons. Similar considerations apply to γ-ray photons, electrons and other emanations from radioactive sources. Since these arise from disintegration of individual nuclei, and since the disintegrations occur randomly, fluctuations in the rate of emission are expected.

Common experience abounds with situations where fluctuations occur. Thus a marksman does not hit the bull's-eye every time, but only for a certain fraction of the time. From seeds that look alike grow plants that are not identical but show a distribution of height, weight, and so forth. Extreme examples are seen in games of chance. Here the variability is very great; paradoxically, prediction then becomes possible through properties of the variations themselves. For example, attempts to throw a "two" in rolling a die are extremely unsuccessful. The reproducibility of such an experiment (rolling a die) is so poor that we can make the following statement: "In N such experiments the fraction in which the outcome is 'two' will approach (1/6) as N increases." In other words, the probability of "two" is (1/6). Also the probability is the same for "one," "three," or any other of the six numbers on the die. Thus ignorance leads to knowledge: In throwing a die we have no idea which of the six faces will come up, and from this we learn that the probability for any one of them is (1/6).

We are led to interest in making predictions about averages or probabilities for various fluctuating quantities. This is frequently possible; while individual variations occur in an unpredictable manner, statements can often be made about averages over large numbers of experiments or measurements. The "random walk" is an entertaining example of this. Here an inebriated gentleman starts from a lamppost and takes N steps, random in direction but equal in length. It is found that if N is large, his distance from the lamppost is, on the average, proportional to $\sqrt{N}$. Also the probability of his reaching a given distance can be calculated. Hence a definite prediction is made about average results of a large number of experiments (steps) in spite of (or because of) ignorance about the outcome of an individual experiment. The random walk offers an analogy to the Brownian motion of particles in suspension, a topic to which we have already referred.

In other situations to which the random walk presents no obvious analogy a result reminiscent of the "$\sqrt{N}$ dependence" is nevertheless found. For example suppose N determinations are made of a given quantity such as length or weight, and that a measure of the error in each determination is σ. Then an equivalent measure of the error in the average of the N determinations is found to be just $\sigma/\sqrt{N}$.

In 1927 Heisenberg stated an uncertainty principle that applies to fluctuations of a very fundamental nature. Profound questions have been raised about the origin, or philosophical implications of, the fluctuations to which the principle applies. However, no exceptions to its applicability have been found in the physical realm. In one of its forms the principle states that if Δp is a measure of the uncertainty of the momentum of a given particle along a given direction, and Δx is the uncertainty of its corresponding position, then the product of Δp and Δx is of the order of h or greater, where h is Planck's constant. (A more precise statement can be made if we are more specific about the definitions of Δp and Δx in terms of probabilities.) This principle is important in biology, for example in considering the possibility of specifying positions and motions of particles in suspension.

The subject of fluctuations and probabilities is a basic one for science generally. For our purposes in this book we now take up, in this chapter, certain aspects of the subject that have special relevance to biology. It is helpful to start by considering a laboratory procedure for determining the number of particles in a suspension. Such procedures are much used in biology and medicine. For example, the "red cell count" gives the number of red cells per unit volume in the blood and is an important medical index. Simple cell counting provides a good example for learning basic concepts underlying probability theory.

8.3 CELL COUNTING

Statistical reasoning can be applied to obtain useful information about various systems of particles, such as biological cells or macromolecules in suspension. For illustration we take up in this section a sampling problem which often arises. We consider a suspension of volume V containing N particles (Fig. 8.2) by the average number density c of the suspension we mean

$$c = \frac{N}{V} , \tag{8.4}$$

the units of c being particles per unit volume. A common method of determining the number density is to count the particles contained in a selected small region in the suspension, which we call the sampling space.

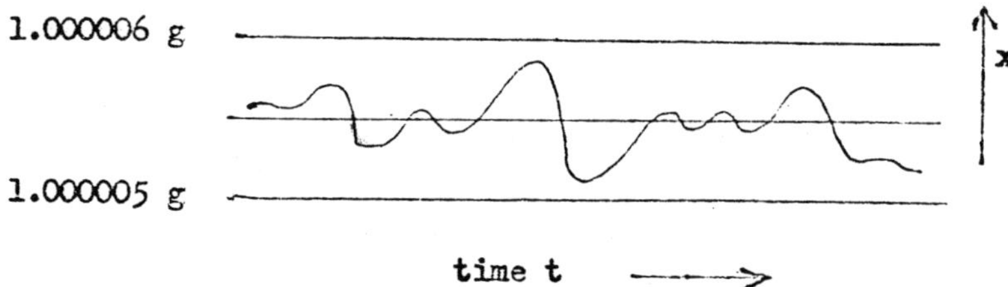

Fig. 8.1 Recording of balance position x versus time t for a very sensitive balance, showing fluctuations.

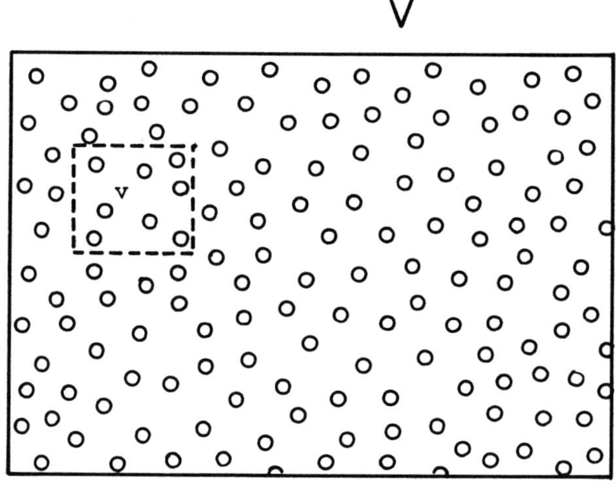

Fig. 8.2 Schematic for sampling problem. There are a fixed number N of particles in a (relatively large) region of fixed volume V. In the sampling space (indicated by dashed lines) of fixed volume v the number n of particles varies from time to time, or from sample to sample.

If n particles are found in a sample of volume v the ratio n/v gives a typical value for the number density. But n/v will seldom (or never) be exactly equal to c. If it were, the sample count n would always have the same value vc, and this is not found to be true.

Indeed we do not expect this since, for one thing, the actual count n is, of course, an integer, while vc would usually not be. On the assumption of a uniform distribution of particles we might perhaps expect n always to be either of two integers, the one just above, the other just below the number vc (if nonintegral). Thus if vc proved to be 33.74 perhaps the observed n would always be either 33 or 34. But this proves to be far from true. In a series of determinations of the particle count n_i, each made on an independent sample of the suspension, a surprising range of values for n_i is found; for example, the readings 28, 34, 32, 38, 33, etc., might be obtained for n_i. No amount of care in sampling procedures or in stirring the suspension to ensure against systematic variations in particle density, will reduce the fluctuations in n_i below a basic value. This basic variation is, in fact, fundamental to the sampling process and we shall see that it is, to an extent, predictable.

It is necessary to introduce some terminology. Let $\bar{n}$ be the mean of a series of k determinations n_1, n_2, etc., of the particle count in v; by definition we then have

$$\bar{n} = \frac{(n_1 + n_2 + - - - n_k)}{k} . \tag{8.5}$$

It is sometimes convenient to use a simpler symbol for $\bar{n}$; thus we let

$$a = \bar{n} \tag{8.6}$$

and use either symbol a or $\bar{n}$ as we see fit. We find experimentally that the mean value $\bar{n}$ for a set of independent observations approaches more and more nearly the value vc as the number k of the observations increases. Let δ_n be the deviation of any given observation of n from the mean; by definition we have

$$\delta_n = n - a . \tag{8.7}$$

It can be shown (Prob. 8.4) that the mean value of δ_n is exactly zero;

thus

$$\bar{\delta}_n = 0 \; . \tag{8.8}$$

We let σ^2 be the mean square of the deviation; thus for a set of k observations we have

$$\sigma^2 = \frac{(\delta_1{}^2 + \delta_2{}^2 + ---\delta_k{}^2)}{k} \; . \tag{8.9}$$

The quantity σ is known as the root-mean-square deviation from the mean, or simply the <u>rms deviation</u> or <u>standard deviation</u>. It is possible to show (see, for example, Section C.8) that when $\bar{n} \ll N$ the expected value of σ^2 is approximately

$$\sigma^2 = \bar{n} = a \; . \tag{8.10}$$

Thus although individual observations of n fluctuate in a random manner, the mean value $\bar{n}$ approaches a definite number vc as k increases, and the rms deviation σ approaches $\sqrt{a}$.

8.4 THE PROBABILITY p

For insight into the nature and origin of fluctuations in the count n we refer to certain concepts and results of probability theory. Basic definitions and principles are reviewed in Appendix C; those who want to pursue the subject further may consult any of a number of available texts, some of which are listed in the References.

We refer again to a suspension of N particles in a container whose total volume is V. Consider first the probability p that <u>any one</u> of them is to be found in a given sampling space of volume v. It is helpful to employ an <u>operational</u> definition of p: this is a definition wherein we describe experimental means (not necessarily practical) for determining p.[+] For

---

[+]A requirement that scientists speak only in terms of concepts that can be operationally defined is somewhat stern, but leads to clarity in our language by eliminating words which may be meaningless. Arguments for operational terminology have been put forth particularly strongly by Bridgman (1936).

this purpose we consider a large number M of identical but independent experiments in each of which the observation is made whether or not the given particle is in v. (To make the experiments independent we assume that the suspension is well stirred between successive determinations.) Using the terminology of Section C.1 this is an experiment in which only two possibilities or Events are possible: Event A, the particle is found in v; and Event B, it is not. The quantity p gives the probability for Event A; let q be the corresponding probability for Event B. Let M_A by the number of experiments in which Event A occurs and M_B the corresponding number for Event B; then M is the sum $(M_A + M_B)$. We now can give operational definitions for p and q:

$$p = M_A/M; \quad q = M_B/M; \quad M = M_A + M_B \ . \tag{8.11}$$

It is clear from Eqs. (8.11) that

$$p + q = 1 \ . \tag{8.12}$$

From the definition of Events A and B it is clearly a certainty that one and only one of these will occur. Hence Eq. (8.12) is in accord with a general principle developed in Sect. C.2 (see Corollary to Theorem I): If it is certain that one and only one of a set of events will occur, then the sum of probabilities for the individual events is unity.

For a given situation it is often possible to assign explicit values to the probabilities p and q without need for such experiments as are implied in the operational definitions. These so-called <u>a priori</u> values tend to be arrived at intuitively; though they may be based ultimately on experience, the latter is often subtle and indirect. For example, in our situation it is reasonable to assume that

$$p = v/V \ , \quad q = (V-v)/V \ ; \tag{8.13}$$

it is clear that these <u>a priori</u> values satisfy the condition of Eq. (8.12).

8.5 PRESCRIBED SETS

8.5.1 The Probability P_n^N

Let us now calculate a <u>joint probability</u> for a given set of "single particle" experiments: that is, let us determine the probability that each experiment in the set will turn out in a prescribed way. As an example consider a set of ten experiments, all involving a given particle for which the results are as in Table 8.1 Here "Yes" means for the experiment in question that the particle was found in the sampling space, and "No" means that it was not. Let P_6^{10} be the probability that a set of ten experiments will yield precisely the results in the table, specifically that a prescribed six (Nos. 1,2,4,5,6 and 8) of the ten yield "Yes". In calculating P_6^{10} we remember that the probability of any given experiment yielding "Yes" is p while the similar probability of a "No" is q. Furthermore, we recall (see, for example, Section C.3) that the probability of occurrence of a number of independent events is just the product of the separate probabilities. Hence we find that

$$P_6^{10} = p^6 q^4 \ . \tag{8.14}$$

In general let P_n^N be the probability that in a set of N independent experiments, a prescribed set of n of these will have "Yes" as a result. Then we see that

$$P_n^N = p^n q^{(N-n)} \ . \tag{8.15}$$

8.5.2 Alternative Interpretation of P_n^N

In our discussion so far we have taken "Yes" or "No" to mean that a particular one of the particles in suspension would or would not be found in the sample space; the given particle might just as well have been the only particle in the entire volume V. We may now reinterpret P_n^N so that it leads to information about an arbitrary number N of particles which coexist in the volume V. For this purpose we employ our imagination in another way: we suppose the individual particles can be identified and thus (as in a Social Security system) assign each individual a number. Assuming the particles are numbered 1 to N, we now consider a new kind of observation: here we stir the suspension well, then observe at a given instant which of the selected individuals is in the sampling space.

As an example, suppose just ten particles are being considered (i.e., N=10) and that in a given observation the results for the ten individuals are as in Table 8.2. Here "Yes" indicates that the individual is found in the sampling space, and "No" otherwise. For ease in comparing with the situation of Table 8.1 we have supposed that the results are similar.

The probability that the results should be as in Table 8.2 can readily be seen to be the joint probability p_6^{10} arrived at earlier, Eq. (8.14), in connection with Table 8.1. Thus the probability of a "Yes" result for a given individual in Table 8.2 is just p (given by Eqs. (8.11) or (8.13), and the probability of a "No" result is q. Clearly then the probability that results for ten individuals in a given observation on a suspension should be as in Table 8.2 is the same as the probability that the results of ten one-particle experiments in a given set should be as in Table 8.1.

Generalizing from the above discussion we see that P_n^N as given in Eq. (8.15) can be interpreted as the probability that in a given sampling of a suspension of N numbered particles a prescribed set of n particles will be found in the sampling space.

8.6 DISTRIBUTIONS

8.6.1 The Probability W_n^N

Consider now another set of results on an observation of ten particles as shown in Table 8.3; as in Table 8.2 there are 6 Yes's and 4 No's, but the Yes group is composed of different individuals than before. It is clear that P_n^N is unaltered by a change of this kind. Similarly there are other possible sets of ten-particle results such that for each set six particles are found in the sampling space, but where the six differ in their individual make-up. In fact it has been shown (see, for example

Section C.6) that there are just C_6^{10} such sets, where

$$C_6^{10} = \frac{10!}{6!4!} . \tag{8.16}$$

Generalizing, in a total of N particles there are C_n^N possible sets for each of which there are n "Yes" particles and (N-n) "No" particles but which differ in the individual makeup of the "Yes" and "No" groups; here

$$C_n^N = \frac{N!}{n!(N-n)!} . \tag{8.17}$$

It is obviously not usually practical to distinguish between individuals of a suspension. Supposing again that N is 10 we now consider the probability that in a given observation exactly 6 of the particles will be found in the sampling space, without regard to the identity of individual particles. We call this probability W_6^{10}. From previous discussion we see that there are C_6^{10} "ways" of accomplishing this, in the sense that there are C_6^{10} possible sets of observations for each of which there are 6 prescribed "Yes" particles. For each of these prescribed sets the probability is P_6^{10} . It follows from basic principles (see Appendix C) that the probability W_6^{10} for a set which will have 6 Yes's and 4 No's (individuals unprescribed) is just the product of the number of prescribed sets of this kind times the probability per (prescribed) set; that is,

$$W_6^{10} = C_6^{10} P_6^{10} .$$

In general, the probability that in a suspension of N particles n of them will be found in the sampling space is

$$W_n^N = C_n^N P_n^N , \tag{8.18}$$

where C_n^N is given by Eq. (8.17) and P_n^N by Eq. (8.15). We have thus finally arrived at a result that can be applied to a real suspension. An important property of the W_n^N is that

$$\sum_{n=0}^{N} W_n^N = 1 \ . \tag{8.19}$$

Equation (8.19) can be readily confirmed (see, for example, Prob. 8.2) and also is expected from general considerations (See Sect. C.1, Corollary to Theorem I).

8.6.2 Illustrative YES-NO Experiment

With the intention of clarifying ideas about the concepts and results discussed above, let us see what they mean in other specific situations. We consider further experiments of the "Yes-No" kind, that is, experiments in which the result is always one of two alternatives. There are many examples of such experiments. We have noted that the cell-counting situation is in this category; here an object either appears in a certain space or does not. Another example comes from tests of a definite effect (such as death, or a mutation) caused by some agent on an organism; here either the agent causes the effect in question, or does not. Still another such situation arises in counting nuclear disintegrations; a specified nucleus in a radioactive sample will either disintegrate (in a specified time interval) or it will not.

For our present purposes we choose still another example, and suppose we are testing the hearing response of a student volunteer. The subject is seated in a soundproofed room in which a barely audible sound of known level is produced at intervals. When asked if he hears the sound, the subject presses an appropriate button indicating "Yes" if he does, and "No" otherwise. Of course, to avoid giving cues to the subject, the question is asked at times when the sound is "off," as well as when it is "on."

Suppose that in 100 questions asked when the sound is "on" at a given level the subject replies "Yes" 60 times and "No" 40 times. Then if p is the probability of "Yes" and q that of "No" we obtain from the experiment fairly good approximations to p and q, namely,

$$p = 0.60 \quad \text{and} \quad q = 0.40 .$$

Suppose now that for some reason the questions are asked in sets of <u>three</u>; the situation then offers a simple example for which we can calculate various quantities from probability theory which have been discussed in this chapter.

In Table 8.4 are listed all possible sets of replies in a three-question experiment. There are 8 of these sets; for there are two ways of answering each question, and hence the number of ways of answering three questions is 2^3. Also given in Table 8.4 are the relevant values or expressions for n, P_n^3, C_n^3, and W_n^3, where n is the number of Yes's. There is only one set each for n = 3 and n = 0; there are three sets each for n = 2 and n = 1. Consider, for example, n = 2; the joint probability P_2^3 is the same (p^2q) for all sets; W_2^3 gives $C_2^3\ P_2^3$ or $3P_2^3$ and is just the sum of the P_2^3 for the 3 sets in which n = 2. It can be verified that

$$W_3^3 + W_2^3 + W_1^3 + W_0^3 = 1 , \tag{8.20}$$

as expected from Eq. (8.19). Clearly, it is then also true that the sum of the P_n^3 for all eight sets is equal to unity.

Of course, numerical values for the P_n^3 and W_n^3 may be obtained by substituting the expected values p = 0.6 and q = 0.4.

8.6.3 Use of W_n^N for Calculating Averages

In a large number of observations of the count n, the average value $\bar{n}$ is given by

$$\bar{n} = \sum_{n=o}^{N} n\ W_n^N . \tag{8.21}$$

TABLE 8.1
RESULTS OF TEN EXPERIMENTS ON A SINGLE PARTICLE

Expt. No.	Result
1	Yes
2	Yes
3	No
4	Yes
5	Yes
6	Yes
7	No
8	Yes
9	No
10	No

TABLE 8.2
SINGLE EXPERIMENT ON A SET OF TEN PARTICLES

Indiv. No.	Result
1	Yes
2	Yes
3	No
4	Yes
5	Yes
6	Yes
7	No
8	Yes
9	No
10	No

TABLE 8.3
ANOTHER TEN-PARTICLE EXPERIMENT

Indiv. No.	Result
1	Yes
2	Yes
3	Yes
4	Yes
5	Yes
6	Yes
7	No
8	No
9	No
10	No

TABLE 8.4

EXAMPLE ILLUSTRATING QUANTITIES FROM PROBABILITY THEORY

Set of replies			n	P_n^3	C_n^3	W_n^3
Yes	Yes	Yes	3	p^3	1	p^3
No	Yes	Yes	2	p^2q	3	$3p^2q$
Yes	No	Yes	2	p^2q		
Yes	Yes	No	2	p^2q		
Yes	No	No	1	pq^2	3	$3pq^2$
No	Yes	No	1	pq^2		
No	No	Yes	1	pq^2		
No	No	No	0	q^3	1	q^3

To see this we suppose that the individual counts are designated n_1, n_2, n_3, etc., and that their mean value $\bar{n}$ is defined as in Eq. (8.5), with the total number k of counts assumed very large. Any given value of the count, say "3", may occur a large number of times. From the general meaning of W_n^N we see that

$$W_3^N = \frac{\text{number of "3" counts}}{\text{total number of counts}} ,$$

the denominator on the right being just k. Hence we see that

$$\text{Number of "3" counts} = k\, W_3^N .$$

In determining the total of all counts we may either sum all the individual n_i , or sum these in groups by adding terms of the type $3kW_3^N$. (The latter term is the product of the count value "3" by the number of counts with that value.) Thus we have

$$n_1 + n_2 \text{ --- } n_k = kW_1^N + 2kW_2^N + 3kW_3^N + \text{---} \tag{8.22}$$

From Eq. (8.5) we see then that

$$\bar{n} = W_1^N + 2W_2^N + \text{- - - - -} = \sum_{n=1}^{N} nW_n^N ,$$

as was to be shown. (Of course, the sum on the right of Eq. (8.21) is the same whether the lower limit is "0" or "1".)

Making use of Eq. (8.21) we can derive (Prob. 8.5) the convenient expression

$$\bar{n} = pN . \tag{8.23}$$

With the foregoing discussion as background we can arrive at an expression for averaging any quantity f_n which is a single-valued monotonic function of n. (Such a function, exemplified by n^2, n^3, e^n, etc., depends on n in such a way that the value f_n occurs when and only when the count "n" occurs.) In a large number k of observations the value f_3 occurs kW_3^N times, and so on; the average value $\bar{f}$ is therefore

$$\bar{f} = \frac{(f_o kW_o^N + f_1 kW_1^N + \text{- - -})}{k} ,$$

or,

$$\bar{f} = \sum f_n W_n^N \ . \tag{8.24}$$

An example of f_n already considered is, of course, n itself. Other examples are the deviation $(n - \bar{n})$, and the square of this deviation. For the former it is found (Eq. (8.8)) that the average value is zero; for the latter the average value is approximately $\bar{n}$, provided that $N >> \bar{n}$. (See Eq. (8.10)).

8.6.4 Significance of W_n^N and Related Concepts

As we have seen the quantity W_n^N gives the probability that n objects will be found in a sample of volume v taken from a total population of N objects occupying volume V. Application of W_n^N to a wide variety of situations is possible by imaginative translation of the words "object," "sample" and "volume" to appropriate language for the situation.

In a straightforward application to biology the objects might be cells in a suspension, and the sample a portion of the suspension which is withdrawn by pipette; determination of the count n can be made directly by visual inspection under a microscope. In other situations the objects might be molecules or ions in solution and the "sampling space" might be the interior of a specified cell, or part of a cell; here the count n obviously cannot be determined visually, and often can not be obtained by any direct measurement. One can nevertheless _imagine_ counts being made and draw conclusions on this basis. Letting our imagination range further we can let the "object" be an _event_, such as the disintegration of a radioactive atom used as tracer, or absorption of an Xray photon in a suspension of bacteria. We might know that in a given time period, N such events have occurred in a volume V and wish to know how many took place in a sampling volume v. One application here is to the "target" problem which is so important in radiation biophysics.

Other examples could be cited; the probability W_n^N is applicable to a wide variety of experimental situations. However, it is usually not practical to make calculations from the expression in Eq. (8.18) with exact values of C_n^N and P_n^N. In the remainder of this chapter convenient approximations to W_n^N are arrived at and discussed. In particular the approximate formulae of Poisson and Gauss are obtained. To obtain the Poisson distribution we assume the probabiliity p is small and hence that $\bar{n}$ is much less than N; this assumption applies, for example, to the sampling situation of Fig. 8.2, if the sampling volume v is much less than the total volume V. We are led from the Poisson result to the Gaussian distribution if the typical sample count is large enough, so that $\bar{n} \gg 1$. In still another approximation, where $\bar{n}$ is nearly equal to $\frac{1}{2}N$, we again obtain a Gaussian distribution; this result applies, for example, to coin-tossing experiments where n gives the number of heads in N tosses of the coin. It also applies to the "one-dimensional random walk" which offers an analogy to Brownian motion and diffusion.

8.7 POISSON DISTRIBUTION: SMALL p

Suppose $p \ll 1$ and hence that typical values of n (being of the order-of-magnitude of $\bar{n}$, the average value of n) are much less than N. We may then obtain the approximation from Eq. (8.17) that (Prob. 8.8)

$$C_n^N = N^n/n! \ . \tag{8.25}$$

It is convenient to define a symbol X such that

$$q^{(N-n)} = e^X \ . \tag{8.26}$$

Taking the natural logarithm of both sides of the above equation we obtain

$$X = \ln q^{(N-n)} \tag{8.27}$$

We find, letting $q = (1-p)$, that

$$X = (N - n)\ \ell n(1 - p) \cong -p(N-n) \cong -pN \ ; \tag{8.28}$$

here n has been neglected relative to N and the approximation has been made, valid for p sufficiently small (see Appendix B) that

$$\ln (1 - p) \cong -p \ . \tag{8.29}$$

Letting $\underline{a} = \bar{n}$ be the average value of n we have (Eq. 8.23) that $\underline{a}$ is equal to pN and hence from Eqs. (8.26) and (8.28) that

$$q^{(N - n)} = e^{-a} \ . \tag{8.30}$$

Referring back to Eq. (8.15) we find that we can write

$$P_n^N = p^n \ e^{-a} \ . \tag{8.31}$$

Using Eqs. (8.18), (8.25) and (8.31) together with the relation $\underline{a}$ = pN from Eq. (8.23) we obtain

$$W_n^N = W(n,a) = a^n e^{-a}/n! \ . \tag{8.32}$$

Equation (8.32) is the Poisson distribution law in its common form. The symbol W(n,a) becomes appropriate to replace W_n^N because we see from the right hand member in Eq. (8.32) that the probability no longer depends explicitly on N, but only on n and its average value $\underline{a}$. Thus W(n,a) gives the probability of obtaining the count n in a situation where the average count is $\underline{a}$. Numerical values of this function are given in Table C.2 for choices of n and $\underline{a}$ lying in the ranges 0-25 and 1-12, resp. Special properties of the Poisson functions are taken up as exercises at the end of this chapter.

8.8 GAUSSIAN DISTRIBUTION

8.8.1 Small p, Large n

If typical values of n are sufficiently large compared to unity the expression for W(n,a) from Eq. (8.32) may be put in the form of the Gaussian or normal distribution. To see this we refer (Appendix B) to the Stirling (large n) approximation to n!, namely

$$n! \cong (2\pi)^{\frac{1}{2}} \ n^{(n+\frac{1}{2})} \ e^{-n} \ . \tag{8.33}$$

Using this value for n! in Eq. (8.32) we obtain

$$W(n,a) = (2\pi)^{-\frac{1}{2}} \ a^n n^{-(n+\frac{1}{2})} \ e^{(n-a)} . \tag{8.34}$$

We now assume that observed values of n differ from the average $\underline{a}$ only a relatively small amount; letting δ be (n-a) we state specifically that δ is much less than $\underline{a}$. Nevertheless n and $\underline{a}$ are each so large that for most observations we can expect δ to be much greater than unity. More concisely we assume that

$$n = a + \delta \ , \text{ where } 1 \ll \delta \ll a \ . \tag{8.35}$$

One then obtains (Prob. 8.12) from Eq. (8.34) that

$$W(n,a) = W(\delta,a) = (2\pi a)^{-\frac{1}{2}} \exp\left(-\delta^2/2a\right) \ . \tag{8.36}$$

Equation (8.36) is the desired result, representing a normal curve; we see that W can be expressed in terms of δ and $\underline{a}$ alone. Since Eq. (8.36) is derived from the Poisson distribution, the restriction $\underline{a} \ll N$ still applies. Hence Eq. (8.10) is valid and $\underline{a}$ may be replaced in Eq. (8.36) by σ^2 yielding

$$W(\delta,\sigma) = (2\pi\sigma^2)^{-\frac{1}{2}} \exp\left(-\delta^2/2\sigma^2\right) \ . \tag{8.37}$$

Here $W(\delta,\sigma)$ is the famous Gaussian distribution and gives the probability that a given observation of n will differ from the mean by an amount δ ; the constant σ is such that σ^2 is the mean value of δ^2 in a large number of experiments. We proceed now to show that the same function applies also for somewhat different conditions than those assumed in arriving at Eq. (8.37).

8.8.2 p = q , Large n

Here we abandon the Poisson approximation in which it was required that p should be small compared to unity. Instead we assume p and q are each equal to ½, as would be true, for example, in a coin-tossing experiment. We find that

$$P_n^N = (\tfrac{1}{2})^N \tag{8.38}$$

Under these conditions and for large N, typical values of n lie in the vicinity of ½N. It is then convenient to write

$$n = \tfrac{1}{2}N + \delta; \quad \delta \ll N \ . \tag{8.39}$$

It can be shown (Prob. 8.13) that Eq. (8.17) becomes

$$C_n^N \cong 2^N \left(\frac{2}{\pi N}\right)^{\frac{1}{2}} \exp\left(\frac{-2\delta^2}{N}\right) . \tag{8.40}$$

Hence we have, from Eqs. (8.18), (8.38) and (8.40)

$$W_n^N = C_n^N P_n^N = \left(\frac{2}{\pi N}\right)^{\frac{1}{2}} \exp\left(\frac{-2\delta^2}{N}\right) . \tag{8.41}$$

We see that Eq. (8.41) is of the same form as Eq. (8.37) and that the two are identical if we replace N by $4\sigma^2$, so that

$$\sigma = \tfrac{1}{2}\sqrt{N} . \tag{8.42}$$

We shall be referring back to these results in later chapters.

8.9 PROBABILITY AS A CONTINUOUS FUNCTION

When n is large the probability W_n^N differs only slightly from W_{n+1}^N ; i.e., unit increase of n causes only a small fractional change in W_n^N . Values of W_n^N for consecutive (integral) values of n may appear as shown by vertical solid lines in Fig. 8.3. It is then useful to define a continuous function $W^N(n)$ in such a way that $W^N(n)$ is equal to W_n^N for integral n, and such that $W^N(n)$ is a "smooth" function for all n, integral or nonintegral. See the dashed curve in Fig. 8.3. In principle this definition of $W^N(n)$ leaves it rather arbitrary for nonintegral n. But in practice it is fairly well defined since independent individuals tend to draw about the same "smooth" curve when fitting the values for integral n. With this in mind we realize that for large n each of the probability functions given in Eqs. (8.36), (8.37) and (8.41) can be regarded as a continuous function of δ. This is true in spite of the fact that these functions were derived on the assumption that n is an integer and hence that δ is defined only for a set of discrete values.

Consider now the probability $W[\alpha,\beta]$ that n lies in the range $\alpha < n < \beta$ where α and β are integers. From previous considerations $W[\alpha,\beta]$

is the sum of probabilities for all integral values of n in the range; that is,

$$W[\alpha,\beta] = W_{\alpha}^{N} + W_{\alpha+1}^{N} + \cdots + W_{\beta-1}^{N} + W_{\beta}^{N} = \sum_{n=\alpha}^{\beta} W_{n}^{N} . \tag{8.43}$$

Assuming α is large we can define a "smooth" continuous function $W^N(n)$ (valid in the range $\alpha \leq n \leq \beta$ for all n, integral or nonintegral) which is equal to W_n^N for integral n in the specified range. From principles of integral calculus we can thus proceed from Eq. (8.43) to write approximately

$$W[\alpha,\beta] = \int_{n=\alpha}^{\beta} W^N(n)\, dn . \tag{8.44}$$

Frequently the expression for $W[\alpha,\beta]$ in terms of an integral as in Eq. (8.44) is more convenient than that in terms of a sum, as in Eq. (8.43).

8.10 PROPERTIES OF THE GAUSSIAN DISTRIBUTION

We shall find frequent application for the Gaussian function given by Eq. (8.37). This can be rewritten in the convenient form

$$W(x) = A \exp(-\alpha x^2) , \tag{8.45a}$$

where, in terms of a constant σ ,

$$A = \frac{1}{\sigma\sqrt{2\pi}} ; \qquad \alpha = \frac{1}{2\sigma^2} \tag{8.45b}$$

Here x replaces the symbol δ used in Eq. (8.37), and may represent other quantities than the deviation δ . As in Section 8.9, W(x) is a continuous probability function for the quantity x . Let x_1 be a particular value of x and let dx be a small range of x . Then the probability that x lies between x_1 and x_1 + dx is approximately given by $W(x_1)$ dx . (Of course, the approximation approaches exactitude as dx approaches zero.)

In Table C.3 values of $W(x)$ from Eq. (8.45) are given vs x for $\sigma = 1$. Plots of $W(x)$ based on Eqs. (8.45) are shown in Fig. 8.4 for several values of σ . We see that the function $W(x)$ has a single maximum which lies at $x = 0$. From Eqs. (8.45) we find that at this maximum the probability has the value

$$W(0) = \frac{1}{\sigma\sqrt{2\pi}} . \tag{8.46}$$

The function is symmetrical about $x = 0$, that is, $W(x)$ is equal to $W(-x)$. Also $W(x)$ decreases monotonically with increasing $|x|$. When σ is small the central peak is high and narrow, while for large σ it is low and broad.

Since the probability of x lying between x_1 and x_1+dx is given (for indefinitely small dx) by $W(x_1)\,dx$ it follows that the probability $W[x_1,x_2]$ of x lying between arbitrary values x_1 and x_2 is given by

$$W[x_1,x_2] = \int_{x_1}^{x_2} W(x)\,dx . \tag{8.47}$$

Thus the probability in question is given by the area under the W vs x curve between x_1 and x_2 . It can be shown (Prob. 8.15) that the area under the total curve (i.e., the integral in Eq. (8.47) when x_1 and x_2 are $-\infty$ and $+\infty$, resp.) is unity for any value of σ. This indication of unit probability, or certainty, is as expected if, for example, x represents the deviation δ ; obviously the deviation has <u>some value</u> between $-\infty$ and $+\infty$.

From tables (see Table C.3) the integral in Eq. (8.47) can be evaluated for any values of the limits; if x_1 and x_2 are each expressed in units of σ the results for W [ ] are independent of σ . A few useful values are:

$$\begin{aligned} W[-\sigma, \sigma] &= 0.6826 \\ W[-2\sigma, 2\sigma] &= 0.9544 \\ W[-3\sigma, 3\sigma] &= 0.9974 \\ W[-\infty, \infty] &= 1 \end{aligned} \tag{8.48}$$

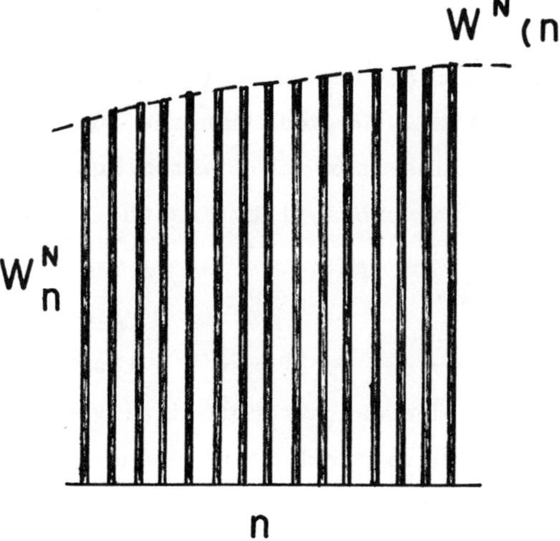

Fig. 8.3 The probability function for consecutive values of integral n (vertical solid lines) and as a continuous function (dashed curve).

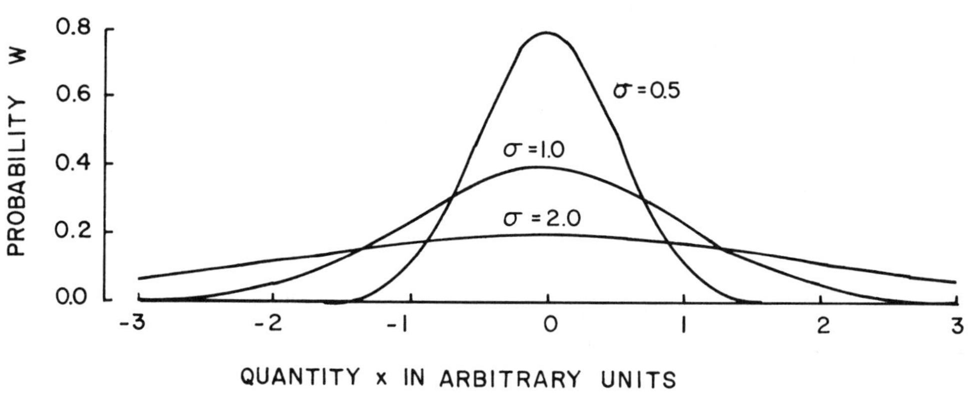

Fig. 8.4 The Gaussian distribution plotted from Eqs. 8.45 for several values of the standard deviation σ . The height of the central maximum is $0.40/\sigma$.

Consider, for example, the result for $[-3\sigma , 3\sigma]$; this means that in a large number of determinations of the quantity x it is expected that 99.74% of them will lie between -3σ and $+3\sigma$, and that 0.26% will lie outside these limits. Because the Gaussian function is symmetrical it is clear that ½(99.74)% will lie in the range $[0,3\sigma]$ and an equal percentage in the range $[-3\sigma,0]$. Similar interpretations can be given to other results in Eqs. (8.48). (Of course, the result for $[-\infty, \infty]$ has already been discussed).

As we have seen before (Section 8.6.3) a probability function can be useful for computing averages. When the Gaussian distribution applies we may determine the average value $\langle f \rangle$ of any function f(x) from the expression

$$\langle f \rangle = \int_{-\infty}^{\infty} f(x)\, W(x)\, dx \; . \tag{8.49}$$

A derivation of Eq. (8.49) will not be given here, but can be arrived at in a fairly straightforward manner by comparing Eq. (8.49) , applicable to the probability function for a continuous variable x , and Eq. (8.24), applicable to the probability function for a variable which takes on only integral values.

Letting f(x) represent, in turn, the functions x, x^2, and $|x|$ each of the following important results can be obtained (Prob. 8.16):

$$\langle x \rangle = 0 \tag{8.50}$$

$$\langle x^2 \rangle = \sigma^2 \tag{8.51}$$

and

$$\langle |x| \rangle = \left(\frac{2}{\pi}\right)^{\frac{1}{2}} \sigma = 0.80\ \sigma \; . \tag{8.52}$$

Equation (8.50) is analogous to Eq. (8.8). Equation (8.51) shows that σ is the root-mean-square value of x, and is analogous to Eq. (8.9). From Eq. (8.52) we obtain a simple relation between σ and the average magnitude of x.

PROBLEMS

8.1 Give operational definitions for P_n^N and W_n^N .

8.2 Show that W_n^N are the terms arising in the binomial expansion $(p + q)^N$ and hence verify Eq. (8.19).

8.3 (a) Make a table similar to Table 8.4 for a set of 4-question experiments of the Yes-No type.
(b) Plot C_n^4 vs n.
(c) Plot W_n^4 vs n assuming p = 0.5 and p = 0.1 , resp.

8.4 Using Eq. (8.24), show that the average value of the deviation δ_n is exactly zero, where δ_n designates $(n - \bar{n})$. In other words, prove Eq. (8.8).

8.5 Starting from Eq. (8.21) show that $\bar{n} = pN$. In other words prove Eq. (8.23). (Hint: Substitute for W_n^N in terms of p and q, factor out pN, then interpret the result.)

8.6 Proceeding from arguments and results in Theorem I, Appendix C, obtain Eq. (C.7) and thus prove the Corollary to Theorem I.

8.7 Proceeding from arguments and results in Theorem IV, Appendix C, obtain Eq. (C.15) and thus prove Theorem V.

8.8 Justify the approximation for C_n^N in Eq. (8.25).

8.9 Plot W_n^N vs n from the Poisson law, Eq. (8.32), for $\underline{a}$ = 1,2 and 3. See Table C.2. What trends are seen in the plots as $\underline{a}$ increases?

8.10 Show by use of the series expansion for e^a (Appendix B) that the Poisson expression for W_n^N, Eq. (8.32), satisfies the condition

$$\sum_{n=o}^{\infty} W_n^N = 1 .$$

Note that this is an approximation to the correct result, Eq. (8.19), according to which the upper limit to the sum is N, not infinity.

8.11 Using the general expression for an average given by Eq. (8.24), and using the approximate expression for W_n^N given by Eq. (8.32), obtain the result

$$\bar{n} = a ,$$

provided that $N >> a$. (You may wish to use a series expression from Appendix B).

8.12 Show that Eqs. (8.34) and (8.35) lead to Eqs. (8.36). This exercise offers experience in the "science of approximations." One approach is to follow these steps: Take the logarithm of both sides of Eq. (8.34); make use of Eq. (8.35) and properties of logarithms to eliminate the term ($n \ln a$); use approximations judiciously, for example, in writing a series expression from Appendix B for the logarithm of $(1 + \frac{\delta}{a})$; reorganize.

8.13 Derive the expression for C_n^N in Eq. (8.40) using the assumption that $|\delta| << N$.

As in Prob. 8.12 a number of steps are required in which approximations are introduced. Suggestion: Introduce Stirling approximations (Appendix B) before or after taking logarithms; make use of Eq. (8.39) and an approximate series expression from Appendix B for the logarithm of $(1 + \frac{\delta}{N})$; reorganize.

8.14 (a) From Eq. (8.18) show that

$$W_o^N = (1 - p)^N.$$

(b) Show that if $p << 1$ this yields

$$W_o^N = \exp(-pN).$$

Since $pN = a$ this is just the Poisson result.

8.15 Show that

$$\int_{-\infty}^{\infty} A \exp(-\alpha x^2)\, dx = 1,$$

when α and A given by Eqs. (8.45b). See Appendix B.

8.16 Verify Eqs. (8.50), (8.51) and (8,52) by using Eq. (8.49) and referring to tables of definite integrals or to Appendix B.

CHAPTER 9

BROWNIAN MOVEMENT AND DIFFUSION

9.1 INTRODUCTION

In Chapter 8 we referred to Brownian movement, the random dancing motion of a particle suspended in fluid. This motion may be thought of as resulting from fluctuating forces on the particle. In turn the forces may be attributed to collisions with the particle by molecules in the surrounding fluid. These collisions occur with great frequency; for example a sphere of radius 1 μm in air at room temperature would be subject to more than 10^{16} collisions per second. The consequences of a given collision between a gaseous molecule and a 1 μm particle could, in principle, be calculated fairly well from classical mechanics if the conditions could be specified. However this is clearly not likely for any collision on an atomic scale, and is certainly not possible for the vast number of collisions to which the particle is subjected. Furthermore in a liquid the encounters of molecules with a suspended particle are not simple collisions, since they are complicated by interactions between neighboring molecules.

From the point of view of classical mechanics we realize that the situation is hopelessly complex for a Brownian particle. Hence we do not expect to be able to predict details of Brownian movements in the sense that the particle velocity is determined as a function in time as in, say, Section 6.3. We find, however, that the enormous complexity of the situation from one point of view actually leads to simplicity from another standpoint. Because the number of collisions, or other kinds of interaction events experienced by a Brownian particle is so large, it proves possible to assume that the events are random. Hence the force on a particle at a given time is as likely to be in one direction as another. Statistical reasoning can thus be applied. We find here as we did for several examples in Chapter 8 that ignorance about details leads to definite statements about averages. In Section 9.2 an equation is derived for one-dimensional Brownian movement. In Sections 9.3 and 9.4 we give our attention to diffusion, a flow of particles in the direction of decreasing concentration which results from Brownian movement. We derive equations for this flow and, in the remainder of this chapter, consider consequences of Brownian movement and diffusion for situations important to biology.

9.2 BROWNIAN MOVEMENT

9.2.1 Theory for Movement Along x.

We follow a relatively simple approach discussed by King (1962) and developed originally by Langevin (1908), directing our attention to the displacement of a particle along an arbitrarily chosen x axis. Consider a particle of mass m acted on by a force X in the x direction, where the magnitude and sign of X vary randomly with time. Let the velocity dx/dt of the particle be designated simply as $\dot{x}$, and its acceleration d^2x/dt^2 as $\ddot{x}$. Assume a frictional force $(-f\dot{x})$ retards the particle, where the frictional constant f is given for a rigid sphere by Eq. (6.15). Then from Newton's second law the equation of motion is

$$X - f\dot{x} = m\ddot{x}. \tag{9.1}$$

Suppose that at t = 0 the particle is at the origin x = 0. We wish to show that as time goes on there is a steady increase in the average distance of the particle from its starting point. Specifically we shall show that the average value of x^2 increases linearly with the time. To do this let $y = x^2$ and consider the derivatives

$$\frac{dy}{dt} = 2x\dot{x}\ , \tag{9.2}$$

and

$$\frac{d^2y}{dt^2} = 2\dot{x}^2 + 2x\ddot{x}. \tag{9.3}$$

Taking the time-average of Eqs. (9.2) and (9.3), and indicating the average of a given term by brackets < > , we obtain

$$\left\langle \frac{dy}{dt} \right\rangle = \langle 2x\dot{x} \rangle , \tag{9.4}$$

and

$$\left\langle \frac{d^2y}{dt^2} \right\rangle = \langle 2\dot{x}^2 \rangle + \langle 2x\ddot{x} \rangle . \tag{9.5}$$

The terms $\langle x\dot{x} \rangle$ and $\langle x\ddot{x} \rangle$ can be found in an equation derived from Eq. (9.1) by multiplying through by x and time-averaging:

$$\langle xX \rangle - f \langle x\dot{x} \rangle = m \langle x\ddot{x} \rangle . \tag{9.6}$$

Combining the last three equations we obtain

$$\langle xX \rangle - \tfrac{1}{2}f \left\langle \frac{dy}{dt} \right\rangle = m\left[\tfrac{1}{2} \left\langle \frac{d^2y}{dt^2} \right\rangle - \langle \dot{x}^2 \rangle\right]. \tag{9.7}$$

Since $\langle \dot{x}^2 \rangle$ appears in Eq. (9.7) we suspect that Eq. (8.2) , representing a basic assumption of kinetic theory, can be used here. We realize that because of symmetry between the three rectangular coordinates $\tfrac{1}{2}m\langle \dot{x}^2 \rangle$ must be just one-third of the total average kinetic energy <K.E.>. Hence we obtain from Eq. (8.2) that

$$\tfrac{1}{2}\, m \langle \dot{x}^2 \rangle = \tfrac{1}{2}\, kT$$

or

$$\langle \dot{x}^2 \rangle = \frac{kT}{m} . \tag{9.8}$$

It is convenient to define a quantity z as

$$z = \left\langle \frac{dy}{dt} \right\rangle . \tag{9.9}$$

One can show (Prob. 9.1) that the operations of averaging and differentiating can be interchanged so that an alternate expression for z is

$$z = \frac{d}{dt} \langle y \rangle . \tag{9.10}$$

Similarly,

$$\frac{dz}{dt} = \frac{d}{dt} < \frac{dy}{dt} > = < \frac{d^2y}{dt^2} > . \tag{9.11}$$

Also, since X varies randomly we can expect that

$$<xX> = 0. \tag{9.12}$$

Combining Eqs. (9.7) through (9.12) we find that

$$m \frac{dz}{dt} + fz = 2kT . \tag{9.13}$$

This equation is of the same type as Eq. (6.25a); a solution is

$$z = \frac{2kT}{f} + A'\exp\left(-\frac{t}{\tau}\right) , \tag{9.14}$$

where τ is given by m/f and A' is an arbitrary constant. Typically τ is very short, as shown in Prob. 6.4. When $t >> \tau$ we have, remembering the definitions of z and y,

$$z = \frac{d}{dt} <x^2> = \frac{2kT}{f} . \tag{9.15}$$

Integrating with respect to time we obtain

$$<x^2> = \frac{2kTt}{f} + B , \tag{9.16}$$

where B is an arbitrary constant. Since x = 0 when t = 0 we see that B = 0 so that

$$<x^2> = \frac{2kTt}{f} . \tag{9.17}$$

This equation is one derived by Einstein and also by Smoluchowski (see References for Chap. 8) in 1906. It will be shown later that kT/f is equal to D, where D is the coefficient of diffusion for the particle-fluid combination. Equation (9.17) is both simple and appealing. It expresses a result which may seem surprising. Thus according to Eq. (9.17) the average square of x (displacement of the particle from an arbitrary starting position) increases steadily with time, in spite of the fact that in any given time interval the particle displacement is as likely to be forward as backward.

We gain some insight on this matter by referring to Section 8.8.2; the analysis given there is applicable to one-dimensional "random walk". Here one imagines that a person carries out a series of N steps along the x axis under the unusual condition that each step is as likely to be backward as forward. Taking p to be the probability of a given step being forward, and q to be the corresponding probability of the step being backward we see that p and q are each "½" as assumed in Section 8.8.2. Let n be the number of forward steps; then (N-n) will be the number of backward steps and the <u>net</u> number of forward steps will be n_o given by

$$n_o = n - (N - n) = 2n - N = 2\delta,$$

where δ is the quantity defined in Eq. (8.39). If the steps are of length b, then bn_o will be the distance from the origin reached by the walker when n out of N steps are forward; we let $x = bn_o$. It follows from Eq. (8.42) that σ^2, the average value of δ^2, is N/4. Since n_o^2 is $4\,\delta^2$ we see that the average value of x^2 is

$$\langle x^2 \rangle = b^2 \langle n_o^2 \rangle = 4b^2 \langle \delta^2 \rangle = Nb^2, \qquad (9.18)$$

where the symbol $\langle\ \rangle$ represents average. If the steps are taken at a constant rate of N' steps per unit time we can replace N by N't to obtain the average value of x^2 in a time interval t:

$$\langle x^2 \rangle = N'b^2 t. \qquad (9.19)$$

We see that Eq. (9.19) is identical in form to the Brownian movement equation given in Eq. (9.17). One may with some imagination consider the dancing Brownian particle to be executing a succession of backward and forward steps, as in a "random walk".

Brownian movement is an example of a phenomenon which is intriguing philosophically, since some aspects can be calculated while others are completely unpredictable, Thus Eq. (9.17) is deterministic since it gives a definite expression for $\langle x^2 \rangle$ as a function of time; furthermore the expression has been tested experimentally and found to be a valid one. On the other hand the direction in which the particle moves cannot be predicted at all; a displacement along +x has the same probability as one along -x.

We introduced Brownian movement by describing it as a motion for which deterministic principles of classical mechanics are not applicable. Some further explanation of the statement is worthwhile. Classical mechanics is used as a basis for Eq. (9.1). Determinism is lost specifically in the step represented by Eq. (9.12); here it is stated that on the average the product xX is zero. This statement is justified if, for any given displacement x which the particle may experience, the force X is as likely to be positive with a given magnitude as negative with the same magnitude. Hence a statement of ignorance about the direction of X is not only consistent with the deterministic Eq. (9.17) but is an important step in arriving at it. In pondering this situation, and such phenomena as Brownian movement in general, there seems to be room for various philosophies. Those inclined toward determinism can meditate on results like Eq. (9.17), while those abhoring the constraints of predictability can take refuge in Eq. (9.12) and other indications of fundamental uncertainty.

9.2.2 Theory for Movement in Three Dimensions

Consider now the movement of a particle with respect to Cartesian axes (x,y,z) and let r measure distance from the origin. Then r^2 is the sum of x^2, y^2, and z^2 and the average value of r^2 is the sum of $\langle x^2 \rangle$, $\langle y^2 \rangle$ and $\langle z^2 \rangle$. Since an equation like Eq. (9.17) holds for each of the coordinates x, y and z it follows that

$$\langle r^2 \rangle = 3\,\langle x^2 \rangle = \frac{6kTt}{f}\,. \tag{9.20}$$

9.2.3 Theory for Rotational Movement

Rotary motion can be treated in a manner analogous to linear motion. Consider a body subject to torque L about a given axis and let the angular displacement about the same axis be θ. In a viscous medium a steady state velocity will be achieved where the applied torque L is equal and opposite to a drag torque proportional to $\dot{\theta}$; for this situation we write, by analogy to Eq. (6.11),

$$L = f_r \dot{\theta} , \tag{9.21}$$

where f_r is the rotational friction constant. A particle suspended in a fluid is subject to rotational fluctuations, and thus takes part in a rotary form of Brownian movement. According to classical kinetic theory each particle will have an average rotational kinetic energy of ½kT about a given axis. Following through analysis similar to that for linear Brownian movement it is found (Prob. 9.2) that the average square of the angular displacement of a particle from its starting orientation is

$$\langle \theta^2 \rangle = \frac{2kTt}{f_r}. \tag{9.22}$$

We now cite a few expressions which have been derived for the rotational friction constant f_r (analogous to those for the linear friction coefficient f given in Section 6.2.) For a sphere of radius R rotating about a diameter in a fluid of viscosity coefficient η , the rotational friction constant has been shown (Lamb, 1945) to be

$$f_r = 8\pi\eta R^3 . \tag{9.23}$$

For a long thin cylinder of radius R and length ℓ rotating about its axis the constant is (Prob. 9.3)

$$f_r = 2\pi\eta R^2 \ell . \tag{9.24}$$

Particles in a variety of shapes can be represented approximately as ellipsoids. Consider, for example, an ellipsoid of revolution whose axis of symmetry is the x axis. Let its semi-axis along x be a_1 and let a_2 be the other semi-axis. It has been shown (see Morawetz, 1965) that the rotational friction constant for motion about an axis perpendicular to x, passing through the center of the ellipsoid, is given by

$$f_r = 8\pi\eta\, a_1 a_2^2\, Y(p) , \tag{9.25}$$

where p is the axial ratio a_1/a_2 and Y (p) is a complicated function plotted in Fig. 9.1. The plot for p greater than unity is for rod-shaped particles, and that for p less than unity is for particles resembling discs. For large p an approximation to the frictional constant is

$$f_r = \frac{8\pi\eta a_1^3}{3.57 + 3 \ln p} . \tag{9.26}$$

It is left as an exercise to determine f_r for a dumbbell (Prob.9.4).

9.2.4 A Laboratory Experiment

A test of the theory for Brownian movement was carried out in a student laboratory at the University of Vermont. (The assistance of Dr. J. A. Rooney in developing this experiment is gratefully acknowledged.) Observations were made on suspensions in water of polystyrene spheres (diameter 1.305 microns, standard deviation 0.016 microns, obtained from Dow Chemical Company). These small spheres were viewed with dark field microscopy, a technique by which the small particles are made to stand out as luminous points against a dark background. A sample of the suspension was contained in a cell mounted in the microscope stage. (Fig. 9.2).

This simple cell was constructed from a standard glass microscope slide A (about 1.2 mm thick) and two cover slides B and C (about 0.15 mm thick) of which B is plastic. There is a central hole of about 6 mm diameter in B, and both sides of B are thinly coated with grease (such as Cello Grease, Fisher Scientific Company). When B is pressed into contact with A the hole forms a well into which a few drops of the desired suspension can be pipetted before an experiment. Slide C is then carefully placed in contact with B, so as to cover the well without trapping bubbles. (If the well is not completely filled with suspension or if the grease on both sides of B does not make a good seal, convection currents may occur

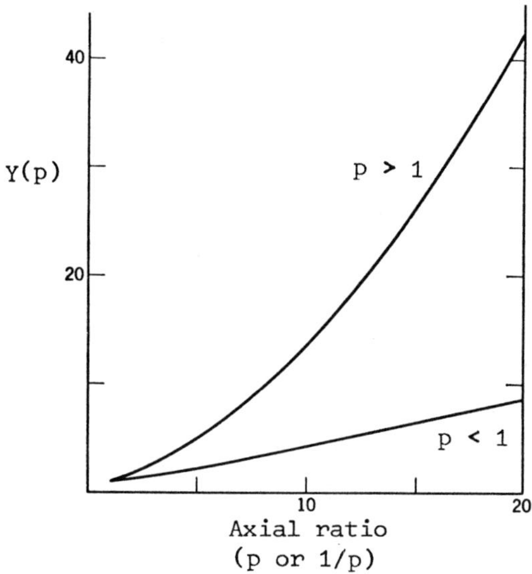

Fig. 9.1 Plot of function Y(p) which appears in Eq. (9.25). For the upper plot p is greater than unity and the abscissa is p; for the lower one p is less than unity and the abscissa is 1/p. (From Morawetz, Ref. 9.4, Fig. VI.6)

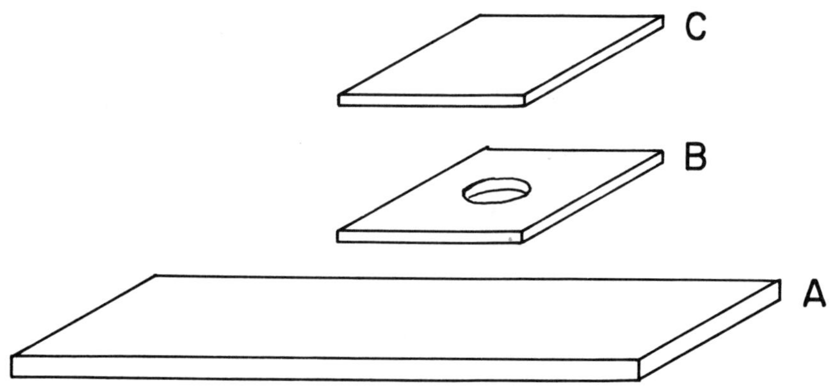

Fig. 9.2 Construction of cell for observations of Brownian movement.

because of evaporation and other processes.) The pieces A, B and C now form a sandwich comprising the desired cell and containing the suspension of interest. Since A is a standard slide the cell may be adjusted horizontally on the microscope stage by conventional means.

Light from an incandescent source will usually cause thermal convection in the suspension. This can be minimized by filtering out the infrared component (by passing the light through a sufficient length of $CuSO_4$ solution, for example).

To study displacements of the particles a ruled grid or scale is used in the microscope, calibrated by viewing a standard scale (stage micrometer) on the microscope stage. In Brownian movement the particle follows a zig-zag path in three dimensions; in this experiment the observer concerns himself only with progress in the "x direction," the latter chosen for convenience relative to the microscope grid. In making measurements the observer selects a particle which is in a suitable part of the field and observes it for a period T of the order of minutes. The x-distance traversed in the time T is recorded. Then the same particle, or another one, is selected for a second observation, etc. Table 9.1 shows results of ten individual observations by one student; here T was 60 seconds. Both x and x^2 are shown. The unit for x in Table 9.1

TABLE 9.1

PARTICLE DISPLACEMENTS ALONG x IN 60 SECONDS

Units of x are 3.1 μm

x	-3	0	3	0	2	-2	-3	0	4	-3
x^2	9	0	9	0	4	4	9	0	16	9

is one division of the microscope scale which was found to be 3.1 μm . Randomness of the motion is suggested by the fact that positive and negative displacements occur about equally frequently. We find from the data that $< x^2 >$ is 5.8×10^{-7} cm^2 .

Similar data were obtained for time intervals of 120 sec and 180 sec, yielding for $< x^2 >$ the values 10.5×10^{-7} and 18.1×10^{-7} cm^2 , resp. Analysis of these data are taken up in an exercise. (Prob.9.5)

9.3 DIFFUSION EQUATIONS

In a suspension all particles are engaged in Brownian movement. A consequence of this is the tendency for flow of particles from regions of high concentration to regions of low concentration. This flow, or diffusion, obviously has the effect of making the concentration more uniform throughout the region occupied by the suspension. Diffusion is important in biology as one of the processes by which movement or "transport" occurs of oxygen, elementary foodstuffs and waste products within and between cells. The laws of diffusion can be arrived at by starting from equations for Brownian movement, such as Eq. (9.17), then determining the net flow by statistical methods.

Instead we shall proceed here from laws developed independently for the process of diffusion, based on experiments where flow rates were measured under various circumstances. Basic to the approach is a principle often known as Fick's First Law. We shall state this for a situation where the concentration c is independent of the coordinates y and z, and is a function only of x and the time t ; here the diffusion is along the x direction. Let J be the "particle current density;" specifically J gives the number of suspended particles per unit time crossing unit area of a given plane of constant x, in the positive x direction. For this situation Fick's First Law is

$$J = -D \, \partial c/\partial x \, , \tag{9.27}$$

where D is the coefficient of diffusion for the particles and suspending liquid under consideration. The quantity J can be eliminated to yield a differential equation involving only c as a dependent variable, by an argument based on conservation of matter. The approach is similar to one we have encountered several times previously, in continuum mechanics. We make a physical statement relative to a suitable volume element of the medium, then obtain a differential expression as the volume of the element becomes indefinitely small.

Here it is appropriate to choose the volume element in the form of a slab contained between the planes $x = x_1$ and $x = x_2$. See Fig. 9.3.

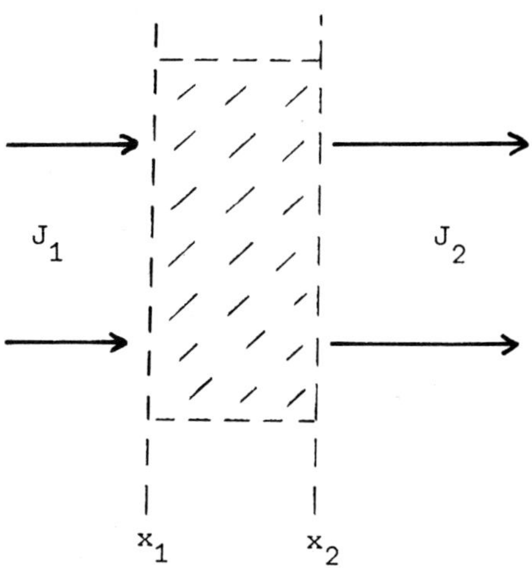

Fig. 9.3 Sketch for derivation of continuity equation. Volume element shown by shading.

Let the cross-sectional area of the element (in any plane of constant x) be unity. Let the particle current density at x_1 be J_1 and that at x_2 be J_2 . Then in unit time J_1 particles will enter the element at x_1 and J_2 particles will leave at x_2 . If J_1 differs from J_2 the number N of particles contained in the element will change with time such that

$$J_1 - J_2 = \frac{dN}{dt} . \tag{9.28}$$

If $\bar{c}$ is the average concentration of particles in the element we see that

$$N = (x_2 - x_1)\, \bar{c} , \tag{9.29}$$

since $(x_2 - x_1)$ is the volume of the element. From Eq. (9.29) we obtain

$$\frac{dN}{dt} = (x_2 - x_1) \frac{d\bar{c}}{dt} . \tag{9.30}$$

Combining Eqs. (9.28) and (9.30) we eliminate dN/dt:

$$J_1 - J_2 = (x_2 - x_1) \frac{d\bar{c}}{dt} . \tag{9.31}$$

If we divide both sides of Eq. (9.31) by $(x_1 - x_2)$ and then let x_2 approach x_1 we arrive at the equality:

$$\lim_{x_2 \to x_1} \frac{(J_2 - J_1)}{(x_2 - x_1)} = \lim_{x_2 \to x_1} \frac{-d\bar{c}}{dt} \tag{9.32}$$

By definition both sides of Eq. (9.32) represent derivatives; we shall now need to write them as partial derivatives, recognizing that J and c are functions of both x and t. The left hand side of Eq. (9.32) is just $\partial J/\partial x$ and the right hand side is $-\partial c/\partial t$; both derivatives are to be evaluated at x_1 at the same instant. But x_1 is really arbitrary and will henceforth not be distinguished from the general point x. From Eq. (9.32) we therefore obtain the partial differential equation

$$\frac{\partial J}{\partial x} = - \frac{\partial c}{\partial t} . \tag{9.33}$$

Equation (9.33) is sometimes called the <u>continuity equation</u>: from its derivation we realize that it amounts to a statement of "conservation of particles." By introducing the expression for J from Fick's First Law, Eq. (9.27), into Eq. (9.33) we obtain

$$D \frac{\partial^2 c}{\partial x^2} = \frac{\partial c}{\partial t} , \tag{9.34}$$

which is the desired diffusion equation in which c is the only dependent variable. Equation (9.34) is sometimes called Fick's Second Law.

9.4 SOLUTIONS OF DIFFUSION EQUATIONS

9.4.1 Steady-State Diffusion Along x

We consider first steady-state diffusion, when $\partial c/\partial t$ is equal to zero. Equation (9.34) then becomes

$$\frac{\partial^2 c}{\partial x^2} = 0, \tag{9.35}$$

of which a general solution is

$$c = a'x + a'' , \tag{9.36}$$

where a' and a" are constants. When Eq. (9.36) applies we have from Fick's First Law, Eq. (9.27), that

$$J = -D \frac{\partial c}{\partial x} = -Da' , \tag{9.37}$$

from which we see that J is a constant. If the diffusion constant D and the current J are known we can determine the constant a' from Eq. (9.37); the constant a" can then be found from Eq. (9.36) if c is known at some particular value of x. Thus if $c = c_o$ when $x = 0$ we see that a" is just c_o .

On the other hand if J is now known but c is known at two values of x, both a' and a" can be determined from this information. Specifically if c has the values c_1 and c_2, respectively, at $x = x_1$ and $x = x_2$ we find (Prob. 9.6) that

$$a' = \frac{c_2 - c_1}{x_2 - x_1} \; ; \qquad a'' = \frac{c_1 x_2 - c_2 x_1}{x_2 - x_1} . \tag{9.38}$$

9.4.2 Time-Dependent Diffusion: Gaussian Distribution

Other important solutions of the differential equation, Eq. (9.34), apply to time-dependent situations. An especially interesting solution c (x,t) is given by the function (Prob. 9.7)

$$c(x,t) = \frac{N}{(4\pi Dt)^{\frac{1}{2}}} \exp\left[-\frac{x^2}{4Dt}\right]. \tag{9.39}$$

This solution can be thought of as representing a situation where at t = 0 a fluid is free of particles except at the plane x = 0. Thus at t = 0 one has from Eq. (9.39) that c = 0 except at x = 0 and that at this singular point c is infinite. This unusual distribution for t = 0 is sometimes called a _delta function_ distribution.

At any value of t the expression for c as a function of x in Eq. (9.39), when divided by N, is identical with the Gaussian distribution of Eqs. (8.45) if σ^2 is replaced by 2Dt. Hence from properties of the Gaussian distribution (Section 8.10) we know that the integral of c over all space is just N, so that N represents the total number of particles in the suspension (per unit area). It follows also that the plots in Fig. 8.4 of W _vs_ x for several values of σ can be used to represent the expression for c(x,t) in Eq. (9.39) . (Prob. 9.8) .

In arriving at Eq. (9.39) our method was, first, to derive the diffusion equation, Eq. (9.34), and then to obtain c(x,t) as a solution of this equation. In deriving Eq.(9.34) we started from Fick's First Law; this law is an empirical one, that is, it was arrived at directly as a "best fit" to experimental data. Alternatively, derivations of Eq. (9.39) have been given (for example, by Einstein, 1956) proceeding directly from basic assumptions such as were used in deriving Eq. (9.17) for Brownian movement. The close connection between Brownian movement and diffusion is seen by comparing Eq. (9.17) with a similar one which can be obtained from Eq. (9.39). Since $c(x,t)$ in the latter equation has the form of the Gaussian distribution of Eq. (8.45) we recognize that (since $\sigma^2 = \langle x^2 \rangle$ the average value of x^2 after time t is

$$\langle x^2 \rangle = 2\,Dt\ . \tag{9.40}$$

Comparing Eqs. (9.17) and (9.40) we see that they are identical if

$$D = \frac{kT}{f}\ . \tag{9.41}$$

Evidently Brownian movement and diffusion are just two aspects of the same phenomenon. One speaks of Brownian movement in connection with individual particles, but of diffusion when referring to the behavior of large numbers of particles. For a sphere of radius 1 μm in a liquid where η is 1 cP the diffusion coefficient at room temperature is

$$D = \frac{kT}{6\pi\eta R} = 2.2 \times 10^{-9}\ \mathrm{cm}^2/\mathrm{sec}.$$

A measure of the distance a particle is displaced in a given time by Brownian movement is the rms (root-mean-square) displacement x_{rms}, that is, the square root of $\langle x^2 \rangle$, the latter being given by Eq. (9.40). For the 1 μm sphere just considered the rms displacement in 1 sec is

$$x_{rms} = \sqrt{2Dt} = 0.66\ \mu\mathrm{m}\ .$$

9.4.3 Time-Dependent Diffusion: Error-Function Distribution.

Another important solution of Eq. (9.34) is

$$c(x,t) = \frac{2}{\sqrt{\pi}} \int_0^z \exp(-y^2)\, dy = \operatorname{erf}(z) , \tag{9.42}$$

where $z = x/\sqrt{4Dt}$. Here erf (z) is known as the error function and is discussed in texts and handbooks on mathematical functions; see, for example, W. Gautschi (1965). Recalling Section 8.10 we see that erf(z) is closely related, but not identical, to $W[0,x_2]$ as defined in Eq. (8.47). From rules of calculus the derivative of c with respect to x is

$$\frac{\partial c}{\partial x} = \frac{dc}{dz} \cdot \frac{\partial z}{\partial x} = \frac{2}{\sqrt{4\pi Dt}} \exp\left(\frac{-x^2}{4Dt}\right). \tag{9.43}$$

We see that the expression for $\partial c/\partial x$ in Eq. (9.43) is identical to the function c(x,t) in Eq. (9.39) for N=2 ; this expression is therefore a solution of the diffusion equation, Eq. (9.34). From this it follows that (as anticipated) the expression for c(x,t) in Eq. (9.42) is also a solution of Eq. (9.34) . (Prob. 9.9) .

A short table of erf (z) is given in Appendix Table C.4 ; the function increases monotonically from zero at z = 0 to a limiting value of unity as z increases indefinitely. Using such a table plots can be made of c <u>vs</u> x for various values of D and t according to the expression in Eq. (9.42); see Fig. 9.4. For example, if D is 10^{-5} cm^2/sec and t is 2.5 sec we see that z = 100 x ; the concentration c for x = 10^{-3} cm is then obtained from Table C.4 as the value of erf (z) for z = 0.1 . These plots represent distributions of concentration at various times for the situation of a suspension initially with unit concentration everywhere; at time t = 0 the suspension is placed in contact with a "particle sink" at x = 0. This sink is such that a particle striking the plane x = 0 is absorbed, or undergoes a reaction, so that it never returns to the suspension. After t = 0 the concentration therefore always remains zero at x = 0. As would be expected the effects of the sink are felt at greater distance as time goes on. For example, we see from Table C.4 that the value of x , call it x_h , at which the concentration is reduced by half (i.e., to 0.5) corresponds approximately to z = 0.48 ; hence

$$x_h = 0.48\sqrt{4Dt} = 0.96\sqrt{Dt} \; . \tag{9.44}$$

The "half-concentration" distance x_h increases proportionally to $\sqrt{Dt}$.

At x = 0 the concentration gradient gradually decreases with time, being given from Eq. (9.43) by

$$\left(\frac{\partial c}{\partial x}\right)_{x=o} = \frac{1}{\sqrt{\pi Dt}} \; . \tag{9.45}$$

We notice the importance in each of the two preceding equations of the quantity $\sqrt{Dt}$. This quantity has physical meaning for us from the analysis in Section 9.4.2. We saw there that the rms value of distance x_{rms} which a particle migrates along x in Brownian movement is just equal to $\sqrt{2Dt}$.

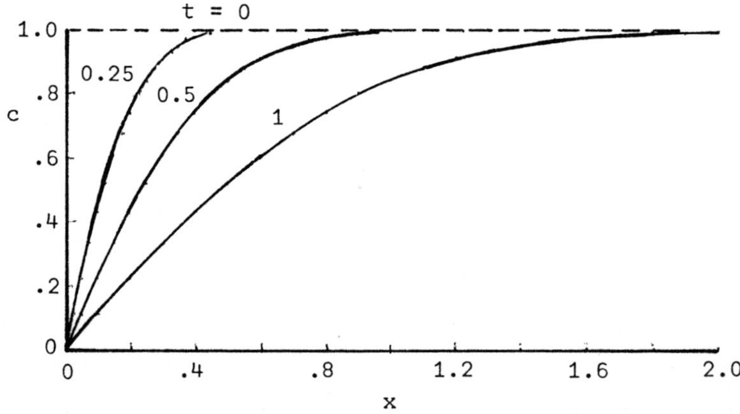

Fig. 9.4 Plots of c vs x based on Eq. (9.42). Numbers attached to curves give $\sqrt{4Dt}$ in cm.

The error-function distribution is a useful one. We shall refer to it again in connection with a technique for measurements of diffusion in gels. (Section 11.6.)

8.4.4 Steady-State Diffusion: Spherical Symmetry

Another situation of considerable interest concerns steady state diffusion when the particles are distributed with spherical symmetry about some point. Such a situation might exist for example, when gases in a solution diffuse toward or away from a biological cell. Instead of Eq. (9.27) we now write for Fick's First Law:

$$J = -D \frac{\partial c}{\partial r} . \tag{9.46}$$

We let Q be the rate at which particles flow outward across a spherical surface of radius r . Then

$$Q = 4\pi r^2 J = -4\pi r^2 D \frac{\partial c}{\partial r} . \tag{9.47}$$

In steady state flow Q must be independent of r . Hence

$$r^2 \frac{\partial c}{\partial r} = A , \tag{9.48a}$$

where A is a constant given by

$$A = - \frac{Q}{4\pi D} . \tag{9.48b}$$

Equation (9.48a) is a simple differential equation of which a solution is

$$c = -Ar^{-1} + B , \tag{9.49}$$

where B is a constant. If Q is known, A is determined by Eq. (9.48b); B may then be determined from knowledge of c at some value of r . For example, if $c = c_\infty$ when r is infinite, then $B = c_\infty$. In the situation involving diffusion toward or from a biological cell r will be effectively infinite for most purposes when it is equal to ten or more cell diameters; the concentration c_∞ is therefore that in the main body of the suspension, where the concentration is essentially uniform. If Q is not known the constants A and B can be determined from knowledge of c at two values of r . Thus for diffusion relative to a cell of radius a we might know the concentration c_o at r = a as well as the concentration c_∞ at "infinite" values of r . We then find (Prob. 9.10) that

$$A = a(c_\infty - c_o) \; ; \; B = c_\infty \; ; \tag{9.50}$$

from Eq. (9.49) we have therefore that

$$c = (c_o - c_\infty)\,\frac{a}{r} + c_\infty \; . \tag{9.51}$$

Using the expression for A from Eq. (9.50) in Eq. (9.48b) we find that

$$Q = 4\pi Da\,(c_o - c_\infty) \; . \tag{9.52}$$

9.5 EQUILIBRIUM DISTRIBUTION OF PARTICLES IN FORCE FIELDS

In previous sections of this chapter we discussed Brownian movement and diffusion as they would occur in the absence of gravitational, electrical and other force fields of external origin. The assumption was also implicit that our coordinate system was an <u>inertial</u> one; that is our coordinate system was not rotating (as it might be if we were discussing centrifugation) or accelerating in any other way. These assumptions are evident in the equation of motion, Eq. (9.1), assumed in treating Brownian movement. For if the coordinate system were not inertial Newton's second law, on which the equation is based, would not be applicable. And if steady force fields were present further terms would need to be added to Eq. (9.1). We recall that the force X which appears in Eq. (9.1) is not a steady force but represents effects of fluctuating intermolecular interactions.

When steady force fields are present the particle motion consists of two parts: a steady part brought about by the force field and a fluctuating part, namely, the Brownian movement. We consider now the first part. If a force field is applied at $t = 0$ there will be a very short time interval during which the velocity of each particle rises to its steady state value; the time required is characterized by the constant τ defined in Section 6.3. After this short transient each particle travels with a velocity given in part by F/f, where f is the frictional constant defined in Section 6.2 and F is the force, assumed here to be along x. If there are c particles per unit volume a particle flow will result such that J, the particle current density along x (see definition of J in Section (9.3)), will be

$$J = \frac{cF}{f} . \tag{9.53}$$

This flow will give rise to gradients of the particle concentration c. When concentration gradients exist another component of net particle flow occurs, because of Brownian movement; this is treated conveniently in terms of laws for diffusion. From Fick's First Law, Eq. (9.27), the particle current density J' for diffusion is $-D\, \partial c/\partial x$.

When equilibrium is established the total current must be zero; that is

$$J + J' = \frac{cF}{f} - D\frac{\partial c}{\partial x} = 0 . \tag{9.54}$$

It is convenient to use Eq. (9.41) and replace D by kT/f; after rearrangement we then have

$$\frac{1}{c}\frac{\partial c}{\partial x} = \frac{F}{kT} \tag{9.55}$$

or

$$\frac{\partial}{\partial x} \ln c = \frac{F}{kT} . \tag{9.56}$$

We now integrate both sides of Eq. (9.56) from zero to x^*, the latter an arbitrary value of x, bearing in mind the possibility that F may be a function of x. We write the result in the form

$$\ln\left(\frac{c}{c_o}\right) = \frac{1}{kT} \int_o^{x^*} F(x)\, dx , \tag{9.57}$$

where c_o is a constant of integration, equal to the concentration c at $x^* = 0$. But we recall from elementary physics that the potential energy (or increase in potential energy) $U(x^*)$ of a body which has been brought from x=0 to x=x* in a field where it is acted on by a force F(x) is just

$$U(x^*) = -\int_o^{x^*} F(x)\, dx \ . \tag{9.58}$$

Comparing the above two equations we see that Eq. (9.57) can be rewritten as

$$\ell n\left(\frac{c}{c_o}\right) = -\frac{U}{kT} \ .$$

Taking the antilogarithm of both sides of this equation we obtain the important result,

$$\frac{c}{c_o} = \exp\left(\frac{-U}{kT}\right). \tag{9.59}$$

The expression on the right hand side of this equation is often called a <u>Boltzmann factor</u>. While we derived Eq. (9.59) for the special situation where the force is along x and dependent only on x , it is applicable much more generally. It follows from this equation that in a suspension (or gas) the concentration c is relatively low in regions when the potential energy is high and, conversely, c has its greatest values at minima of U.

As an example suppose the potential energy U arises from the gravitational field near the earth. Let us assume the x direction is upward, that the particle has volume V and density ρ_o and that the suspending fluid has density ρ . Then the force F along x on each particle is the difference between the buoyant force ρgV and the weight $\rho_o gV$. From Eq. (9.58), replacing x* by h, U is then given at any height h by

$$U = -\int_o^h (\rho gV - \rho_o gV)\, dx = (\rho_o - \rho)\, gVh \ . \tag{9.60}$$

Using this expression for U in Eq. (9.59) we obtain the distribution

$$\frac{c}{c_o} = \exp(-\beta h) , \tag{9.61a}$$

where

$$\beta = \frac{(\rho_o - \rho)gV}{kT} . \tag{9.61b}$$

It is interesting that Eqs. (9.61) represent a variation of concentration with height similar to the variation of pressure with height exhibited in Eq. (5.18) for an isothermal gaseous atmosphere. This similarity reflects the idea that particles in suspension behave like a gas. In Eqs. (9.61) we see that the concentration is c_o at $h = 0$; this is a consequence of the choice of arbitrary constant in writing Eq. (9.57). As the height h increases the concentration c decreases such that c is reduced by a factor of e or 2.72 when h is increased by the distance β^{-1}; this quantity β^{-1} may be called the <u>diffusion layer thickness</u>. For a molecule of hemoglobin Hb in water we let $(\rho_o-\rho)$ be 0.3 gm/cm^3 and V be about 10^{-19} cm^3 ; then β^{-1} becomes 1430 cm. Obviously we expect little settling of Hb in a stationary test tube of water! For red cells in saline $(\rho-\rho_o)$ might be one-third as great and the volume at least 10^8 times greater, leading to a β^{-1} value of about 0.4 μm or less. These cells will clearly settle to the bottom of a suspension if left undisturbed. In the blood of our circulatory system this settling is evidently prevented by convection in the circulation.

We see from Eq. (9.61b) that the gravitational constant g is an important factor in determining the diffusion layer thickness β^{-1} . In outer space, at distances well away from the earth and other massive bodies g is small and hence β^{-1} is relatively large, a fact which may have some significance. On the other hand, in terrestrial laboratories the possibility exists of simulating a considerable increase in g by use of centrifugation. We saw in Section 6.3.2 that for a given particle in a given liquid the sedimentation velocity is greater in a rotating vessel of angular velocity ω than in a static gravitational field by the g-factor $\omega^2 x/g$. Here x is the distance of the particle from the axis of rotation. If the derivation leading to Eqs. (9.61) is examined it will be seen that these same equations are approximately valid for a suspension of particles in a centrifuge tube provided that g is replaced by $\omega^2 x$. In this approximation it is assumed that the distance from the axis of rotation is nearly constant in the region of interest. It follows that the values of β^{-1} in a centrifuge are smaller than those in a gravitational field by the g-factor; this factor can be as large as 300,000. Reduction of a diffusion layer thickness by centrifugation is important in biophysical techniques for studying macromolecules.

9.6 DIFFUSION COEFFICIENTS

9.6.1 Determination of D for a Gas in Solution

A variety of experimental methods have been used to determine the diffusion coefficient D for molecules in solution. For oxygen molecules in water and aqueous solutions Wise and Houghton (1969) measured D by observing the shrinkage of oxygen gas bubbles during diffusion. For a spherical bubble the theory of Section 9.4.4 is applicable, provided the diffusion has been occurring for a time long enough to reach a steady state condition; a criterion for this is that the typical diffusion distance $\sqrt{Dt}$ should be much greater than the bubble radius. For a bubble of radius $\underline{a}$ diffusing into gas-free water Eq. (9.52) is applicable with $c_\infty = 0$; Q then represents the rate at which O_2 molecules diffuse outward from the bubble. The number of O_2 molecules contained in a bubble of radius $\underline{a}$ is given by the volume $4\pi a^3/3$ multiplied by the number density n ; the latter is 2.52×10^{19} molecules per cubic centimeter at $20^\circ C$. As diffusion occurs the molecules leave the bubble at the rate Q ; of course, Q is also equal to the rate of decrease of the number of molecules contained in the bubble. Hence

$$-Q = n \frac{d}{dt}\left(\frac{4\pi a^3}{3}\right) = 4\pi a^2 n \frac{da}{dt} . \quad (9.62)$$

Combining this result with Eq. (9.52) (with $c_\infty = 0$) we obtain

$$D = -\frac{n}{c_o} a \frac{da}{dt} . \quad (9.63)$$

The concentration c_o is the number density of O_2 molecules just outside the bubble surface at $r = a$; this is considerable less than the number density n inside the molecule, being limited by the solubility of oxygen in water. At 20°C the ratio c_o/n is found by Wise and Houghton to be 0.031 . We see that measuring the radius a and its rate of change da/dt yields enough information to determine the diffusion coefficient from Eq. (9.63) . By this means Wise and Houghton obtained data on D for the O_2 molecule in water, as cited in Section 9.6.5 .

9.6.2 Molecular Weight Determination from D and s .

For a macromolecule the diffusion coefficient D is an important measureable characteristic. Together with information on the sedimentation coefficient s and the relative density, knowledge of D leads to an evaluation of the macromolecule mass m. For a combination of Eqs. (6.30) and (9.41) leads (Prob. 9.11) to

$$m = \frac{kTs}{D[1 - (\rho/\rho_o)]} ; \quad (9.64)$$

here ρ_o is the density of the macromolecule and ρ is that of the suspending liquid; m is equal to $\rho_o V$ where V is the molecular volume. To obtain the molecular weight M in daltons we multiply m by Avogadro's number N_a, the number of molecules per mole. Instead of the density ρ_o its reciprocal, the specific volume $\bar{v}$, is often used. Equation (9.64) is frequently written

$$M = \frac{R_g Ts}{D(1-\rho\bar{v})} , \quad (9.65)$$

where R_g is the molar gas constant, equal to kN_a . Equation (9.65) is used for calculating molecular weights of biological macromolecules from measured values of s, D and $\bar{v}$. (Prob. 9.12).

9.6.3 Sedimentation-Equilibrium Method for Measuring D

One method of measuring diffusion coefficients for macromolecules is the technique of sedimentation equilibrium, based on equations similar to Eqs. (9.61). Referring to Eqs. (9.61b), (6.30) and (9.41) we see that the static diffusion layer thickness β^{-1} can be written

$$\beta^{-1} = \frac{D}{sg} . \tag{9.66}$$

As noted in Section 9.5 the length β^{-1} for suspensions of macromolecules can be reduced to convenient values by using centrifugation. When the test sample of suspension in a centrifuge is so small that all parts can be assumed to be at approximately the same distance x from the axis of rotation Eq. (9.66) becomes

$$\beta^{-1} = \frac{D}{s\omega^2 x} , \tag{9.67}$$

where ω is the angular velocity of the centrifuge. In the technique of sedimentation equilibrium a suspension of the macromolecules of interest, placed in a suitable analytical cell, is subjected to continuous rotation in a centrifuge until equilibrium is established. The length β^{-1} is then measured by optical means. If the sedimentation coefficient s is also known for the macromolecule its diffusion coefficient D can be calculated from Eq. (9.67). A disadvantage of this technique is that it is rather slow. The time required for equilibration may be of the order of ℓ^2/D , where ℓ is the depth of the suspension; if ℓ is 0.1 cm and D is 10^{-6} cm^2/sec the time ℓ^2/D is nearly three hours.

9.6.4 Measurements of D with Nonequilibrium Methods

The more widely used methods for measuring diffusion are based on time-dependent solutions of the diffusion equation, such as the error-function solution given in Eq. (9.42). In one situation a layer of uniform suspension in a given liquid is very carefully covered with a layer of pure liquid. The concentration then varies with height according to a step function, as shown for $t = 0$ in Fig. 9.5. As time goes on diffusion occurs and the boundary becomes less sharp. The distribution can be expressed as in Eq. (9.42). Below the original boundary, say, at $x = 0$, the concentration is given by $1 + \text{erf}(-z)$ and above the boundary by $1 - \text{erf}(z)$ where, as in Eq. (9.42), z is $x/\sqrt{4Dt}$. At $x = 0$ the concentration gradient $\partial c/\partial x$, which we write as $(\partial c/\partial x)_0$, is given by Eq. (9.45) and involves D. Hence one method of determining D is to measure the gradient $(\partial c/\partial x)_0$ as a function of time. (Such measurements are conveniently made by a Schlieren optical method; see Schachman (1959)). According to Eq. (9.45) a plot against time t of the reciprocal of the square of $(\partial c/\partial x)_0$ will be a straight line with slope πD. Hence from a measurement of said slope the diffusion coefficient D can be determined. A different, but related, method for analyzing data for this situation is discussed in an instruction manual. (Chervenka, 1970).

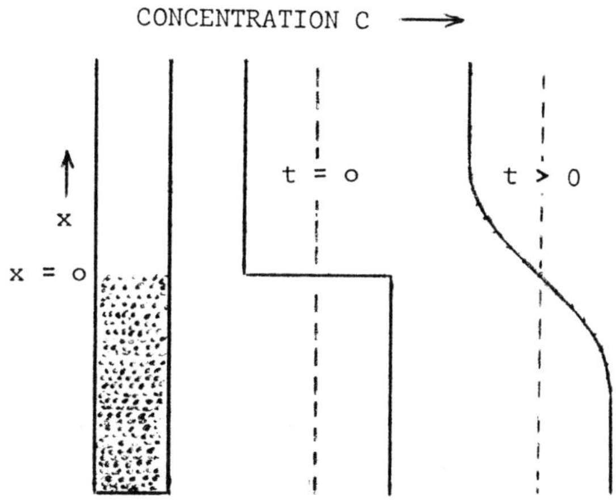

Fig. 9.5 Diffusion starting from a step-function distribution.

Still another method for measuring diffusion coefficients of macromolecules is called the Archibald method. (Schachman, 1959). This method is based on the fact that Eq. (9.54) applies (i.e., the total particle current is zero) at all times at both the top and bottom of a sedimenting cell, even though equilibrium has not been established throughout the cell. Hence Eq. (9.55) applies at all times at these two points. In a gravitational field we replace F by fsg, where s is as defined in Section 6.3.1; in a centrifugal field we let F be approximately given by $fs\omega^2x$, if x is nearly constant throughout the sample. Also replacing kT by Df we have from Eq. (9.55), for the centrifuge,

$$\frac{1}{c}\frac{\partial c}{\partial x} = \frac{s\omega^2 x}{D}, \tag{9.68}$$

valid at either end of the centrifuge cell. Until equilibrium is established both $\partial c/\partial x$ and c vary with time, but in such a way that their ratio, left hand side of Eq. (9.68), is independent of time. A simultaneous measurement of both c and $\partial c/\partial x$ at either end of the tube suffices to determine $s\omega^2x/D$. If s is known from other measurements (and the instrumental parameters ω and x are known) the diffusion coefficient D can be calculated.

9.6.5 Values of D for Small Molecules and Ions

We develop a feeling for the significance of diffusion by considering actual numerical values of the diffusion coefficient D for ions, atoms, and molecules of biological interest. For our unit we use the <u>fick</u>, defined as 10^{-7} cm^2/sec. No diffusant is more important to vital function than the oxygen molecule. Animal cells require oxygen in carrying out biochemical reactions needed for growth and repair and also reactions which provide energy for movement. In a cell suspension oxygen molecules dissolved in the liquid travel to the boundary of a cell, permeate its membrane or wall, and then travel to mitochondria or other sites of biochemical activity. Diffusion, along with convection and other processes still not understood, is involved in transport of the oxygen along the various steps in its path from solution to reaction site. In an organism as complex as the human body extra steps are involved, in connection with the circulatory system. Here oxygen must be transported from the lung to the blood, then pass through the red cell membranes to be adsorbed on Hb molecules. The red cells then travel with blood flow to distant parts of the body. At a destination oxygen is released from the Hb, passes out through the red cell membrane, through the blood fluid to a capillary wall, then through this wall to a cell where it is utilized. Again diffusion is an important process; it is involved in various parts of the journey from lung to cell.

In measuring the diffusion coefficient D for O_2 molecules in water Wise and Houghton used the bubble method described in Section 9.6.1. Making measurements at different temperatures they obtained the results in Table 9.2.

TABLE 9.2

DIFFUSION COEFFICIENTS FOR OXYGEN MOLECULES IN WATER

From Wise and Houghton (1969)

T in $^{\circ}$C. D in ficks.

T	10°	20°	30°	40°	50°	60°
D	182	238	297	393	431	583

Accurate theory does not exist for predicting the values of D for oxygen or other small molecules in solution. However a very useful expression is

$$D = \frac{kT}{6\pi\eta R} , \tag{9.69}$$

which comes from Eq. (9.41) using for f the Stokes expression from Eq. (6.15). The order of magnitude for D is given correctly by this expression; also the temperature dependence of D is accounted for in terms of the temperature dependence of η ; see Table 7.1. (Prob. 9.13).

For most other gases in aqueous solution D has comparable values. For ammonia, bromine, chlorine, carbon dioxide and nitrogen the ratio of D to that for oxygen is about 0.8, 0.5, 0.6, 0.8 and 0.8 , respectively. For hydrogen, however, D is greater, being more than twice the value for oxygen. (Engineering Manual, 1967).

For ions in solution D is often calculated from measurements of electrical conductance. Some of the small ions have diffusion coefficients comparable to those of the oxygen molecule. Thus for the sodium, potassium and chloride univalent ions in aqueous solution at 25°C, the D-values are 133, 196 and 203 ficks, respectively. However, the hydrogen ion H^+ is most unusual; under the same conditions D is 931 ficks for this ion. (Longsworth, 1972).

9.6.6 Values of D for Molecules of Intermediate Size

Among larger molecules, the amino acids are of especially great biological importance. These are the monomer units which link together to form protein macromolecules or protein subunits. There are about 20 different amino acids found in proteins; they vary in molecular weight from 75 daltons for glycine to 777 daltons for thyroxine. In our body diffusion is important for transport of these monomers along paths from the digestive tract to the interior of a cell. In a cell diffusion is involved in protein synthesis; here amino acids are selected and put in ordered arrays, apparently according to instructions from DNA macromolecules.

Careful measurements of diffusion coefficients for a number of amino acids have been carried out by L. G. Longsworth (1953).

His method is one of those discussed in Section 9.6.4 and is illustrated in Fig. 9.4. An approximate step function of concentration is established initially by forming a sharp boundary between suspension and pure liquid. Spreading of the boundary in time is followed by an optical method, specifically, with Rayleigh interference fringes. Results for nine amino acids are given in Table 9.3 for two temperatures, $1^{\circ}C$ and $25^{\circ}C$. A measure of the molecular size for any amino acid is the volume V occupied by one mole of the amino acid in solution. The volume occupied by one molecule is, of course, V/N_a where N_a is Avogadro's number. Values of V are listed in Table 9.3.

TABLE 9.3

DIFFUSION COEFFICIENTS FOR AMINO ACIDS AND SUGARS

From Longsworth; see References. Molecular weight MW in daltons; molar volume V in $cm^3/mole$; diffusion coefficient D in ficks; $DV^{1/3}$ in fick-cm.

MW	Amino Acids	V	$D(1^{\circ}C)$	$D(25^{\circ}C)$	$DV^{1/3}$
75	Glycine	43.5	51.5	105.5	371
89	Alanine ( )	60.6	43.2	91.0	357
115	Proline	81.0	41.9	87.9	380
131	Hydroxyproline	84.4	39.3	82.6	362
155	Histidine	99.3	34.5	73.3	339
165	Phenylalanine	121.3	32.4	70.5	349
204	Tryptophan	144.1	30.4	65.9	346
	SUGARS				
180	Glucose	111.9	31.4	67.3	324
342	Sucrose	209.9	24.1	52.1	310

An interesting comparison was made of these data for D, with those which would be expected if Eq. (9.69) were valid. If the molecules were perfectly spherical V would be proportional to the cube of the radius R; hence D would be inversely proportional to $V^{1/3}$. We see by inspection of Table 9.3 that D does indeed decrease regularly as V increases. A quantitative test is obtained by calculating $DV^{1/3}$, which should be the same for all molecules if Eq. (9.69) were strictly applicable. We see that this is not true with high accuracy, but is valid in a rough way.

Another expectation from Eq. (9.69) is that the diffusion coefficient at 25°C should be greater by a factor of 1.95 than that at 1°C, since the viscosity for water varies (inversely) in this ratio. (Table 7.1). This is found to be roughly correct but the ratio is actually about 2.1.

Also listed in Table 9.3 are data for two sugars: glucose, a monosaccharide of molecular weight 180 daltons, and glucose, a disaccharide (formed by a chemical joining of two monosaccharide units) of molecular weight 342 daltons. We see that Eq.9.69 is valid to about the same extent as for amino acids: the quantity $DV^{1/3}$ is just a little less than for the amino acids, while the viscosity ratios between the two temperatures are each about 2.1.

9.6.7 Values of D and Related Parameters for Macromolecules

For macromolecules measurements of D are made partly for the purpose of gaining information on the size and shape of the macromolecules. We saw in Eq. (9.65) that the diffusion coefficient is used in this important equation for calculating molecular weight. Also from Eq. (9.41) the friction constant f at a given temperature can be calculated from knowledge of D. If a molecule is known to be spherical its radius R can be calculated from knowledge of D at a given temperature in a medium of know viscosity, using

Eq. (9.69). Information on the shape of a molecule is obtained by calculating the frictional coefficient f_o which would exist if the molecule were spherical, and comparing this with the measured coefficient f . To obtain f_o we note that the volume of one molecule is $M\bar{v}/N_a$, where M is the molecular weight, N_a is Avogadro's number and $\bar{v}$ is the specific volume (volume per unit mass) for the molecular substance. For a sphere of the same volume the radius must be R_e such that

$$\frac{4\pi R_e^3}{3} = \frac{M\bar{v}}{N_a} . \tag{9.70}$$

We may call R_e the effective radius of the molecule. The constant f_o is then calculated from the Stokes expression in Eq. (6.15) as

$$f_o = 6\pi\eta R_e , \tag{9.71}$$

where R is obtained from knowledge of $\bar{v}$ and M, according to Eq. (9.70).

In Table 9.4 are data for a few proteins on s, D, $\bar{v}$, M and f/f_o . Here s, D and $\bar{v}$ are measured quantities while M, f and f_o are calculated. Listed first are albumins from various sources, including two human sources. Albumin is the primary protein constituent of human plasma, its concentration being about 4 gm per 100 ml of plasma. It is important in maintaining plasma viscosity and osmotic pressure. By examining data on albumin for the two human sources we get an impression of the extent to which values for s, D and $\bar{v}$ are reproducible for a given species. By comparing data from human, horse and other sources we see the extent to which values are species-dependent. Since f/f_o differs from unity the albumin molecules are not accurately spherical but may be approximately ellipsoidal. For an ellipsoid an f/f_o ratio of 1.3 has been shown to represent a ratio of about 7 in the lengths of the major and minor axes. (Cohn and Edsall, 1943).

Another important protein constituent of plasma is γ-globulin, which is present in concentrations of about 1.2 to 1.8 gm per 100 ml of plasma and is important in defense against diseases. When an animal organism is invaded by bacteria the number of γ-globulin units increases considerably. The new units are antibodies; they react with antigens, that is, proteins produced by the bacteria. Clumps are then formed which are subsequently devoured by white cells.

Transferrin is a relatively minor protein constituent of blood, accounting for about 0.25 gm per 100 ml of plasma, which plays an important role in transporting iron.

Fibrinogen (see Section 7.4) is present in the human blood plasma at a concentration of about 0.3 gm per 100 ml. Its high value of f/f_o is consistent with the fact that its molecular length is much greater than its width. Fibrinogen is important in the clotting of blood, an interesting and complicated process which has been discussed by Waugh (1959).

Insulin was the first protein to have its basic monomer structure determined in detail; for this monomer the molecular weight is about 6000 daltons. The unit whose properties are given in Table 9.4 is apparently an aggregate of 4 monomers. The molecular weight of insulin is also frequently quoted as 12,000 daltons, namely, a unit comprised of 2 monomers. For a protein such as this one, with a strong tendency to aggregate, there can be difficulty in arriving at a "true" molecular weight. The 6,000 dalton monomer is rather like a two-stranded rope, one of whose strands is shorter than the other. Each strand is a polypeptide chain, meaning that it consists of a series of amino acids chemically bonded together. One of the chains contains 21, and the other 30 amino acids. Insulin is a hormone produced in the pancreas; it has an important function in regulating the glucose level in the blood.

TABLE 9.4

PARAMETERS OF MACROMOLECULES

From Handbook of Biochemistry (1970) except starred entry from Wise and Houghton (1969). Sedimentation constant s in svedbergs; diffusion coefficient D in ficks; specific volume $\bar{v}$ in cm^3/gm; all at 20°C. Molecular weight MW in daltons.

	s	D	$\bar{v}$	M	f/f_o
Human serum albumin	4.60	6.10	0.733	68,500	1.29
Human serum albumin	4.67	5.93	0.736	72,300	1.30
Horse serum albumin	4.58	6.42	0.748	68,600	1.22
Canine serum albumin	4.84	6.19	0.729	70.000	1.26
Rat serum albumin	4.41	6.0	0.720	63,650	1.35
γ globulin	7	4.0	0.739	160,000	1.38
Transferrin	5.5	5.0	0.725	90,000	1.37
Human fibrinogen	7.63	1.97	0.723	341,000	2.34
Insulin	1.95	7.30	0.735	24,400	1.52
Human hemoglobin	4.20				
	4.31				
	4.48				
	4.5				
	4.55				
	4.56			64,450	
	4.76			76,000	
	5.3	6.7			
		5.7*			
Horse hemoglobin			0.749		

9.7 ROTARY DIFFUSION

We saw in Section 9.2.3 that rotary Brownian movement occurs and is quite analogous to linear Brownian motion. It is to be expected then that diffusion equations similar to those taken up in Section 9.3 should apply to rotational motion. To see the form these take let us consider a particle in suspension; suppose that it is a body of revolution (i.e., symmetrical about an axis) and that it is centered at 0. We define Cartesian axes (x, y, z) with origin at 0 and suppose the axis PP of the particle lies in the xy plane making an angle θ with the x axis as it rotates about 0. (See Fig. 9.6). To discuss a suspension we refer to a quantity w representing concentration in θ-space, analogous to the quantity c which represents concentration in (x, y, z) space. In defining

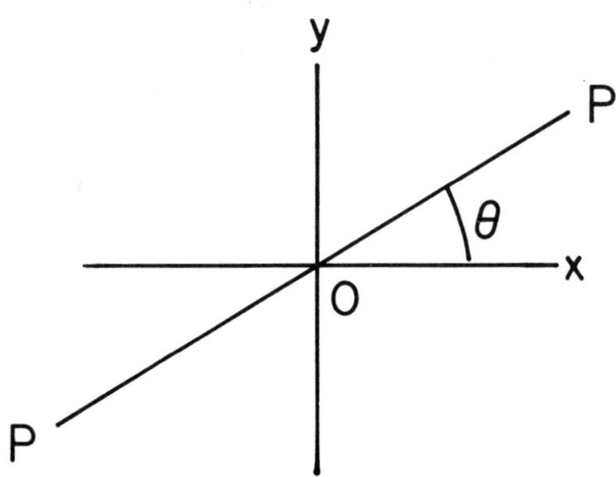

Fig. 9.6 The line PP represents the axis of symmetry for a rod-shaped particle which rotates in the (x,y) plane about its center at 0.

w we let n_{12} represent the number of particles per unit volume for which the angle θ lies between θ_1 and θ_2, and define w at θ_1 by a limiting procedure:

$$w(\theta_1) = \lim_{\theta_2 \to \theta_1} \frac{n_{12}}{\theta_2 - \theta_1} \quad . \tag{9.72}$$

Since θ_1 is an arbitrary point we see that the quantity w for any angle θ represents the number of particles in the immediate vicinity of θ per unit volume per unit angle. If the particles are randomly oriented $w(\theta)$ will be essentially constant, independent of θ . But in a force field where the potential energy of a particle depends on θ , or in a flow field where velocity gradients exist, the particles will tend to be oriented; that is, $w(\theta)$ will exhibit a maximum in a favored direction. From the definition of $w(\theta)$ we can always write

$$\int_0^{2\pi} w \, d\theta = c \ , \tag{9.73}$$

where c (as before) is the particle density, that is, the number of particles per unit volume. In general w will be a function of θ and, because of this, rotary diffusion will occur analogous to the linear diffusion which takes place in a concentration gradient $\partial c/\partial x$. In fact, analogous to Fick's First Law of Eq. (9.27), we can write

$$J_r = - D_r \frac{\partial w}{\partial \theta} \ , \tag{9.74}$$

where J_r represents rotary diffusion flow; that is, J_r gives the number of particles per second per unit volume whose angle increases past a given value θ . By arguments similar to those used in Section 9.3 we arrive at a diffusion equation analogous to Eq. (9.34), namely,

$$D_r \frac{\partial^2 w}{\partial^2 \theta} = \frac{\partial w}{\partial t} \quad . \tag{9.75}$$

For special problems the solutions taken up in Section 9.4 can be applied with suitable translation of symbols. For example, a Gaussian distribution analogous to that in Eq. (9.39) is useful with c replaced by w, D by D_r and x by θ . It applies to a situation where the particles are all perfectly aligned by an external force field, which is then removed at $t = 0$. We now recall Eq. (9.41) and realize that because of the close analogy between linear and rotary diffusion we can write

$$D_r = \frac{kT}{f_r} , \tag{9.76}$$

where f_r is the rotational friction constant defined in Section 9.2.3. Instead of Eq. (9.22), by analogy to Eq. (9.40), we now can write

$$\langle \theta^2 \rangle = 2D_r t , \tag{9.77}$$

where θ is the angular displacement from an initial orientation at $t = 0$. For a sphere of 1 μm radius in a fluid where η is 1 cP we obtain, using the expression for f_r in Eq. (9.23),

$$D_r = \frac{kT}{8\pi\eta R^3} = 0.167 \text{ radian}^2/\text{sec} ;$$

the rms angular displacement in 1 second is

$$\theta_{rms} = \sqrt{2D_r t} = 0.58 \text{ radians} = 33^\circ .$$

A good method for measuring D_r is a suspension of particles involves (1) applying an electric field to align them, then (2) cutting off the field so that particles are free to take part in Brownian movement and thus to diffuse away from the initial orientation. We see from Eq. (9.77) that if $\theta = 0$ initially for all particles the time t required to achieve a specific value of $\langle \theta^2 \rangle$ is inversely proportional to D_r . Somewhat related to this, it has been shown that the time required for the average value of $\cos \Theta$ to decrease from 1.0 to 1/e (or 0.368) is

$$\tau = \frac{1}{6D_r} \quad . \tag{9.78}$$

The characteristic time τ given by this equation is called the relaxation time.

Further analogies to linear diffusion exist. In a force field where the potential energy U of the particle is a function of θ one may follow through reasoning similar to that in Section (9.5) and show that the Boltzmann-type equilibrium distribution of Eq. (9.59) holds also in θ-space, with c and c_o replaced by w and w_o . This result is especially relevant to electrically polarized molecules in electric fields.

Electro-optical methods have been used to obtain values of D_r for biomacromolecules and viruses. O'Konski and Haltner (1956) used a method involving transient electric birefringence to study the basic units of tobacco mosaic virus, TMV, in solution. An electric field was used to align the TMV units; then the field was cut off and the resulting de-orientation of the units followed by an optical birefringence method. (See Section 9.9.2) . The relaxation time τ was found to be about 0.5 msec; from Eq. (9.78) this yields a value for D_r of 330 sec^{-1} . From this value of D_r information on the size and shape of the TMV particles was confirmed. Using Eq. (9.76) they calculated f_r ; then using an expression for f_r suitable for a rod of diameter 150 A and unknown length they calculated the latter to be 3400 A. Calculating the volume of the rod and multiplying by the known density 1.37 gm/cc they obtained its mass. Multiplying this by Avogadro's number they obtained 49.8×10^6 daltons as the molecular weight for the basic TMV virus particle in agreement with results obtained with other methods.

For the collagen molecule D_r was measured by Ananthanarayanan and Veis (1972); they found the high value of 811 sec^{-1}, consistent with a length-width ratio greater than 200 for collagen. (Section 7.3.3).

Another highly asymmetric molecule is fibrinogen, the biomolecule required for blood clotting. (Section 7.4). Billick and Ferry (1956) measured D_r for the fibrinogen monomer obtaining a value of 25,000 sec^{-1} for water as solvent.

Higher values of D_r are expected for molecules of comparable size that are more globular in shape. Krause and O'Konski (1959), using an extended electric birefringence method, obtained for bovine serum albumin a value for D_r of 830,000 sec^{-1}. This high value of D_r for serum albumin compared to that for fibrinogen is especially striking in view of the fact that the molecular weights of the two molecules differ by only a factor of five. (Table 9.4).

9.8 ORIENTATION OF PARTICLES IN A VELOCITY-GRADIENT FIELD.

9.8.1 Basic Theory: Dumbbell Particles

Suppose the velocity u in a suspending liquid is along x and given by Gy, as in Eq. (5.41) and in the arrangement of Fig. 5.2. This fluid flow represents simple shear, and the velocity gradient G is sometimes called the <u>rate of shear</u>. As may be seen by referring to Section (4.6), each element of fluid in this field takes part in clockwise rotation with angular velocity G/2. As might be expected, the particles in suspension also rotate. To examine the motion we take a simple example.

Suppose the particle is a dumbbell, axis in the xy plane, as pictured in Fig. (6.4) and discussed in Section 6.4.2. The angular velocity of the dumbbell is $\partial\theta/\partial t$, which we call $\dot{\theta}$. We assume the center 0 of the dumbbell moves with the fluid in its own vicinity. The situation is therefore the same as if 0 were at $y = o$ which, for convenience, we assume to be true. Using the symbols of Section 6.4.2 the fluid velocity near the sphere A is then $G\varepsilon$ in the x direction; the component of this in the θ direction is $-G\varepsilon \sin\theta$. It is reasonable to assume also that a small dumbbell (with density similar to that of the fluid) rotates with angular velocity $\dot{\theta}$ such that the velocity of the sphere A along θ , namely, $b\theta$, is just equal to the fluid velocity along θ in its own vicinity.

For if $\dot{\theta}$ deviates from this condition a drag force will act on sphere A causing torque on the dumbbell along the z axis through 0 ; the direction of this torque will be such as to restore $b\dot{\theta}$ to the value $-G\epsilon \sin\theta$. We suppose the response to a drag torque is instantaneous, an acceptable approximation if the characteristic time τ' for this response (analogous to τ defined for sedimentation in Section 6.3.1) is small compared to the time for complete rotation in the shearing field. This will be true if $G\tau' \ll 1$. (Prob. 9.14). We then have

$$b\dot{\theta} = -G\epsilon \sin\theta = -Gb \sin^2\theta$$

or

$$\dot{\theta} = -G \sin^2\theta \quad . \tag{9.79}$$

We arrive at the same equation by considering the sphere B . We see that as the dumbbell rotates its angular velocity $\dot{\theta}$ changes according to the value of θ; thus $\dot{\theta}$ is small when the axis is nearly aligned with the flow (i.e. when $\theta \cong 0$) and has its greatest value when the axis is perpendicular to the flow ($\theta = \pi/2$). Since the average value of $\sin^2\theta$ is ½ we see that

$$\langle\dot{\theta}\rangle = -\frac{G}{2} \quad , \tag{9.80}$$

where $\langle\ \rangle$ in this instance refers to an average with respect to θ . It is interesting that, according to Eq. (9.80), the average angular velocity of the dumbbell is equal in magnitude and direction to the angular velocity of an element of fluid. A similar result has been found for a particle in the shape of a thin ellipsoid, that is, a needle-shaped particle. (Jeffery, 1922-23; Cohn and Edsall, 1943).

In the above analysis we have ignored effects of Brownian movement. Because of these effects the dumbbell will experience random angular displacements which are superposed on the motion predicted in Eq. (9.79). Since we are concerned with a large number of particles we can represent the random motion by diffusion theory. Let us consider the total "rotational particle current" J_{tot}, that is, the number of particles for which θ increases past a given value in one second. The total current consists of two parts, one of which is caused by diffusion and is given by Eq. (9.74). The other part is the current which would exist in the absence of diffusion, namely, $w\dot{\theta}_o$, where $\dot{\theta}_o$ has the value $-G \sin^2\theta$ according to Eq. (9.79). For the total current we then have

$$J_{tot} = -D_r \frac{\partial w}{\partial \theta} - wG \sin^2\theta \quad . \tag{9.81}$$

It is clear that J_{tot} must be a constant, independent of θ , in an equilibrium distribution. For suppose the current across θ_1 , for example, was greater than that across a somewhat greater angle θ_2 ; then the number of particles with orientation between θ_1 and θ_2 would increase, in violation of the stated equilibrium condition. For convenience we now divide through by D_r in Eq. (9.81) and introduce symbols α and j defined by

$$\alpha = \frac{G}{D_r} \quad ; \qquad j = \frac{J_{tot}}{D_r} \quad . \tag{9.82}$$

Then Eq. (9.81) reduces to

$$\frac{\partial w}{\partial \theta} + \alpha w \sin^2\theta + j = 0 \quad . \tag{9.83}$$

When there is no flow, G and α are zero and there is no preferred direction for the dumbbells; hence w is independent of θ . Letting w be w_o when $\alpha = 0$, and using Eq. (9.73), we find that $2\pi w_o$ is equal to c, or

$$w_o = \frac{c}{2\pi} \tag{9.84}$$

From Eq. (9.83) it is clear that when $\alpha = 0$ and $w = w_o$ the quantity j must be zero. This means that the current J_{tot} is zero, as is expected in the absence of flow. When α is nonzero, but small, the following approximate solutions of Eq. (9.83) can be found:

$$j = -\frac{\alpha c}{4\pi} \quad , \tag{9.85}$$

and

$$w = \frac{c}{2\pi}\left[1 + \frac{\alpha}{4}\sin 2\theta - \frac{\alpha^2}{8}\sin^4\theta\right] . \tag{9.86}$$

Equations (9.85) and (9.86) are approximately valid when $\alpha \ll 1$; terms involving α^3 , α^4 , and higher powers of α have been neglected. It is not hard to verify that if j and w from the above equations are substituted into Eq. (9.83) the latter is satisfied if terms proportional to α^3, α^4, etc. are neglected.

9.8.2 Solving the Flow-Orientation Equation; Small α.

We now outline a method for arriving at the above expressions for j and w, leaving details for an exercise. (Prob. 9.15). Suppose j and w are of the form

$$j = a_1\alpha + a_2\alpha^2 + \ldots$$
$$w = w_o + b_1\alpha + b_2\alpha^2 + \ldots \quad ; \tag{9.87}$$

here the a_i are constants (independent of θ and α) while the b_i may be functions of θ, but are independent of α; w_o is given by Eq. (9.84); we assume $\alpha \ll 1$. To determine the a_i and b_i, we substitute these series expressions for j and w into Eq. (9.83), then group terms according to powers of α as shown:

$$[\quad]\,\alpha + [\quad]\,\alpha^2 = 0 \,, \tag{9.88}$$

where each bracket [ ] may be a function of θ, but is independent of α; we neglect powers of α higher than α^2. (It is found that terms independent of α cancel out.) Now this equation must be true for arbitrary α (provided that $\alpha \ll 1$) ; this is possible only if each of the brackets [ ] is separately equal to zero. Equating the brackets to zero yields

$$\frac{\partial b_1}{\partial\theta} + w_o \sin^2\theta + a_1 = 0 \,, \tag{9.89a}$$

and

$$\frac{\partial b_2}{\partial\theta} + b_1 \sin^2\theta + a_2 = 0 \,. \tag{9.89b}$$

Further relationships are obtained by integrating all terms in Eqs. (9.89) with respect to θ in the range 0 to 2π. From the first equation we obtain

$$a_1 = -\tfrac{1}{2} w_o \,, \tag{9.90}$$

and from the second equation,

$$2\pi a_2 = -\int_o^{2\pi} b_1 \sin^2\theta \, d\theta \quad . \tag{9.91}$$

After the above expression for a_1 is substituted into Eq. (9.89a) the latter can be integrated to yield

$$b_1 = \frac{c}{8\pi} \sin 2\theta \ . \tag{9.92}$$

With this expression for b_1 the integration in Eq. (9.91) can be carried out, and it is found that

$$a_2 = 0 \ . \tag{9.93}$$

With b_1 and a_2 known, Eq. (9.89b) can now be integrated with the result

$$b_2 = -\frac{c}{16\pi} \sin^4 \theta \ . \tag{9.94}$$

When expressions obtained above for the a_i and b_i are substituted into Eqs. (9.87) we see that the expressions for j and w are indeed those anticipated in Eqs. (9.85) and (9.86).

9.8.3 Discussion of Theoretical Results

Examining Eq. (9.85) we realize that by virtue of Eqs. (9.82) and (9.84) this can be rewritten as

$$J_{tot} = -\frac{Gw_o}{2} \ . \tag{9.95}$$

The quantity $-\frac{1}{2}G$ is just the angular velocity of an element of fluid in the shearing flow and also, by Eq. (9.80), it is the average angular velocity of a dumbbell in the absence of diffusion or Brownian movement. Thus J_{tot} in Eq. (9.95) is just the rotational current that would exist if the angular distribution function w had the value w_o (as if there was no flow), while the average angular velocity had the value corresponding to "no diffusion."

From Eq. (9.86) we see that when $\alpha = 0$ the expression for w reduces to w_o given in Eq. (9.84), as expected. When α is not zero but small enough so that the α^2 term can be neglected the function w has a maximum when θ is $\pi/4$, that is 45° ; at this maximum w takes on the value w_m given by

$$w_m = (1 + \frac{\alpha}{4}) \ w_o \ . \tag{9.96}$$

For example, if $\alpha = 0.5$ the quantity w_m is equal to $1.125\ w_o$. When the α^2 term on the right of Eq. (9.86) is taken into account the position of the maximum is shifted somewhat. Continuing with the assumption that α is small we find (Prob. 9.16) that the maximum occurs at an angle $\theta = \theta_m$ such that

$$\theta_m = \frac{\pi}{4} - \frac{\alpha}{8} \quad . \tag{9.97}$$

Equation (9.97) represents a useful approximation when the velocity gradient G is relatively small; it shows that the angle θ_m corresponding to the most probable orientation decreases as α (and hence the velocity gradient) increases.

The results of Sect. 9.8.2 and those discussed so far in this section illustrate important features of the flow orientation processes. However they are based on simplifying assumptions and, in particular, are valid only for small α . More sophisticated theory was developed for long thin particles by Boeder (1932) and discussed by Morawetz (1965) and by Cohn and Edsall (1943). Table 9.5 shows Boeder's results for θ_m , the value of θ at which maximum W occurs, for various values of α . We see that as α increases from zero the preferred angle θ_m decreases from 45° , approaching 0° as α approaches infinity. In Fig. 9.7 the probability distribution W is shown as a function of θ for several values of α . As α increases the probability curve is characterized by a peak which is increasingly sharp as the particles become more nearly aligned with the flow.

TABLE 9.5

ANGLE θ_M OF MOST PROBABLE ORIENTATION FOR LONG THIN PARTICLES IN A SHEARING FIELD

$\alpha = G/D_r$ where G is velocity gradient and D_r is rotary diffusion coefficient. From Boeder (1932)

α	0.0	0.5	1.0	2.0	3.5	5.0	7.5	10.0	∞
θ_m	45°	$41^o27'$	$38^o15'$	$32^o13'$	$25^o50'$	22°	$17^o30'$	15°	0°

9.9 CONSEQUENCES OF PARTICLE ALIGNMENT

9.9.1 Dependence of Viscosity on Rate of Shear

It was seen in Chapter 7 that for biological suspensions which are caused to flow the shear viscosity coefficient η is not always constant, but instead tends to decrease with increasing velocity gradient G ; this was seen especially for blood and synovial fluids. While detailed explanations are not usually available for this non-Newtonian behavior some insight can be obtained from the ideas discussed in Section 9.8. There it was shown that asymmetric particles tend to become aligned in a velocity gradient field; since the contribution of each particle to the viscosity depends on its orientation we have here the basis for a dependence of η on the velocity gradient G.

To examine possibilities somewhat more specifically we return to the treatment of dumbbell-particles in Section 6.4.2. Here it is shown, Eq. (6.49), that the extra heat generated by the presence of such particles in a field of simple shear is proportional to $\sin^2 \theta$. Following through steps similar to those in Eqs. (6.51) - (6.54) we then find (Prob. 6.9) that κ is proportional to $\sin^2 \theta$. It follows that the intrinsic viscosity $[\eta]$ is also proportional to $\sin^2 \theta$. But in Section 9.8 it is shown for asymmetric particles that as the velocity gradient G increases from zero the most probable angle θ_m decreases monotonically from 45° , approaching zero as G becomes indefinitely large. Hence, as expected, the viscosity is a decreasing function of G. A result valid for relatively small values of G was obtained by Kuhn and Kuhn (1945), by integrating over all orientations in an equilibrium distribution. Their specific expression is

$$\kappa_o = \kappa_{oo} \left[1 - \frac{1}{32} \alpha^2\right] ; \qquad (9.98)$$

here κ_{oo} is the limiting value of κ_o as the velocity gradient G approaches zero. (The quantity κ_{oo} is thus obtained as a result of two limiting processes, since κ_o is defined in Section 6.4 as the

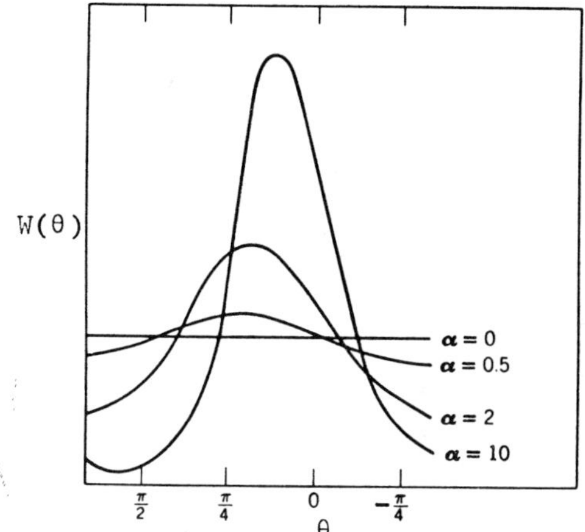

Fig. 9.7 Plots of probability W vs orientation angle θ for dumbbells or long thin ellipsoids. $\alpha = G/D_r$. See text. (From Morawetz, Reference 9.4, Fig. VI.8)

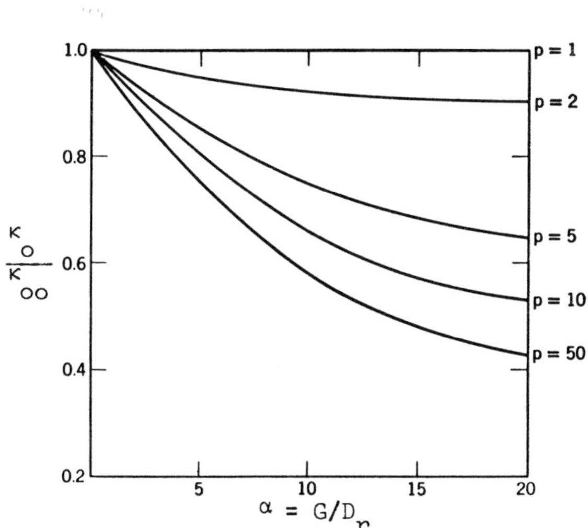

Fig. 9.8 Viscosity ratio κ_o/κ_{oo} as a function of parameter α for prolate ellipsoids with axial ratio p . Symbols κ_o , κ_{oo} and α are defined in text. (From Morawetz, Reference 9.4, Fig. VI.13)

limiting value of κ as the concentration approaches zero). Equation (9.98) applies only for small values of α. Analogous results have been developed by Scheraga (1955) for prolate ellipsoids under a wide range of conditions; results are plotted in Fig. 9.8. Note that these plots can hardly be compared with Eq. (9.98) since the latter is valid only for $\alpha < 1$.

9.9.2 Birefringence and Dichroism of Suspensions

Orientation of particles in a suspension is often determined by optical techniques in which plane polarized light is transmitted through the suspension. Suppose the light propagates in the z direction; then its electric field vector $\underline{E}$ will lie in the xy plane. Suppose also that all the particles are aligned in the x direction. Then, in general, the propagation characteristics of the light will depend on the angle θ between $\underline{E}$ and the x axis. If the particles are *birefringent* the velocity of light will be a function of θ; if they are *dichroic* the attenuation of the light will vary with θ. We shall not discuss details of the techniques but simply note that the properties of birefringence and dichroism are the bases of methods in use for determining the orientation of particles in suspension. In techniques based on *flow birefringence* and *flow dichroism* determinations are made of the orientation of particles in velocity gradient fields. In techniques based on *electric birefringence* and *electric dichroism* the particle orientation is brought about by an electrical field. Both methods yield information on the rotational diffusion coefficient D_r.

PROBLEMS

9.1 In the derivation discussed in Section 9.2 use is made repeatedly of the idea that when a quantity such as y is subjected to the operations of (a) averaging with respect to time and (b) differentiation with respect to time, the order in which the operations are carried out is immaterial. Thus in proceeding from Eq. (9.9) to (9.10) it is assumed that

$$\left< \frac{dy}{dt} \right> = \frac{d}{dt} \left< y \right> .$$

Justify this assumption by considering carefully what the operations mean.

9.2 Derive Eq. (9.22) in a manner analogous to the development used in deriving Eq. (9.17).

9.3 When a long thin rod of circular cross section rotates steadily about its axis the surrounding fluid is set into steady motion with cylindrical symmetry, such that all particles travel in circles about the axis. Let u be the velocity at a distance r from the axis. Consider a region of fluid in the shape of a cylindrical shell, bounded by the surfaces $r = r_1$ and $r = r_2$.

(a) Show that torque L_2 is exerted on the shell by the fluid outside it, and that the magnitude of L_2 is

$$L_2 = 2\pi\eta r_2^{\,2} \; (\partial u/\partial r)_2$$

per unit length. Here η is the viscosity coefficient for the fluid and the subscript "2" means that $\partial u/\partial r$ is to be evaluated at r_2 .

(b) Show similarly that torque L_1 in the <u>opposite direction</u> is exerted on the shell by fluid inside it, and determine the magnitude of L_1 .

(c) Explain why the magnitudes of L_1 and L_2 must be equal, in steady flow.

(d) Show then that, since R_1 and R_2 are arbitrary, the velocity u must be inversely proportional to r , and that for a rod of radius R rotating with angular velocity $\dot{\theta}$ the velocity must be given by

$$u = \frac{\dot{\theta}R^2}{r} .$$

(e) Making use of these results justify the expression for f_r in Eq. (9.24).

9.4 Consider the dumbbell described in Section 6.4.2; centered at 0, and rotating about an axis; the latter is perpendicular to the dumbbell axis and passes through 0 . Show that the rotational friction constant is given by

$$f_r = 12\pi\eta b^2 R .$$

9.5 Carry out an analysis of the laboratory data given in Table 9.1 , and cited in the associated text. (Section (9.2.4). Determine $< x^2 >$ and calculate kT/f , assuming Eq. (9.17) holds. Then calculate f , given that the temperature is 20° C. Compare with the value of f expected for a sphere of 2.30 μm diameter in water.

9.6 Verify the values for a' and a" in Eq. (9.38).

9.7 Show that the expression for c(x,t) in Eq. (9.39) is a solution of Eq. (9.34).

9.8 Make a rough copy of the plots of Fig. 8.4, labelling ordinates, abscissae and parameters so that the plots represent concentration c versus distance x for various values of the time t .

9.9 Given that the expression for $\partial c/\partial x$ in Eq. (9.43) is a solution of Eq. (9.34) explain why the expression for c(x,t) in Eq. (9.42) is also a solution.

9.10 Verify the values for A and B given in Eqs. (9.50).

9.11 Carry out the indicated steps to derive Eq. (9.64).

9.12 Choosing one of the proteins for which data on s, D, $\bar{v}$, etc. are given in Table 9.4; test the consistency of these data with formulae given in Section 9.6.2.

9.13 Test the applicability of Eq. (9.69) to the oxygen molecule. For example, at some temperature use numerical values for D, T, k and η to determine the effective radius r of the oxygen molecule; then compare this with information available from handbooks or other sources. (Or, alternatively, choose a value for r and calculate D.) Also test the extent to which Eq. (9.69) correctly gives the temperature dependence of D.

9.14

(a) Proceeding as in Section 6.3.1, derive a differential equation (analogous to Eqs. (6.25)) for the rotation of a sphere in a viscous medium, in response to applied torque.

(b) Considering a solution analogous to Eq. (6.26), show that the characteristic response time τ' is given by I/f_r, where I is the moment of inertia of the sphere about a diameter and f_r is the rotational friction constant (see Section 9.2.3). Using expressions for I and f_r show that

$$\tau' = \rho_o R^2/2\eta \ .$$

(c) Evaluate τ' for spheres of macromolecular dimensions and use your results to discuss the assumption made in Section 9.8.1 (applied there to dumbbells), that $G\tau' \ll 1$.

9.15 Fill in steps of the development outlined in Section 9.8.2. Thus obtain, in turn, each of Eqs. (9.89) through (9.94).

9.16

(a) Referring to Eq. (9.86), show that the maximum of w occurs at $\theta = \theta_m$, where

$$\alpha = \frac{2 \operatorname{ctn} 2\theta_m}{\sin^2 \theta_m} .$$

(b) As α approaches zero, θ_m approaches $\pi/4$. Hence assume that for α nonzero but small,

$$\theta_m = \frac{\pi}{4} + \delta ,$$

where $\delta << 1$. Introduce this expression into the equation of (a), make suitable approximations and thus arrive at Eq. (9.97).

9.17 It has been estimated that a fully oxygenated single red cell contains roughly 10^9 molecules of oxygen (four for every H_b molecule). With this information make a rough calculation of the time required for a red cell to receive its complement of oxygen as it passes through the lung. For this purpose assume the cell is a sphere of radius a = 3 μm; also that the surrounding liquid is similar to water and saturated with oxygen; at 20° C the O_2 concentration in the liquid would then be about 3×10^{17} molecules/cm^3 . As an extreme assumption suppose that until it is saturated the red cell binds O_2 as soon as it arrives. Equation (9.52) can then be used, it being implicit that $c_o = 0$, that is c = 0 at r = a.

Note: This calculation gives a lower limit to the time required as far as diffusion is concerned in the absence of stirring. In the actual case where c_o is not zero the diffusion occurs more slowly. On the other hand, when stirring effects occur because the cell tumbles about, or from other causes, concentration gradients decrease and the rate of oxygen transport increases.

CHAPTER 10

ELEMENTS OF THERMODYNAMICS

10.1 INTRODUCTION

Numerous books have been written on the many aspects and applications of thermodynamics. Our treatment of the subject here is limited to those concepts and principles which we need for discussion of biophysical mechanics in later chapters. A few representative textbooks, recommended for further reading, are listed in the References.

In thermodynamics much use is made of the word system. A system may be, for example, a body of gas in a container, or a fiber stretched between supports, or a series of compartments containing suspensions. Classical thermodynamics applies only to systems that are in an equilibrium state. For a mole of gas its state is specified by giving values of any two of the parameters: pressure, temperature and volume. For a stretched fiber its state can sometimes be specified by giving any two of the quantities: length, tension and temperature. But biological fibers (such as muscle fibers) are not isolated from their surroundings in the way that gas in a closed container can be. As a result more parameters are usually required to specify the state of a fiber; specifically, the state is usually found to depend on the chemical nature of the environment.

In general, the number of parameters required to specify the state of a system can only be determined experimentally. Thus suppose a system is compared at two different times t_1 and t_2. At both times the system is in equilibrium and the parameters P and T (which might represent pressure and temperature) are the same. If by any test whatever a significant difference is found in the system at these two times we know that the states are different, and one or more parameters in addition to P and T are required to specify the state. If no difference is found we conclude that the system is in the same state at times t_1 and t_2 , and that P and T suffice as state parameters. Obviously the latter conclusion is subject to revision if a more sensitive test is found which shows the system to be different at t_1 and t_2 . Our definitions of state and state parameters as described here are of the operational type discussed in Section 8.4 , that is, they are based on experimental tests.

Another thermodynamic concept is that of the state function, or state variable, defined as a property of a system whose value depends only on the state of the system. It is clear that for a mole of gas its pressure, temperature and volume are state functions. Less obvious examples will be taken up later.

10.2 THE FIRST LAW

Most of thermodynamics is based on two Laws which were advanced in the 19th century, and which have been thoroughly tested experimentally. The First Law of Thermodynamics is in part an expression of the idea that heat is a form of energy. A quantity U, the internal energy of a system, is defined as the total energy of all kinds contained within the system. Consider a change of state in which a small amount of heat dQ is added to the system and a small amount of work dW* is done by the system (on its surroundings); the resulting change dU in the internal energy is given by

$$dU = dQ - dW^* \,.^{(1)} \tag{10.1}$$

---

(1) In thermodynamics it is convenient to write expressions in terms of dU, dQ, etc., and treat these as algebraic quantities. In this usage the symbol dU (for example) means a small but finite increment of U. When the same symbol is used in an integral such as $\int dU$ it represents an infinitely small, or infinitesimal, increment. A ratio of increments such as dU/dt becomes a derivative in the limit where the increments become infinitesimal.

The statement in Eq. (10.1) is a form of the First Law. (Note: The symbol W^* is used here for "work" to distinguish it from the symbol W introduced in Chap. 8 for "probability"). Different forms of the law are obtained for different kinds of systems. For a gas under pressure P the work done dW^* in a small increase dV of the volume is P dV , so that the First Law is

$$dU = dQ - PdV \ . \tag{10.2}$$

For a fiber under tension F whose length is increased by a small amount $d\ell$ the work done by the fiber is $(-F\,d\ell)$ and we have

$$dU = dQ + F\,d\ell \ . \tag{10.3}$$

Since the heat dQ added to a system and the work dW^* done by it can both be measured experimentally (at least, in principle) the change dU in internal energy can be determined for a given small change in state. For a series of small changes in state the total change in U is given by a summation of all the increments dU , or by the integral $\int dU$. Suppose the system is carried through a series of small changes, and finally returned to its original state. Then what can be said about the total change in U during the cycle, represented by the integral $\oint dU$? It is found (basically from experiment, although the chain of reasoning is indirect) that this integral around a cycle is always zero, a most appealing result which applies for any choice of the original state, and for any series of changes. From findings of this kind it follows that the internal energy U of a system, being independent of its history, is a function of its state only; in other words, U is a state function. This important property of U adds significance to the statement of the First Law in Eq. (10.1). Thus a re-statement of this Law might be: "For any system a state function U exists such that in any small change in state Eq. (10.1) holds". The First Law is to some extent a statement that "energy is conserved", but its full implications are probably not suggested by this simple statement.

The difference $U_2 - U_1$ in internal energy between any two states "1" and "2" is obtained by considering any set of small changes in state, that is, any path, leading from "1" to "2" ; $U_2 - U_1$ is then given by a summation of the increments dU along this path, or by the integral

$$U_2 - U_1 = \int_1^2 dU .$$

If state "1" is regarded as a reference state, U_1 might be assigned the value zero and U_2 regarded as the absolute value of the internal energy for state "2"; this is an arbitrary procedure, however. We are usually not concerned with the absolute value of the internal energy, but only with the difference in U between two states.

By contrast it is found that in a similar cycle of changes in state the integrals $\int dW^*$ and $\int dQ$ are not, in general, equal to zero. This is shown in Appendix D for a system consisting of ideal gas contained in a cylinder; calculations of $\int dW^*$ can be carried out here by use of the gas law (which is in good agreement with experiment) .

10.3 THE SECOND LAW

We now proceed to the Second Law of Thermodynamics. This law is intimately concerned with another state function, namely, the entropy S. In stating the thermodynamical definition of S we refer to the concepts of "reversibility" and "irreversibility". A change in state for a system is reversible if the system and all of its surroundings can be brought back to their original states; otherwise it is irreversible. Real processes are never completely reversible, but some are nearly so. To approach reversibility a change must be carried out slowly, so that the system remains nearly in equilibrium with its surroundings and so that dissipative processes, such as those that exist in flow of a viscous fluid, are negligible.

Now consider a small reversible change in state for a system maintained at Kelvin temperature T while heat dQ is added to it. In this change the entropy of the system increases by an amount dS, defined as

$$dS = \frac{dQ}{T} . \tag{10.4}$$

Suppose a change from an initial state "1" to an arbitrary final state "2" is carried out in small steps from state to state in a series leading from "1" to "2". The total change of entropy $S_2 - S_1$ is given by a summation of the increments dS for the separate steps, or by the integral

$$S_2 - S_1 = \int_1^2 dS = \int_1^2 \frac{dQ}{T} . \tag{10.5}$$

We refer to a particular choice of states leading from "1" to "2" as the path of integration. Since, as anticipated, the entropy is a state function, the integrals in Eq. (10.5) are independent of the path of integration and depend only on the end states. As remarked in connection with the internal energy U, we might regard state "1" as a reference state; if S_1 is set equal to zero, S_2 can be regarded as an absolute value of the entropy at state "2". But again, our interest is usually in differences rather than absolute values.

The discussion above anticipates the first of two assertions in the following statement of the Second Law of Thermodynamics (Lindsay and Margenau, 1936):

(a) The entropy S (whose increment is defined in Eq. (10.4) is a state variable ; and

(b) Its value for an isolated system never decreases.

There is a considerable background of theory and experiment to support this Law, and there are a number of interesting alternative and equivalent formulations of it. Also there are many important implications of the Second Law. We shall not attempt to discuss this material here in any detail, but instead refer the reader to standard textbooks. In Appendix D an example is worked out to support the statement that S is a state function.

In the foregoing discussion we have defined the entropy S in thermodynamical terms. A statistical definition has also been given which affords valuable insight into the significance of entropy, and of the Second Law. We refer to the quantity W introduced in Chap. 8 and here use it to characterize the state of a system. Specifically, for any given state we let W be the probability that the system is in that state. The statistical entropy S is then defined by the simple equation

$$S = k \ln W \,, \tag{10.6}$$

where k is the Boltzmann constant already referred to in Section 8.1. The thermodynamical and statistical definitions of S appear very different at first, but it is another postulate of thermophysics that they are, in fact, the same. Support for this postulate comes from examples where agreement is found between calculations based on the two definitions of S. An example involving a perfect gas is taken up as an exercise. (Prob. 10.1).

From a philosophical point of view the most interesting aspect of the Second Law is very likely part (b), the statement that for an isolated system the entropy never decreases. According to the statistical definition of entropy, Eq. (10.6), this statement means that the probability of an isolated system never decreases. This is a concise, albeit negative, way of saying that the probability W and entropy S of the system either remain constant or increase. Sometimes it is clear that the entropy should tend toward a maximum. As an example, suppose our system consists of gas in a cylindrical vessel with a partition separating two compartments. See Fig. 10.1 . Compartment "1" has length L_1 and compartment "2" has length L_2, the sum $L_1 + L_2$ being L. Assuming N gas particles altogether, we let W be the probability that there are N/2 particles in each compartment. Then from Chap. 8 we easily obtain

$$W = p^m q^m C_m^N \ ; \tag{10.7}$$

here $m = N/2$ and C_m^N is given by Eq. (8.17) while

$$p = \frac{L_1}{L} \ ; \quad q = \frac{L - L_1}{L} \ . \tag{10.8}$$

We readily find (Prob. 10.2) that the entropy S can be written

$$S = A + mk\,[\ln p + \ln(1 - p)] \ , \tag{10.9}$$

where A is independent of L_1 . The bracket [ ] on the right hand side of Eq. (10.9) is plotted versus $p = L_1/L$ in Fig. 10.2 . We see that the bracket, and hence also S, is symmetric about a maximum at $p = 0.5$. Hence the entropy of the system is maximum when $L = \frac{1}{2}L_1$, that is, when the compartments are of equal volume. Now we realize from our knowledge of the behavior of gases that if the partition is free to move (e.g., if it is a frictionless piston) it will seek the position $L = \frac{1}{2}L_1$, since the pressure on its two faces will then be equal. Hence we conclude from the gas laws, together with the expression for S in Eq. (10.9), that if $L \neq \frac{1}{2}L_1$ the system will tend to adjust itself (by motion of the partition) in such a way that its entropy increases. This conclusion is in agreement with the Second Law.

We shall see later (Section12.2) that thermodynamical principles also yield a method from which, given the expression for S in Eq. (10.9), we can calculate the net force on the partition if $L \neq \frac{1}{2}L_1$.

In a suspension of particles, in the absence of force fields, the most probable state is one of uniform concentration. In Chap. 9 we considered the process of diffusion which occurs when there are spatial variations in the concentration. Since diffusion tends to produce uniform concentration it causes the entropy S (of the total suspension) to increase, and provides an example of the Second Law "at work".

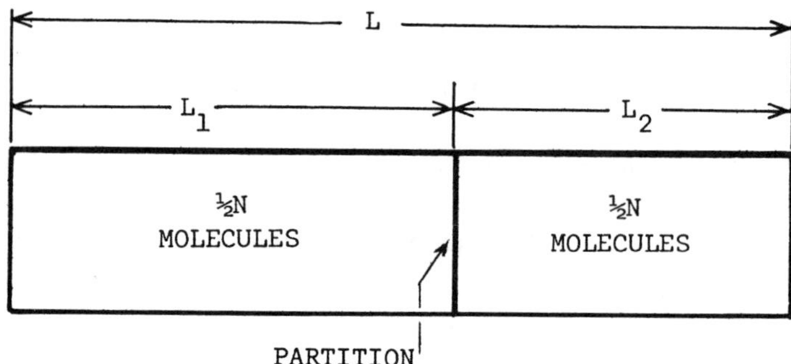

Fig. 10.1 Two-compartment system referred to in Section 10.3 . Partition may be moveable.

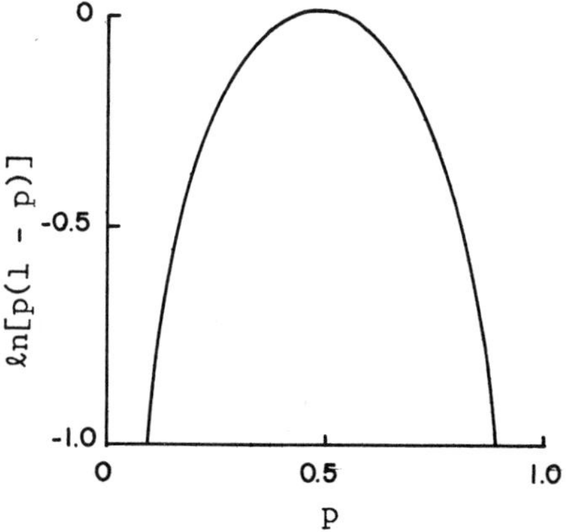

Fig. 10.2 Plot vs p of the bracket [ ] which appears in Eq. (10.9).

The fact that entropy tends to increase with time has led to the idea, interesting philosophically, that the direction of time and the direction of entropy change are closely related. Further, it has been argued that biological evolution is a consequence of the Second Law. (Blum, 1962). On the other hand, this Law is also the basis for statements that "the Universe is running down", that is, its randomness is increasing.

Some have argued that since life continues to be generated, and since a living system is highly improbable (from the point of view of statistical mechanics) the Second Law does not apply to life processes. The question of the ultimate applicability of thermodynamics to living systems, or to the Universe as a whole, is difficult and perhaps unanswerable. For the Second Law applies only to equilibrium situations and only to isolated systems with prescribed components. These restrictions pose problems, both for living systems and for the Universe as a whole. Nevertheless there are aspects of life processes, and countless aspects of the physical universe, for which the thermodynamical principles apply very well indeed. That profound insight on life processes can be gained from the Second Law has been shown compellingly by Schrödinger. (1946)

10.4 GIBBS FREE ENERGY G AND THE CHEMICAL POTENTIAL μ

Another state function, or state variable, which finds much application in biology is the Gibbs free energy, for which we use the symbol G. Consider a system of volume V subject to a pressure P at a Kelvin temperature T. Then G is defined as

$$G = U + PV - TS \,, \tag{10.10}$$

where U is the internal energy and S the entropy of the system. For each state of the system there is a unique set of values for the measureable parameters P, V and T; hence these parameters are state variables. In addition we have seen that U and S are state variables, so it follows that G is also.

Most applications of the Gibbs function G are to processes which occur at constant pressure and temperature; suppose our system undergoes some kind of a small change under these conditions with resulting small increments dU in the internal energy, dV in the volume and dS in the entropy. Then the free energy G is increased by dG, where

$$dG = dU + PdV - TdS \ . \quad \text{(Constant P and T).} \qquad (10.11)$$

From the definition of dS, Eq. (10.4), we realize that TdS is just dQ . Combining Eq. (10.11) with the First Law, Eq. (10.1), then yields

$$-dG = dW^* - P\,dV \ . \ \text{(Constant P and T)} \ . \qquad (10.12)$$

From this equation we gain some insight on the significance of G : the negative increment -dG of this function during a process at constant temperature and pressure is equal to the work done, dW* , minus the work of expansion p dV . When an isolated strip of soft tissue is stretched its cross-section typically decreases in such a way that its volume remains essentially constant, that is, dV is nearly zero. If the tissue strip is stretched at constant pressure and temperature, the decrease in its Gibbs free energy -dG is approximately equal to the work done by it.

The quantity dW* can represent work of various kinds. If a "solute" particle (such as an ion, a molecule or a larger object) exists in a fluid, the "solvent", there will usually be an alteration of intermolecular forces in and near the region occupied by the solute particle. Thus energy, either positive or negative, is required when solute is accomodated in the solvent interior. This is true even when the "solute" added is only more of the solvent itself. In a change of state involving addition of solute and/or solvent to a solution the work dW* includes the "energy of solution" in addition to the expansion work P dV; but the P dV term cancels in Eq. (10.12) and the negative increment -dG is just the energy of solution.

We now consider a few thermodynamical relations which apply to a system relevant to biology, consisting of an aqueous solution or suspension . We refer to the solute elements (ions, molecules, etc.) as "particles" and to the solvent as "water" . The state of the system depends on the pressure P, the temperature T, the total volume V and the particle concentrations. In a small change of state suppose dn identical particles and dn_w water molecules are added to the system, and that other state parameters are also varied. Considering the energy of solution we let the work dW* done by the system in this change of state be

$$dW^* = P\ dV - \mu\ dn - \mu_w\ dn_w\ , \tag{10.13}$$

where $(-\mu)$ is the energy required to dissolve one particle and $(-\mu_w)$ is the corresponding energy for a water molecule. The quantities μ and μ_w are called chemical potentials and are discussed further in the next section. (The symbol μ is the same as that used earlier for the shear modulus of elasticity; there should be few occasions for confusion.)

We determine the change dU in the internal energy for the change of state in question by combining Eqs. (10.1) and (10.13) while remembering that dQ = T dS; the result is an equation given by Gibbs:

$$dU = T\ dS - P\ dV + \mu\ dn + \mu_w\ dn_w\ . \tag{10.14}$$

We now return to the Gibbs free energy G, which is given in general by Eq. (10.10). In any small change of state the increment of G is

$$dG = dU + P\ dV + V\ dP - T\ dS - S\ dT\ . \tag{10.15}$$

For the situation under consideration dU is given by Eq. (10.14) and we obtain

$$dG = \mu_w \, dn_w + \mu \, dn + V \, dP - S \, dT \, . \qquad (10.16)$$

This equation is useful in that it allows us to write expressions for a number of partial derivatives of interest. Consider, for example, $\partial G/\partial n_w$; by this we mean the ratio of dG to dn_w under conditions where there is no change in the other variables n, P and T, whose increments appear on the right of Eq. (10.16). This ratio is to be taken in the limit where the increments become infinitesimals. (See footnote in Section 10.2.) When n, P and T are constant dG is equal to $\mu_w \, dn_w$ and (in the indicated limit) $\partial G/\partial n_w$ becomes μ_w . Proceeding similarly with the other possible partial derivatives we obtain

$$\frac{\partial G}{\partial n_w} = \mu_w \; ; \quad \frac{\partial G}{\partial n} = \mu$$
$$\frac{\partial G}{\partial P} = V \; ; \quad \frac{\partial G}{\partial T} = -S \, . \qquad (10.17)$$

It should be mentioned that in a more specific notation the derivatives would be written in the form

$$\left(\frac{\partial G}{\partial n_w}\right)_{n,\,P,\,T} \, ,$$

where the subscripts list the variables that are held constant in the differentiation. For our purposes here we simplify the notation by omitting the subscripts, but then must remember the information which these would contain.

10.5 CHEMICAL POTENTIALS IN SOLUTIONS

We now proceed to obtain expressions relating to the chemical potentials μ_w and μ , which apply to water molecules and solute particles in a suspension or solution. We expect these potentials to depend on the temperature T and the pressure P as well as on the solute concentration c.

To determine how μ_w depends on the pressure P we consider the partial derivative $\partial\mu_w/\partial P$; this derivative represents the ratio of $d\mu_w$ to dP for fixed values of the other variables T and c . Using relations in Eqs. (10.17) we have

$$\frac{\partial\mu_w}{\partial P} = \frac{\partial}{\partial P}\left[\frac{\partial G}{\partial n_w}\right] = \frac{\partial}{\partial n_w}\left[\frac{\partial G}{\partial P}\right] = \frac{\partial V}{\partial n_w} \quad . \tag{10.18}$$

Here we have made use of the fact that, since G is a well-behaved function, the order of differentiating G with respect to P and n_w can be interchanged. Now V/n_w (the increase in volume of the system resulting from the addition of one water molecule) is essentially the volume $\bar{v}_w$ of a water molecule. Proceeding similarly for μ and letting the volume of a solute molecule be $\bar{v}$ we obtain the results:

$$\frac{\partial\mu_w}{\partial P} = \bar{v}_w \; ; \quad \frac{\partial\mu}{\partial P} = \bar{v} \quad . \tag{10.19}$$

Considering in a similar way how μ and μ_w vary with temperature we obtain the equations (Prob. 10.3):

$$\frac{\partial\mu_w}{\partial T} = -\frac{\partial S}{\partial n_w} \; ; \quad \frac{\partial\mu}{\partial T} = -\frac{\partial S}{\partial n} \quad . \tag{10.20}$$

By definition $\partial S/\partial n_w$ is the increase in entropy of the system which results from adding one water molecule to the solution; it can, in this sense, be regarded as the "entropy per water molecule". In the same sense $\partial S/\partial n$ is the entropy per solute particle.

We shall wait until Section 11.3 to analyze the dependence of μ and μ_w on the particle concentration c. From the information in Eqs. (10.19) and (10.20) we proceed to the conclusion that in a small change in state the increments in μ and μ_w are given by

$$d\mu = \bar{v}\, dP - \frac{\partial S}{\partial n} dT + d\mu^c$$
$$d\mu_w = \bar{v}_w\, dP - \frac{\partial S}{\partial n_w} dT + d\mu_w^c \, , \tag{10.21}$$

where $d\mu^c$ and $d\mu_w^c$ are concentration-dependent contributions to $d\mu$ and $d\mu_w$. It will be shown later, in Section 11.3, that they are given by

$$d\mu^c = \frac{kT\, dc}{c} = kT\, d\,(\ell n\, c)$$
$$d\mu_w^c = - \bar{v}_w\, kT\, dc \, . \tag{10.22}$$

Equations (10.21) contain the information expressed in Eqs. (10.19) and (10.20). For example, when dT and dc are both zero we see (from the first of Eqs. (10.21)) that $d\mu = \bar{v}\, dP$ and hence obtain the first of Eqs. (10.19)

10.6 THE GIBBS-DUHEM EQUATION

Other useful relationships are derived by making use of the Gibbs Equation Eq. (10.14), together with certain general considerations. In Eq. (10.14) a number of thermodynamic variables appear which apply to a system of interest; these can be classified according to how they depend on the size of the system. The variables V, S, n and n_w increase if the system enlarge and are called extrinsic variables. On the other hand the characteristics T, P, μ and μ_w which apply to a solution are not dependent on size of the system, and are intrinsic variables.

Consider now the dependence of U on the extrinsic variables. Suppose each of V, S, n and n_w is multiplied by the same arbitrary factor λ. Then we expect U also to be multiplied by the same factor. To express this mathematically we set V, S, etc., equal respectively to λV_o, λS_o, etc. and write

$$U(V\, , S\, , n\, , n_w) = \lambda U_o \, , \tag{10.23}$$

where U_o is just $U(V_o, S_o, n_o, n_{wo})$. Differentiating with respect to λ (using the "chain rule" for the left hand side) we obtain from Eq. (10.23):

$$\left(\frac{\partial U}{\partial V}\right)^{*} \frac{\partial V}{\partial \lambda} + \left(\frac{\partial U}{\partial S}\right)^{*} \frac{\partial S}{\partial \lambda} + \ldots\ldots = U_o \ . \qquad (10.24a)$$

Here the symbol $(\partial U/\partial V)^{*}$ denotes the partial derivative of U with respect to V with S, n and n_w held constant at the respective values λS_o, λn_o , λn_{wo} , and evaluated at $V = \lambda V_o$; similar meanings apply to $(\partial U/\partial S)^{*}$, etc. Since $V = \lambda V_o$ the derivative $\partial V/\partial \lambda$ yields V_o ; similarly $\partial S/\partial \lambda$ is S_o , etc. Hence Eq. (10.24a) becomes

$$U_o = V_o \left(\frac{\partial U}{\partial V}\right)^{*} + S_o \left(\frac{\partial U}{\partial S}\right)^{*} + \ldots \qquad (10.24b)$$

When $\lambda > 1$ the derivatives $(\partial U/\partial V)^{*}$, etc., in this equation apply to a system λ times larger than that to which V_o, S_o, etc., apply. However, we need only the result for $\lambda = 1$; the concept of arbitrary λ has now served its purpose. When $\lambda = 1$ all quantities in Eq. (10.24b) apply to the same system; we may therefore drop the subscripts and asterisks, writing simply:

$$U = V\frac{\partial U}{\partial V} + S\frac{\partial U}{\partial S} + n\frac{\partial U}{\partial n} + n_w\frac{\partial U}{\partial n_w} \ . \qquad (10.25)$$

From the Gibbs Equation, Eq. (10.14), we can now evaluate $\partial U/\partial V$, etc., and obtain

$$U = -PV + TS + n\mu + n_w \mu_w \ . \qquad (10.26)$$

We see that this integrated form of U is related in a very simple way to Gibbs Equation, Eq. (10.14). Nevertheless it would hardly be possible to justify Eq. (10.26) without the intervening steps represented by Eqs. (10.23) to (10.25), or their equivalent.

An important result is now derived (Prob. 10.4) by using Eq. (10.26) to form a total differential of U, then combining this with the Gibbs Equation, Eq. (10.14); we obtain

$$-V\,dP + S\,dT + n\,d\mu + n_w\,d\mu_w = 0 \,. \tag{10.27}$$

This is the Gibbs-Duhem Equation and will prove useful in subsequent discussion .

10.7 CONCEPTS OF NONEQUILIBRIUM THERMODYNAMICS

10.7.1 Introduction

As stated in previous sections of this chapter, the classical thermodynamics treated there deals only with equilibrium situations. While classical thermodynamics is important to biology, its applicability is limited, since (by any usual definition of life) it is obvious that a system in equilibrium cannot be alive. (Prob. 10.5). An extension to thermodynamics has been developed for dealing with processes important to biology such as diffusion, heat flow, fluid flow, and transport of ions and molecules in force fields. Any process of this kind occurs in response to nonequilibrium conditions, but proceeds in the direction of equilibrium; its rate tends to zero as equilibrium is approached. Because of its unidirectional character it is frequently called an irreversible process.

This new theory is referred to as nonequilibrium thermodynamics or as irreversible thermodynamics. It is based on postulates advanced in a general form by Onsager (1931). Before stating these we need to define certain concepts.

10.7.2 Spatial Variations of μ ; Thermodynamic Forces.

We have discussed the chemical potential μ in Section 10.5 . This quantity has the dimensions of energy and, in fact, plays a role analogous to potential energy functions (such as that referred to in Section 7.6). In solutions or suspensions where the pressure, temperature or concentration vary with spatial coordinates (x, y, z) the chemical potential can also be regarded as a function of (x, y, z). Consider Eqs. (10.21) and (10.22) from which we find that the increment $d\mu$ for a particle in suspension is given by

$$d\mu = \bar{v}\, dP - \frac{\partial S}{\partial n}\, dT + kT\, \frac{dc}{c} \; . \tag{10.28}$$

In the derivation of Eq. (10.28) the quantity $d\mu$ was defined as a change in μ brought about in a given system. An alternative definition of $d\mu$ proves convenient for a solution or suspension where any or all of P, T and c vary continuously with spatial coordinates x, y and z ; here $d\mu$ is the difference in μ at two separated points in the solution. Let the pressure at two points A and B be represented by P_A, P_B, resp., and let dP be the difference (P_B-P_A) . Let dT and dc similarly represent the differences in T and c, resp., at points A and B. Define a system A as a small volume of solution in the vicinity of point A. Then define a similar system B near point B . Specifically, choose B so that is is identical in properties with the system which would result from A if we were to carry out on it the small changes dP, dT and dc.

Now let μ_A and μ_B be the chemical potential applicable at points A and B, resp. Then $d\mu$ in Eq. (10.28) is just the difference $(\mu_B-\mu_A)$. This statement is, of course, restricted to situations where $d\mu$ is small. It is also required that the solution everywhere be "almost" in equilibrium. We can now define spatial derivatives of μ by choosing A and B as "neighboring" points. For example, suppose A and B are on the x axis and have the coordinates x_A and $(x_A+\Delta x)$, resp. Then the derivative $\partial\mu/\partial x$ at A is given by

$$\left(\frac{\partial\mu}{\partial x}\right)_A = \lim_{\Delta x \to o} \frac{\mu_B - \mu_A}{\Delta x} \; . \tag{10.29}$$

But $(\mu_B-\mu_A)$ is given by $d\mu$ in Eq. (10.28) when dP is (P_B-P_A), etc. Hence $\partial\mu/\partial x$ can be obtained by dividing each side of Eq. (10.28) by Δx and passing to the limit. (To agree with usual terminology in calculus the differentials $d\mu$, dP , etc. in Eq. (10.28) should be replaced by $\Delta\mu$, ΔP, etc., before passing to the limit). We then obtain

$$\frac{\partial\mu}{\partial x} = \bar{v}\frac{\partial P}{\partial x} - \frac{\partial S}{\partial n}\frac{\partial T}{\partial x} + \frac{kT}{c}\frac{\partial c}{\partial x} . \tag{10.30}$$

In the language of nonequilibrium thermodynamics, spatial variations of the chemical potential μ give rise to thermodynamic forces. The component of force in, say, the x direction is

$$X = -\frac{\partial\mu}{\partial x} . \tag{10.31}$$

Equation (10.31) is similar to analogous equations relating force to potential functions in other areas of physics. A familiar result is obtained when the temperature and concentration are independent of x . For then Eqs. (10.30) and (10.31) yield simply

$$X = -\bar{v}\frac{\partial P}{\partial x} ; \tag{10.32}$$

this is the expected result (see Section 6.1) for the force on a body of volume $\bar{v}$ in a pressure gradient.

For another example suppose the temperature and pressure are both constant, but that the concentration varies with x . Then we obtain from Eqs. (10.30) and (10.31) that the thermodynamic force on a particle is

$$X = -\frac{kT}{c}\frac{\partial c}{\partial x} . \tag{10.33}$$

Assuming the force causes the particle to move along x with steady velocity U_{rel} relative to a Newtonian fluid it will be acted on by a viscous drag force fU_{rel} equal and opposite to X . Then

$$U_{rel} = \frac{X}{f} = -\frac{kT}{cf}\frac{\partial c}{\partial x} . \tag{10.34}$$

Since there are c particles per unit volume the particle current J along x is cU_{rel} , or

$$J = -D\frac{\partial c}{\partial x} , \tag{10.35}$$

where D is the diffusion coefficient, given by kT/f according to Eq. (9.41). This application of the chemical potential has therefore led us to Fick's First Law for diffusion. (See Section 9.3).

10.7.3 Entropy Production

As a particle moves through a viscous medium along the x direction heat is generated at a rate equal to the product of its velocity $\dot{x}$ and the force X. If there are c identical particles per unit volume, all moving together, the rate of heat generation per unit volume Φ_v, called the (volume) dissipation function, is given by

$$\Phi_v = c\dot{x}X . \tag{10.36}$$

Since the particle current J is given by $c\dot{x}$ we obtain

$$\Phi_v = JX = \text{"current" times "force"} . \tag{10.37}$$

Assume the medium is at constant Kelvin temperature T, and further that the particle motion produces heat at such a slow rate that the temperature is little affected, even near the particle. The entropy per unit volume S_v will then increase at the rate Φ/T; that is

$$S_v = \frac{JX}{T} . \tag{10.38}$$

Equation (10.38) applies only in a special situation. According to the theory of non-equilibrium thermodynamics (Katchalsky and Curran 1967) the entropy production quantity S_v for a process occurring at temperature T can be written more generally as a sum of products $T^{-1} J_i X_i$, that is in the form

$$S_v = T^{-1} \sum_i J_i X_i . \tag{10.39}$$

In this generalization of Eq. (10.38) the choice of quantities J_i to be designated as "flows" and X_i as "forces" is somewhat arbitrary except, of course, that each product $J_i X_i$ must have the same units, those of energy (heat) per unit volume per unit time.

The dissipation function for this isothermal process is just

$$\Phi_V = \sum_i J_i X_i \tag{10.40}$$

Examples will be taken up in Chap. 11.

10.7.4 Phenomenological Equations and Onsager's Law.

As discussed in Sections 10.2 and 10.3 classical thermodynamics is based primarily on the First and Second Laws. These laws were originally stated as postulates; scientists gained confidence in them as experimental evidence accumulated in agreement with the postulates, with apparently no significant disagreements. Similarly, nonequilibrium thermodynamics is based on postulates which have been tested experimentally. Onsager (1931) suggested that when n flows J_i occur in response to n forces X_i in a system which is almost, but not quite, in equilibrium the forces and flows are related in a linear fashion, so that

$$\begin{aligned} J_1 &= L_{11} X_1 + L_{12} X_2 + \dots\dots\dots\dots L_{1n} X_n \\ J_2 &= L_{21} X_1 + L_{22} X_2 + \dots\dots\dots\dots L_{2n} X_n \\ &\dots\dots\dots\dots\dots\dots\dots\dots\dots\dots\dots\dots \\ J_n &= L_{n1} X_1 + L_{n2} X_2 + \dots\dots\dots\dots L_{nn} X_n \, . \end{aligned} \tag{10.41a}$$

These can be written succinctly as

$$J_i = \sum_{j=1}^{n} L_{ij} X_j \, , \tag{10.41b}$$

where the index i successively takes on the integral values 1 through n and the L_{ij} are constants. Equations (10.41) can be solved for the X_i , yielding

$$X_i = \sum_{j=1}^{n} R_{ij} X_j \, , \tag{10.42}$$

where the R_{ij} are constants. The above linear relationships between the X_i and J_i are called the phenomenological equations. (In general, phenomenological equations are relations between quantities which are macroscopic and directly observable, such as particle currents and pressure gradients, in contrast to submicroscopic quantities such as atomic and subatomic properties.) It was further stated by Onsager that all coefficients L_{ij} satisfy the reciprocity condition

$$L_{ij} = L_{ji} . \qquad (10.43)$$

It follows that the R_{ij} satisfy a similar condition. Equation (10.43) is called Onsager's Law. The constants L_{11}, L_{22} , etc., for which $i = j$ are called the "straight coefficients" while those for $i \neq j$ are the "coupling coefficients" or "cross coefficients". It can be shown that the product of any two straight coefficients is equal to, or greater than, the square of the associated coupling coefficient; that is, for any i and j ,

$$L_{ii} L_{jj} \geq L_{ij}^2 . \qquad (10.44)$$

PROBLEMS

10.1 Consider the system sketched in Fig. 10.1, consisting of gas in a compartmented cylinder. Starting from the condition $L_1 = \frac{1}{2}L$ assume the partition is displaced isothermally to an arbitrary value of L_1. Referring to results in Appendix D write expressions for:
(a) The heat added to each compartment during the displacement;
(b) The corresponding increase in entropy for each compartment, using the thermodynamic definition for change of entropy given in Eqs. (10.4) and (10.5); and
(c) The total increase in entropy for the two-compartment system. Compare this last expression with Eq. (10.9), which was derived from the probabilistic definition of entropy, Eq. (10.6).

10.2 Using the statistical definition of entropy S in Eq. (10.6) and expressions in Eqs. (10.7) and (10.8), derive the result for S in Eq. (10.9), which applies to the two-compartment system of Fig. 10.1.

10.3 Derive the expressions in Eqs. (10.20) for the partial derivatives of μ and μ_w with respect to temperature. You may proceed as in the derivation of Eqs. (10.19).

10.4 Derive the Gibbs-Duhem equation, Eq. (10.27), by proceeding as indicated in the text.

10.5 For various definitions of life, discuss the assertion in the text: "A system in equilibrium cannot be alive."

CHAPTER 11

OSMOSIS AND PASSIVE TRANSPORT

11.1 INTRODUCTION

Heat has been shown to be a form of energy involving random motion at the molecular level. Thermophysics deals with implications of this submicroscopic motion for the behavior of systems as we observe them, that is, at the macroscopic level. For example, in Chap. 9 we treated the subjects of Brownian movement and diffusion; these are observable macroscopic phenomena which can be accounted for in terms of random movements of atoms and molecules. In this chapter and the next we treat other topics in thermophysics which are relevant to biology.

In so doing we shall find application for the principles taken up in Chaps. 8 and 10. Kinetic theory and irreversible thermodynamics provide the basis for understanding osmotic phenomena taken up in Sections 11.2 and 11.3; these phenomena are important, for example, in maintaining normal differences in pressure across cell membranes. Principles of diffusion, of hydrodynamics and of irreversible thermodynamics are useful in understanding the passive movement or <u>transport</u> of molecules and ions across membranes. Topics related to passive transport comprise most of the remainder of Chap. 11. In Chap. 12 principles of thermodynamics and physical statistics are appealed to in explaining some aspects of bioelasticity.

11.2 OSMOTIC PRESSURE

We have commented earlier (Chap. 7) on the importance of water to living systems. But in biological tissue water is probably never in a pure form; instead the typical biological fluid is better described as an aqueous medium in which ions and molecules are dissolved and macroscopic particles are suspended. Cells and intracellular bodies such as mitochondria and nuclei are often sensitive to changes in the concentration of solute in their immediate fluid environment. For example, as we saw in Section 2.7, when a red cell is suspended in solutions of NaCl the shape and size of the cell depend on the NaCl concentration c . When c is less than about 0.9% the solution is "hypotonic", and water flows into the cell causing an increase of its volume. While the cell is changing shape the increase of volume does not necessarily lead to strain in the membrane which forms its envelope. But when the spherical shape is arrived at, further increase in volume causes stretching of the membrane. When c becomes as low as 0.4% to 0.5% the cell bursts, releasing its hemoglobin contents into the surrounding medium.

On the other hand, if the concentration c is increased above 0.9% the solution is <u>hypertonic</u>; the red cell shrinks somewhat, acquires a crinkled appearance and is said to be <u>crenated</u>.

In the human body the various cells suspended in the blood, or bathed by it, are sensitive to changes in concentration of NaCl and other solutes in the blood. Complex regulatory systems carry out the important task of maintaining blood solute concentrations at normal levels.

On the other hand, in laboratory situations, it is sometimes useful to manipulate cells by adjusting solute concentrations in their environment. For example, it is often desirable to have quantities of cellular contents available for scientific study, or for practical purposes. A common method for rupturing cellular envelopes, and thus releasing their contents, is to subject the cells to "osmotic shock" by suspending them in distilled water or other media of low solute concentration.

Such effects as we have just described, resulting from suspended ions, molecules and other particles, are <u>osmotic</u> phenomena, and can be accounted for in thermophysical terms. We shall first treat the subject by using simple kinetic theory, an approach which has the advantage of giving physical insight. (A disadvantage is that this theory applies only under restricted conditions; results of more general validity are obtained from thermodynamics.) We idealize the situation by supposing that both inside and outside the cell there are suspensions in which a large number of particles dance about in Brownian movement. As a result of this motion there is a continual gentle bombardment of the bounding membrane by the particles, whose effect is to cause an average pressure on the membrane. It has been shown by experiments (Perrin, Ref. 8.5) that when the particle concentration is low, so that there is little interaction between particles, the results of kinetic theory can be applied as if the particles formed an ideal gas. Letting the pressure resulting from the particles be p_p we then have from Eq. (8.3) that

$$p_p = ckT , \tag{11.1}$$

where c is the number of particles per unit volume. Also each of the H_2O molecules can itself be regarded as a Brownian particle with average kinetic energy given by 3 kT/2 ; see Eq. (8.2) . Because of this motion the H_2O molecules exert an average partial pressure p_w on the membrane. However, we would not expect Eq. (11.1) (with c giving the number of H_2O molecules per unit volume) to be a good approximation for p_w ; since the H_2O molecules are very closely spaced the results for a dilute "particle gas" cannot be used. Fortunately for our purposes here it is unnecessary to obtain a specific expression for the partial pressure of water. In the following reasoning only differences in this quantity need to be evaluated.

We shall now let c, p_p , and p_w apply to the suspension inside the cell; the pressures p_p and p_w then act outward on the cell membrane. If the concentration of particles in the outer suspension is c' they will exert an inward partial pressure p_p' on the membrane, given by

$$p_p' = c' kT , \tag{11.2}$$

assuming there is no temperature difference between the outer and inner media. Also the outer water molecules will exert an inward partial pressure p_w' on the membrane.

In explaining osmotic effects we suppose for the simplest situation that the membrane is freely permeable to water but will not allow passage of suspended particles, either those inside the cell or those outside. To visualize this we may think of the membrane as a sieve whose holes are large enough to pass H_2O molecules but not the larger particles. If the partial pressure of water on one side of the membrane is momentarily different from that on the other, water will flow in the direction of the smaller partial pressure until p_w is equal to p_w'. The total internal pressure is equal to $(p_w + p_p)$ while the total external pressure is $(p_w' + p_p')$; hence in equilibrium (for which p_w is equal to p_w') the total pressure inside will exceed that outside by the difference between p_p and p_p'. This difference is called the <u>osmotic pressure</u> difference; representing it by π_o, we see from the above reasoning, and from Eqs. (11.1) and (11.2), that

$$\pi_o = p_p - p_p' = (c - c')\, kT \,. \tag{11.3}$$

This equation is known as the van't Hoff law. To gain some understanding of its significance, suppose that initially identical suspensions of concentration c exist inside and outside a model cell of the kind under consideration. Then at a given time suppose a few particles are removed from the outer suspension, reducing its concentration to c'. This will not cause any significant change in the total external pressure $(p_p' + p_w')$ which is typically essentially equal to atmospheric pressure; however, there will evidently result an increase in p_w' equal to the decrease in p_p', the latter being $(c-c')kT$. Water will then flow into the cell until the interior partial pressure of water p_w is equal to p_w'. We can see that at equilibrium the total pressure in the cell will be greater than that outside by the amount π_o given in the van't Hoff law, Eq. (11.3) .

If the membrane is not completely impermeable to the particles in suspension the van't Hoff law will not be valid. For such a leaky membrane it is found that the osmotic pressure developed is less than the value π_o given in Eq. (11.3); instead it is given by $\sigma \pi_o$, where σ is called the reflection coefficient. The magnitude of σ varies between 0 and 1, and depends on the effective diameter of the pore relative to that of the particle. The reflection coefficient is discussed further in Sections 11.5.2 and 11.5.3.

11.3 CONCENTRATION DEPENDENCE OF CHEMICAL POTENTIAL

We can now proceed to confirm the expressions for $d\mu^c$ and $d\mu_w^c$ in Eqs. (10.22). Let μ and μ_w represent the chemical potential for a particle and for a water molecule, resp., inside the cell. Suppose that initially the pressure, temperature and particle concentration are the same inside and outside the cell; then the chemical potentials (assumed to depend only on those variables) are also the same inside and out. Now imagine a small change of state in the cell involving changes of the inner concentration and pressure, without any change in temperature and without any change in the external concentration or pressure; the outer chemical potentials are then unaffected. Also, since the membrane is permeable we expect that as the change in state occurs water will flow into or out from the cell as required so that the chemical potential remains equal on the two sides of the membrane. It follows that after equilibrium is achieved μ_w will be unaffected by the change in state, that is, that

$$d\mu_w = 0 \,. \tag{11.4}$$

From this simple statement we obtain from Eq. (10.21) that

$$d\mu_w^c = - \bar{v}_w \; dP \,. \tag{11.5}$$

Here $d\mu_w^c$ is the change in the chemical potential of a water molecule inside the cell resulting from a change dc of the inner particle concentration; dP is the associated change of inner pressure. Hence dP can be identified with the osmotic pressure discussed earlier; from the van't Hoff law, Eq. (11.3), we realize that dP is just equal to kT dc so that

Eq. (11.5) becomes

$$d\mu_w^c = - \bar{v}_w \, kT \, dc \ , \tag{11.6}$$

as anticipated in Eq. (10.22). To obtain an expression for $d\mu^c$ we refer to the Gibbs-Duhem equation, Eq. (10.27), which in an isothermal change becomes simply

$$-V \, dP + n \, d\mu = 0 \ , \tag{11.7}$$

since $d\mu_w$ is zero. The total volume V of the system can be written

$$V = n\bar{v} + n_w \bar{v}_w \ . \tag{11.8}$$

Combining the above two equations with the first of Eqs. (10.21), and using the van't Hoff value kT dc for dP, we obtain (Prob. 11.1)

$$d\mu^c = kT \frac{dc}{c} \ , \tag{11.9}$$

valid for dilute suspensions (for which $n\bar{v} << V$). This is the expression anticipated in Eqs. (10.22).

11.4 THEORY OF PASSIVE TRANSPORT: HYDRODYNAMICS AND DIFFUSION

11.4.1 Introduction

For a cell to be alive and active there must be continual traffic of small molecules and ions across the membrane which separates the cell interior from its environment. Molecular raw materials for generating energy and synthesizing biomolecules must be brought to the cells, while manufactured products and wastes must be carried away. In the remainder of this chapter we take up aspects of transport processes; in so doing we find application for principles of fluid flow and diffusion taken up earlier, especially in Chaps. 5, 9 and 10. But in dealing with these processes on the molecular scale appropriate for biomembranes, we shall see that new aspects appear which require special consideration.

It should be mentioned that while we confine ourselves here to "passive" processes (i.e., flow and diffusion) biological cells also have unique means for "active transport" or "pumping" of ions from one side of a membrane to the other. Biological pumps are not yet understood; but they are the object of much research and it has been shown that they are driven by chemical energy. Specifically the energy for active transport of Na^+ across the red cell membrane (from the interior to the exterior) has been shown to depend on ATP molecules, molecules which we saw are also important to muscle contraction (Section 1.6).

In this and the next section we acquaint ourselves with theoretical approaches to passive transport, applied to rather simple models. First we take up theory which is based primarily on principles of hydrodynamics and diffusion. Then, in Section 11.5, we see how theory of nonequilibrium thermodynamics contributes to the subject. Experimental results for biological media are treated in Section 11.6.

11.4.2 The Membrane as a Liquid Layer

In taking up passive transport we first consider an important model suggested by the characteristics of a phospholipid bilayer (Section 7.9 and Fig. 7.13). In this model the membrane is represented simply as a layer or film of viscous liquid. (The film is perhaps supported by a network of protein macromolecules whose presence is ignored for the present.) Molecules traverse from one side of the layer to the other by (1) dissolving into the layer on the first side, (2) passing through by diffusion, then (3) coming out of solution on the opposite side.

To treat the process quantitatively we suppose the layer is bounded by planes at $x = 0$ and $x = L$. (Fig. 11.1 .) Outside the layer on either side is an aqueous solution containing the diffusant molecules in question, for example, molecules of the sugar glucose. In the solution just outside the layer the concentration of glucose is c_o at $x = 0$ and c_L at $x = L$. In the layer, just inside the boundaries, the glucose concentration is less than in solution by a factor depending on the solubility of glucose in the phospholipid material. We let K be the ratio of the concentration of glucose just inside the layer to that just outside; this constant K is sometimes called the partition coefficient. (It is similar to the solubility ratio c_o/n referred to in Section 9.6.1 .) Then the glucose concentration in the layer is Kc_o at $x = 0$ and Kc_L at $x = L$. In steady state diffusion we expect that in the layer the concentration c will vary linearly with x, as in Section 9.4.1 . The concentration gradient $\partial c/\partial x$ will be constant and given in magnitude by the ratio of the concentration difference $K(c_o - c_L)$ to the layer thickness L . For the particle current J along x we then have

$$J = -D \frac{\partial c}{\partial x} = \frac{DK}{L} \Delta c \ ; \quad \Delta c = c_o - c_L \ ; \tag{11.10}$$

here D is the diffusion coefficient for glucose in the phospholipid substance. In obtaining the sign of J we realize that $\partial c/\partial x$ is negative if $c_o > c_L$, and the particle current is then in the positive x direction. The diffusion permeability P_d for an experimental situation is sometimes defined as the ratio of the observed current J to the measured concentration difference Δc :

$$P_d \equiv \frac{J}{\Delta c} \ . \tag{11.11}$$

We find that P_d has the units of velocity. A physical interpretation may be useful: for a hypothetical liquid in which the particle concentration is Δc a uniform motion of velocity P_d will yield the observed particle current J . When Eq. (11.10) applies we obtain

$$P_d = \frac{DK}{L} \ . \tag{11.12}$$

Another definition is sometimes given for diffusion permeability; the conventional symbol for this alternative coefficient is ω , and it is defined by

$$\omega = \frac{J}{kT\ \Delta c} = \frac{P_d}{kT} \ . \tag{11.13}$$

From Section 11.2 (e.g., Eq. (11.1)) we recall that $kT\ \Delta c$ has the units of pressure. We find that the unit for ω is the reciprocal of the product: force x time .

Applicability of the liquid model to actual membranes is considered in Section 11.6 .

11.4.3 Membrane with Macroscopic Pores

We now take up another simple model, to which we can again apply principles taken up earlier. In this model the cell membrane is regarded as a sheet of thickness L penetrated with a number of right cylindrical holes or pores, each of radius R, as suggested in Fig. 11.2. Aqueous solution exists on both sides of the membrane and also fills the pores. In this section we assume the pore radius R is large compared to the diameter of water molecules, and any other molecules whose passage through the membrane is of interest. We can therefore use principles of fluid flow and diffusion which we took up in Chapters 5 and 9.

We first consider diffusion across this membrane, requiring that the diffusant molecules must travel through the pores. As was done for the liquid-layer model (Section 11.4.2) we let the boundaries be at $x = 0$ and $x = L$ and assume the concentrations in the solutions at these planes (i.e., just outside the membrane) are c_o and c_L , resp.. Since the medium in the pore is an aqueous solution the concentration c is continuous across each of the planes $x = 0$ and $x = L$. (This is unlike the situation for the liquid membrane considered in the previous section. There a jump in concentration from c_o to Kc_o occurred across the plane $x = 0$, for example.) When steady state diffusion is set up the concentration inside a pore varies linearly with x ; the diffusion current density j in the pore is then

$$j = \frac{D}{L}(c_o - c_L) , \tag{11.14}$$

which may be compared with Eq. (11.10); of course the coefficient D is for the diffusant molecules relative to the aqueous medium in the pore. Let the pore area be A_p per unit area of membrane; then A_p gives the

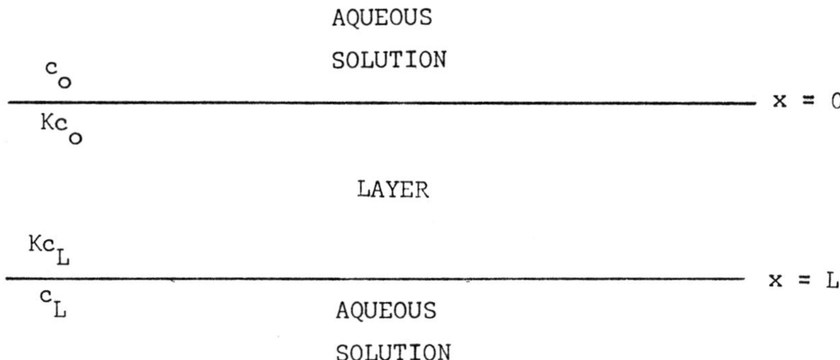

Fig. 11.1 Liquid-layer model for transport of solute through a biomembrane. See Section 11.4.2 .

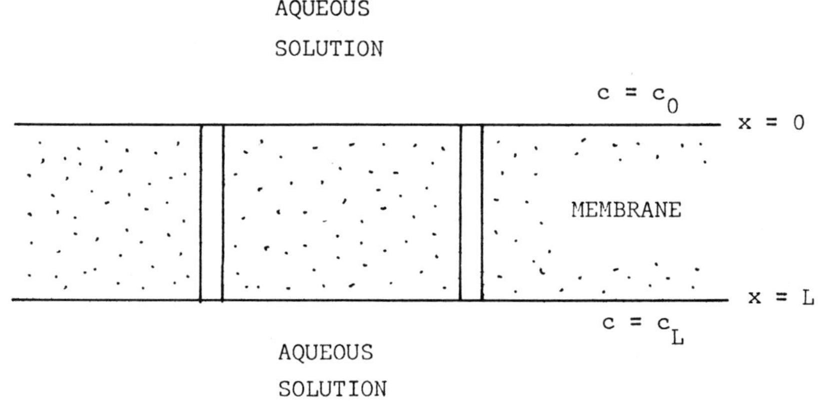

Fig. 11.2 Model of membrane with macroscopic pores.

fractional area occupied by pores. The diffusion current J per unit area of membrane will be jA_p or

$$J = \frac{DA_p}{L} \Delta c \ ; \ \Delta c = c_o - c_L \ . \tag{11.15}$$

This equation is very similar to Eq. (11.10), but has the ratio A_p appearing instead of the ratio K . Defining the permeability P_d again as the magnitude of $J/\Delta c$ we obtain

$$P_d = \frac{DA_p}{L} \ , \tag{11.16}$$

which may be compared with Eq. (11.12).

Secondly, we consider flow of liquid across the membrane, requiring it to flow through the pores which are now taken to be right circular cylinders of radius R. Assuming laminar viscous flow occurs through each pore we use Poiseuille's law, assuming a pressure difference Δp is applied across the membrane. An expression for the volume rate of flow through each pore is given by V/t_o in Eq. (5.63), where G' is $\Delta p/L$. In the area A_p occupied by pores in unit area of membrane there are $A_p/\pi R^2$ pores of radius R . Hence the total volume flow rate per unit area of membrane is J_f , where

$$J_f = \frac{A_p}{\pi R^2} \ \frac{\pi R^4 \ \Delta p}{8\eta L} = P_f \ \Delta p \ ; \tag{11.17}$$

here P_f is the flow permeability and is given by

$$P_f = \frac{R^2 A_p}{8\eta L} \ . \tag{11.18}$$

The quantity J_f has units of velocity, and hence the units of P_f are: velocity divided by pressure. In membranes with small pores their size and number are not usually measured directly. Instead this information is obtained from measurements of diffusion and flow. Consider a membrane for which P_d and P_f have been measured. If the membrane is one for which Eqs. (11.16) and (11.18) apply we see, by combining these equations, that

$$R^2 = 8\eta D \frac{P_f}{P_d} \ . \tag{11.19}$$

If η and D are known or can be estimated, a calculation of R can then be made from Eq. (11.19). (Prob. 11.2) Once R has been found A_p can be determined from Eq. (11.18) if L can be estimated. Alternatively, A_p can be determined from Eq. (11.16) if D and L are known. After A_p has been determined the number of pores per unit area of membrane can be calculated since this is just $A_p/\pi R^2$.

The simple procedures just described are indicative of those which have been followed in calculating pore parameters for membranes. However, corrections are required for the very small pores which are characteristic of biomembranes. These have been discussed by Solomon (1968, Ref. 11.1) and by Anderson and Quinn (1974).

11.5 THEORY OF PASSIVE TRANSPORT: NONEQUILIBRIUM THERMODYNAMICS

11.5.1 Flows and Forces

We now take up another approach to the transport problem. This approach is more abstract and, at the same time, more general than the theory taken up in Section 11.4 . We treat the situation represented in Fig. 11.3 . A membrane separates two solutions or suspensions, A and B. Various choices might be made for the solvent and solute, but for definiteness we take these to be water and glucose, resp. We suppose that the temperature T is the same on both sides while there may be differences in the pressure P , in the glucose concentration c , and in the chemical potentials μ and μ_w for glucose and water, resp . We use unprimed quantities for medium A and primed quantities μ' , etc., for medium B . In considering flow of water and glucose across the membrane we may think of the latter as a simple liquid layer (as was considered in Section 11.4.2) or as a membrane penetrated by pores filled with solution (as considered in Section 11.4.3). Other models of the membrane may also be imagined; an advantage of the thermodynamic method is that statements can be made without regard to a specific model.

In approaching the membrane problem by means of nonequilibrium thermodynamics we consider currents of solvent and solute passing through the membrane from medium B to medium A . As a molecule of water passes from B to A its chemical potential reduces from μ_w' to μ_w ; assuming no change in potential energy or kinetic energy this implies generation of heat in the amount $(\mu_w' - \mu_w)$. Similarly in each passage from B to A of a glucose molecule $(\mu'-\mu)$ units of chemical energy are converted into heat. If the particle currents of water and glucose across the membrane from medium B to medium A are J_w and J , resp., the heat produced per unit area per second is Φ given by

$$\Phi = (\mu_w' - \mu_w) J_w + (\mu' - \mu)J . \qquad (11.20)$$

Here a "particle current" gives the number of molecules crossing unit area of the membrane per second. We now refer to the expressions for $d\mu$ and $d\mu_w$ in Eqs. (10.21) and (10.22). Consistent with the discussion subsequent to Eq. (10.28) in Section 10.7.2 , we interpret

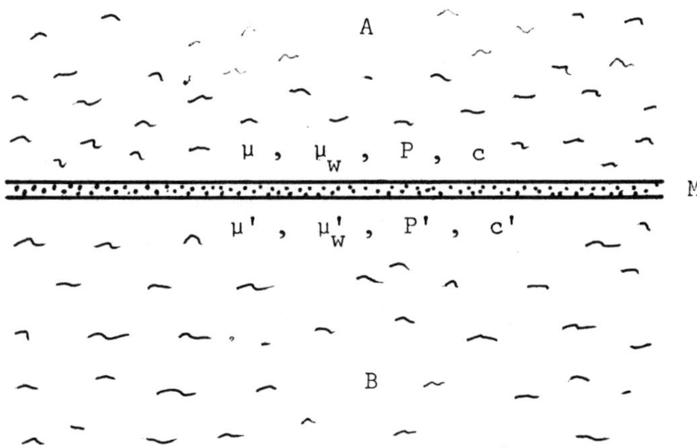

Fig. 11.3 Membrane M separates two media, such as solutions of glucose in water. Unprimed quantities μ , etc , apply to medium A and primed quantities μ' , etc , to medium B .

$d\mu$ (for example) as the difference in μ between two points in space. Specifically, we interpret $d\mu$ for our situation as $(\mu' - \mu)$, and $d\mu_w$ correspondingly as $(\mu_w' - \mu_w)$. Then dP in Eq. (10.28) is interpreted as $(P' - P)$, and dc as $(c' - c)$, with the understanding that these quantities are small (but finite). Further, we follow common usage and let $dP = \Delta P$ and $dc = \Delta c$ so that

$$\begin{aligned} dP &= \Delta P = P' - P \\ dc &= \Delta c = c' - c \end{aligned} \tag{11.21}$$

From Eqs. (10.21) and (10.22) (assuming $dT = 0$) we then can rewrite Eq. (11.20) as

$$\Phi = \vec{v}_w \, [\Delta P - kT\,\Delta c] J_w + [\bar{v}\,\Delta P + kT\,\frac{\Delta c}{c}]\,J \, . \tag{11.22}$$

Following Kedem and Katchalsky (1958, Reference 11.3) we now introduce the symbols J_v and J_D where

$$J_v = \bar{v}_w J_w + \bar{v} J \; ; \quad J_D = \frac{J}{c} - \frac{J_w}{c_w} \; ; \tag{11.23}$$

here c_w is the number of water molecules per unit volume and, in a dilute solution, is approximately equal to the reciprocal of $\bar{v}_w$. Equation (11.22) then becomes

$$\Phi = [J_v][\Delta P] + [J_D][kT\,\Delta c] \, . \tag{11.24}$$

We notice that J_v and J_D both have the units of velocity (and thus differ from the currents J and J_w) . Also both Δp and $kT\,\Delta c$ have the units of force per unit area. In the language of nonequilibrium thermodynamics the quantities J_v and J_D can be regarded as <u>flows</u>, while the quantities p and $kT\,\Delta c$ are then <u>forces</u>. The general theory of nonequilibrium thermodynamics is based on the assumption that when these forces are small, each of the flows is a linear function of all the forces. Stated mathematically this basic assumption is, for the situation under consideration,

$$\begin{aligned} J_v &= L_p\,\Delta p + L_{pD}\,kT\,\Delta c \\ J_D &= L_{Dp}\,\Delta p + L_D\,kT\,\Delta c \, . \end{aligned} \tag{11.25}$$

Another basic assumption of nonequilibrium thermodynamics is the Onsager reciprocal relationship which in this instance states that

$$L_{Dp} = L_{pD} . \qquad (11.26)$$

It can be shown that L_p and L_D are positive while L_{pD} may have either sign; it is also found that

$$L_p L_D > L_{pD}^2 . \qquad (11.27)$$

Equations (11.25) - (11.27) are the principal results of nonequilibrium thermodynamics for the membrane problem under consideration. The development of an expression for Φ , the heat generated per unit area, was carried out primarily for the purpose of identifying quantities to be selected as flows and forces. This selection is somewhat arbitrary except that for an isothermal process the flow-force products must appear in an expression for Φ as illustrated in Eq. (11.24). In a more general situation the expression required is one for the production of entropy; for isothermal processes at temperature T the rate of entropy production is just Φ/T .

Equations (11.23) - (11.27) are rather abstract as yet. The quantities J_v, J_D, L_p, etc., take on physical significance when we see their meanings under special conditions. For a glucose solution in water the flow J_v represents the total volume flow of water and glucose through unit area of membrane; its units are volume per unit area per second or simply velocity. When $J = 0$, as is true if $c = 0$ or if the membrane is impermeable to glucose, the quantities J_v and J_w differ only by the constant factor $\bar{v}_w$. This constant is approximately equal to the volume of a water molecule, about 3.0×10^{-23} cm^3 .

For J_D the units are the same as for J_v , but there are important differences in these flow quantities. Especially obvious is the sign preceding the J_w term, which is "+" for J_v but "-" for J_D . Examining the quantities J_w/c_w and J/c we realize that these are precisely the velocities of the molecules of water and glucose, respectively. Hence J_D has a simple meaning: it gives the velocity of the glucose relative to the water. In a dilute solution the coefficients ($\bar{v}_w$ and $1/c_w$) of J_w are nearly the same in the expressions for J_v and J_D, but the coefficients ($\bar{v}$ and $1/c$) of J are very different. When J is zero in a dilute solution we see that J_D is approximately equal to $-\bar{v}_w J_w$ and is hence nearly equal to $-J_v$.

It is common to refer to J_v as a <u>volume flow</u> and to J_D as a <u>relative velocity</u> or <u>exchange flow</u>.

11.5.2 The Coefficients L_p, L_{pD} and L_D

To get a feeling for the meaning of L_p, etc., we consider experiments by means of which they can be measured for a given membrane. In imagining an experiment for measuring L_p we suppose pure water exists on both sides of the membrane and that a pressure difference Δp is applied, causing water to flow. Then J is zero, of course, and we have, from Eqs. (11.23) and (11.25) that the volume flow J_v is

$$J_v = J_w \bar{v}_w = L_p \Delta p \,. \tag{11.28}$$

Hence the "particle current" for water is J_w, given by

$$J_w = \frac{L_p}{\bar{v}_w} \Delta p \,, \tag{11.29}$$

where $\bar{v}_w$ is equal to 3.0×10^{-23} cm^3. The quantity L_p is called the <u>filtration coefficient</u>. For a given membrane its value can be determined from a water-flow experiment of the kind just considered, in which J_w is measured for given Δp. Comparing Eq. (11.29) with Eq. (11.17) and identifying $J_w \bar{v}_w$ with J_f we see that (when flow of solvent-only is involved) the permeability P_f and the filtration coefficient L_p are the same:

$$L_p = P_f \,. \tag{11.30}$$

We might hope to determine L_D for a given combination of membrane and solution, in a manner analogous to that for determining L_p . However, this is not usually feasible, for then we should need to arrange conditions for observing the passage of solute only; but transfer of solute by a membrane will usually be accompanied by a flow of solvent. Instead we proceed more indirectly and first consider an experiment in which L_{pD} is determined. Here it turns out that an appropriate experiment is one in which the pressure difference Δp and the concentration difference Δc are so adjusted that the volume flow J_v is zero. Defining σ as $-L_{pD}/L_p$ we obtain from Eqs. (11.25) that the relation between Δp and Δc is then

$$\Delta p = \sigma kT\,\Delta c \;; \quad \sigma = -\frac{L_{pD}}{L_p} . \tag{11.31}$$

The nondimensional ratio σ is called the reflection coefficient, which is the quantity referred to in Section 11.2 . We see from Eq. (11.31) that σ can be calculated from measurements of Δp and Δc in the "zero J_v" experiment just described. If L_p is known L_{pD} can then be determined from Eqs. (11.31) .

The pressure difference Δp required to maintain zero volume flow is often called the osmotic pressure and represented by the symbol π . From Eq. (11.3) of Section 11.2 (with change of symbols) the theoretical value π_o of the osmotic pressure for an ideal membrane is $kT\,\Delta c$. We may therefore rewrite Eq. (11.31) as

$$\pi = \sigma\pi_o . \tag{11.32}$$

This important result was pointed out by Staverman (1951).

A determination of L_D for a solution can now be made, by carrying out an experiment in which the flow of solute, such as glucose, is measured while the volume flow J_v is maintained zero. Turning to Eqs. (11.23) we assume that the solution is dilute so that $c_w \simeq 1/\bar{v}_w$ and $c\bar{v} \ll 1$; making use of these assumptions we eliminate J_w and obtain:

$$J = (J_v + J_D)\,c .$$

We now set $J_v = 0$ and substitute for J_D from Eq. (11.25), make use of Eqs. (11.26) and (11.31) and obtain (Prob. 11.3)

$$J = [c(L_D - \sigma^2 L_p)][kT\ \Delta c]\ . \tag{11.33}$$

Comparing this equation with Eq. (11.13) we see that the equations are identical if we give the diffusion permeability ω the value:

$$\omega = c(L_D - \sigma^2 L_p)\ . \tag{11.34}$$

In an experiment of the kind just described where J and Δc are measured the permeability ω can be calculated from the ratio of J to $kT\ \Delta c$.

We have now described means for determining L_p, σ and L_{pD} (the latter being equal to L_{Dp}) . If ω has also been determined the coefficient L_D can be calculated from Eq. (11.34) for given c .

From Eqs. (11.25) we see that characterizing the flow properties of a membrane-solute combination requires knowledge of the three constants L_p , L_{pD} and L_D . Another suitable set of three independent constants, more widely used, are L_p , σ and ω . For some experimentally determined values of these constants see Section 11.6 .

11.5.3 The Reflection Coefficient σ

Further insight on the significance of σ is obtained by considering a situation where the solute concentration c is initially the same on both sides of a membrane and flow is produced by a pressure difference Δp applied at $t = 0$. We have then from Eqs. (11.23), (11.25), (11.26) and (11.31) that at $t = 0$

$$\begin{aligned} J_v &= J\bar{v} + J_w \bar{v}_w = L_p\ \Delta p \\ J_D &= \frac{J}{c} - \frac{J_w}{c_w} = -\ \sigma L_p\ \Delta p\ . \end{aligned} \tag{11.35}$$

For a dilute solution $\bar{v}_w = 1/c_w$; adding the above two equations then yields

$$(1 - \sigma)\ L_p\ \Delta p = J(\bar{v} + \frac{1}{c}) \cong \frac{J}{c}\ , \tag{11.36}$$

since $\bar{v} << 1/c$ in a dilute solution. Dividing all terms of Eq. (11.36) by $L_p \Delta p$, which is equal to J_v by Eqs. (11.35), we obtain

$$1 - \sigma = \frac{J}{c\ J_V} . \tag{11.37}$$

We now examine the right hand side of this equation. The quantity J gives the actual particle current of solute (e.g., glucose) through the membrane. On the other hand J_v gives the total volume current of solution (water plus glucose) through unit area of membrane per second. The concentration c is the number of solute particles per unit volume of solution on either side of the membrane at t = 0 . We arrive at an interpretation of the product cJ_v by noticing that *if* the solution had passed through the membrane at concentration c this product would be just the particle current of solute. In general the membrane will prevent free flow of solute and J will be less than cJ_v . The ratio J/cJ_v is sometimes called the *sieve constant* and, from Eq. (11.37), is given by $(1-\sigma)$.

We now consider several special situations. First suppose the membrane conforms to the pore model discussed in Section 11.4.3 , and that the pores are very large; J will then equal cJ_v and we have $\sigma = 0$. Second, suppose again that the membrane is porous but that it is semipermeable, allowing water to pass but no solute. Then J = 0 and $\sigma = 1$. More generally a porous membrane will allow some, but not all, of the solute to pass; thus σ has a value intermediate between 0 and 1 . We can now see why σ is called a reflection coefficient; it is a measure of the extent to which solute molecules are prevented from accompanying water flow through the membrane, as if they were "reflected" at the membrane surface.

In the above examples we illustrate the significance of σ by referring to a porous-membrane model. However the theory of nonequilibrium thermodynamics, in which the concept of σ appears, is valid regardless of model and is, for example, fully applicable to a liquid layer such as was considered in Section 11.4.2 . For such a layer it may be true that the solute travels through the membrane faster than water does;

then $J > cJ_v$ and σ has a negative value. A process for which σ is negative is sometimes called reverse osmosis, and is important in desalination.

11.6 TRANSPORT THROUGH MEMBRANES, GELS AND CYTOPLASM: EXPERIMENTAL RESULTS

Here we examine findings on transport of nonelectrolyte (uncharged) molecules through permeable structures which are either of biological origin or serve as "models" for biostructures. We shall be especially interested in results which serve as tests for theories discussed in previous sections of this chapter.

11.6.1 Phospholipid Films

As was mentioned in Section 11.4.2 the liquid-layer model is particularly relevant to phospholipid bilayers. Methods of preparing these bilayers in films suitable for quantitative studies (starting with lipids extracted from brain white matter) were described by Mueller, Rudin, Tien and Westcott (1962, 1963). Typically about 60-90 A thick, the films are sometimes called black lipid films or membranes, since destructive optical interference makes them appear black by reflected light.

Cass and Finkelstein (1967) measured diffusion across thin lipid membranes using tritiated water THO as the diffusant. They found the diffusion permeability P_d defined in Eq. (11.11) to be approximately 10^{-3} cm/sec. This magnitude of P_d was regarded as support for the idea that THO (and hence also water) traverses the lipid film as if the latter were a liquid. For previous workers (Hanai and Haydon, 1966) had suggested that the hydrocarbon interior of the film (consisting partly of a solvent such as n-decane or n-hexadecane) is such that the diffusion coefficient D for THO is about 10^{-5} cm/sec and the partition coefficient K is about 10^{-4}. Assuming a film thickness L of about 10^{-6} cm Eq. (11.12) yields 10^{-3} cm/sec for P_d. Since the values for D, K and L are rough estimates the agreement between measured and calculated values of P_d is in order-of-magnitude only. Nevertheless it is significant that a simple theory based on the assumption of a liquid layer should be so successful.

Further support for treating the lipid film as a liquid layer comes from experiments with other films made from different lipids, in which it was found that P_d tends to increase as the expected lipid fluidity increases. Such a result is expected from Eq. (11.12) since the diffusion coefficient D for a molecule in a liquid of viscosity η varies inversely with η and hence is proportional to the fluidity η^{-1}.

A fluid-layer model for phospholipid membranes is very appealing because of its simplicity. We have just seen that the model is useful in enabling us to make rough but quantitative predictions about membrane permeability. Also it is plausible that the hydrocarbon interior of a phospholipid film should be liquid-like since similar hydrocarbons in bulk quantities are liquid. Nevertheless it has not really been established that the membrane interior has literally the properties of a Newtonian liquid.

As an alternative model for these membranes a promising concept is that of the liquid crystal. (Brown, 1972) . Here the phospholipid molecules are regarded as being in a semi-crystalline array; such an array is unlike a solid crystal in that the molecules are relatively free to move about. An interesting theory for transport based on this model has been put forward by Trauble (1971). In this theory he points out that many defects or "kinks" are to be expected in the hydrocarbon structure of a lipid film. A kink provides a "hole" or "pocket" large enough for diffusant molecules such as H_2O . Thermal energy makes the holes take part in Brownian movement, that is, diffusion, rather as particles do. Specifically, Trauble assumes that the diffusant molecules "hitchhike" across a membrane by entering kink-holes and travelling with them as they take part in diffusion. A theoretical estimate of 10^{-5} cm^2/sec is arrived at for an effective diffusion coefficient D applicable to the occupied holes. This is the same value of D as was used for the liquid model; hence if the same values of K and L are used one obviously obtains the same result for P_d *via* Eq. (11.12).

11.6.2 Dialysis Membranes

Cellulose-based membranes are important because of their use for dialysis in "artificial kidney" machines; membranes formed from collagen are also being developed for this purpose (Stenzel, et al, 1971). In the dialysis application it is required that the membrane should allow ready passage of urea (molecular weight 60) and other unwanted molecules of relatively small size while retaining the blood proteins and cells. For several kinds of cellulose membranes Renkin (1954) measured (1) the volume flow rate of water for given values of the pressure difference Δp and (2) the diffusion current J_d of tritiated water THO and other solutions for various values of the concentration difference Δc. He analyzed his data by use of "pore theory" like that in Section 11.4.3. Values of the flow permeability P_f were found to be 9.5 and 22.4, resp., for Visking cellulose and DuPont cellophane, each in units of 10^{-12} cm^3/dyne-sec; for these membranes the thickness L was 55 μm and 80 μm, resp. From diffusion measurements with THO, P_d is determined to have the approximate values 4.9×10^{-4} cm/sec and 4.4×10^{-4} cm/sec. Taking D to be 2.6×10^{-5} cm^2/se for THO in water and letting η be 10^{-2} P we obtain from Eq. (11.19) the approximate values of 20 A and 33 A, resp., as the pore radius R for each of the two membranes.

For collagen membranes Stenzel and coworkers found P_d for D_2O to be about 10^{-3} cm/sec, P_f about 2.8×10^{-11} cm^3/dyne-sec, and L about 50 μm . From t kind of theory discussed in Section 11.4.3 these data lead to calculated values of R in the vicinity of 24 A . (Prob. 11.2) .

Actually electron microscopy for cellulose (collodion) membranes (Bugher, 1953) shows that the "pores" are tortuous in shape and of irregular size; they are also interconnected, that is anastomosing. The model of a membrane penetrated by pores consisting of independant right cylinders is clearly a highly idealized one. Neither are uniform right-cylindrical pores to be expected in collagen membranes; Stenzel, et al., refer to R as a "mathematical phenomenon".

Even for straight cylindrical pores one may well question the validity of results from hydrodynamics for situations where the molecular size is comparable to the pore diameter. After all, hydrodynamics is a continuum theory in which inhomogeneity or "graininess", caused by the finite size of molecules, is ignored. Nevertheless it often proves possible to apply continuum mechanics, thus assuming homogeneity, to flow of molecular media. Perhaps this is true because flow measurements are based on averages in which effects of individual molecules are often negligible.

11.6.3 Red Cell Membranes and the Pore Model

Solomon and his associates have shown that pore theory is useful even for membranes of mature erythrocytes, that is, mature red cells. The erythrocyte membrane has been a favorite for study by many investigators because red cells are important physiologically and are conveniently available. Also (mature) red cells are relatively simple to study since they are free from other tissue, and they contain no organelles such as nuclei or mitochondria.

Among studies of molecular transport across the red cell membrane was a diffusion experiment by Paganelli and Solomon (1957). In this experiment tritiated water THO was quickly mixed into a saline solution in which red cells were suspended. Observations were then made at successive short time intervals after the mixing, to determine the rate at which THO molecules had entered the cells. In this way it was possible to determine the diffusion current J corresponding to a given concentration difference Δc between the inside and outside of the cell membrane. From this information they calculated the permeability P_d defined in Eq. (11.11), obtaining the value 5.3×10^{-3} cm/sec for THO, hence also for H_2O. This is a high value for P_d, being even somewhat higher than typical values for thin lipid films (Section 11.6.1). In Table 11.1 values of P_d for water are listed for a number of biomembranes; for the red cell P_d is over 30 times greater than the next largest value in the table and over 250 times greater than the smallest value.

TABLE 11.1

WATER PERMEABILITY CONSTANTS

P_d as defined in Section 11.4.3 , for THO molecules in water, in units of cm/sec when multiplied by 10^{-3} . From Paganelli and Solomon (1957).

P_d	Membrane
5.3	Human red cell
0.025	Amoeba (Chaos chaos)
0.021	Amoeba proteus
0.0483	Frog gastric mucosa
0.128	Frog ovarian egg
0.068	Zebra fish ovarian egg
0.090	Xenopus body cavity egg
0.036	Zebra fish egg, shed, non-developing
0.075	Frog body cavity egg
0.168	Salmon egg, unactivated
0.113	Toad skin
0.073	Frog skin

In another experiment Sidel and Solomon (1957) measured water flow through the red cell membrane resulting from an osmotic pressure difference $\Delta\pi$ between the interior of a human red cell and its exterior. They established this pressure difference by adjusting the NaCl concentration in the bathing solution. The water flow rate was measured by determining the rate of change of cell volume; it was found to be 0.23 cm^3/sec per cm^2 of membrane area (or simply 0.23 cm/sec) for a concentration difference of one "osmole" per cm^3. An osmole is defined as N_a particles (where N_a is Avogadro's number, 6.02×10^{23}), each of which is osmotically independant. Stated more concisely, the water flow rate observed by Sidel and Solomon is 0.23 cm^4/osm-sec. ("Osmole" is abbreviated as "osm"). This total flow rate comes about partly through diffusion (because of a difference of concentration of water molecules on the two sides of the red cell membrane) and partly because of hydrodynamic flow (caused by the difference in pressure across the membrane).

In considering the contribution from diffusion we take the difference in concentration Δc to be 1 osm/cm^3 or N_a particles/cm^3. The particle current is $P_d \Delta c$, and hence $P_d N_a$, in particles per cm^2-sec. Each "particle" is a water molecule of volume $\bar{v}_w$, so the volume of water diffusing through the membrane is $P_d N_a \bar{v}_w$. But $N_a \bar{v}_w$ is just the molar volume of water, equal to 18.0 cm^3. We have, finally, remembering that P_d is 5.3×10^{-3} cm/sec:

$$\text{Diffusion current} = P_d N_a \bar{v}_w = 0.095 \text{ cm}^4/\text{osm-sec} . \tag{11.38}$$

Subtracting the diffusion current from the total water flow we obtain the part resulting from hydrodynamic flow, namely, 0.135 cm^4/osm-sec. From the van't Hoff law, Eq. (11.3), a concentration of 1 osm/cm^3 at temperature T corresponds to an osmotic pressure of $N_a kT$. Hence we can evaluate the hydraulic permeability constant P_f defined, as in Eq. (11.17), as $J_f/\Delta p$. We obtain (Prob. 11.5) the value

$$P_f = 5.4 \times 10^{-12} \text{ cm}^3/\text{dyne-sec} ; \tag{11.39}$$

This is of the same order as P_f for the membranes of cellulose and collagen, discussed previously.

Information on P_d and P_f is useful in itself, in allowing predictions about the behaviour of red cells in different environmental conditions. But also, knowing P_d and P_f, we can proceed to gain insight on the nature of the membrane. Let us consider, as Solomon and associates did, the pore model of Section 11.4.3. Choosing the simplest possibility, suppose the sketch in Fig. 11.2 was literally applicable, and that the pores are right circular cylinders, all with the same radius R. Furthermore assume for simplicity that the analysis of Section 11.4.3 applies. Using Eq. (11.19) we can then calculate the radius R. For η we choose 10^{-2} P, a value typical of water at room temperature. For D we choose 2.6×10^{-5} cm^2/sec, appropriate for a water molecule diffusing through water at room temperature. Finally, we choose for P_d the value 5.3×10^{-3} cm/sec and for P_f the value given in Eq. (11.39). We obtain

$$R^2 = 8\eta D \frac{P_f}{P_d} = 21.2 \times 10^{-16} \text{ cm}^2 \; ; \qquad (11.40)$$

hence the radius R is calculated to be 4.6×10^{-8} cm , that is, 4.6 A .

Such a value for R is most interesting. A pore diameter of about 9A is consistent with the known facts that water molecules (of diameter about 3A) pass easily through the membrane, while hemoglobin molecules (of diameter about 55 A) normally do not escape from the cell. However, there is as yet no evidence from electron microscopy for the existence of pores in the red cell membrane. Any channels that exist may be tortuous rather than straight cylinders. And, speaking of membranes generally, Solomon suggested in 1959 (Reference 11.1) that "one might liken the cellular membrane to a bowl of spagetti which is shaken continually, opening up new pores or passages in one place and closing them somewhere else."

For small pores the use of continuum theory is suspect, as noted in Section 11.6.2 Using corrections discussed by Solomon (1968, Reference 11.1) the radius R is determined to be somewhat smaller than the value obtained from Eq. (11.19). Anderson and Quinn (1974) discuss the corrections further, showing where errors have been made in previous work.

In spite of such problems the simple pore theory is most useful for making estimates and predictions. Thus in Eq. (11.16) we can use previously indicated values for P_d and D and calculate the ratio A_p/L , obtaining 204 cm . Taking L to be 200 A, somewhat greater than thicknesses usually cited for the red cell membrane, we obtain about 4×10^{-4} for A_p . Thus if the assumptions are valid, the pores take up less than 0.1% of the membrane.

Further experiments with red cells are described by Solomon (1968) . In these experiments red cells were initially in a normal environment (e.g., 0.9% NaCl) but were suddenly projected or mixed into various test solutions by means of small jets or other means. In each test solution the solute molecules were of known radius; examples were urea molecules of effective radius 2.7 A and glucose molecules of radius 4.2 A . For each test solute, say, X, the experiment was repeated a number of times with a different value for the concentration of X, and in each instance observations were made on the initial rate of change in cell volume resulting from the change in environment. In this way a characteristic value c_{ec} was obtained , for each solute, such that when its concentration is c_{ec} the initial rate of change of cell volume is zero. As expected (from the discussion in Section 11.2) water tends to flow into the cell when the test solute concentration c_e is less than c_{ec} , and to flow outward if $c_e > c_{ec}$.

Zero rate of change in cell volume corresponds to the condition $J_v = 0$. Hence Eq. (11.31) would apply (1) if only a single solute were involved, and (2) if its particles were free of interactions, i.e., if they formed ideal solutions. When multiple solutes are present, all in ideal solution, a modified form of Eq. (11.31) applies: The excess Δp of the pressure in the cell over that outside is given by a sum of contributions $\sigma_i kT \, \Delta c_i$, where σ_i is the reflection coefficient and Δc_i is the concentration difference for the ith solute. In the experiments described by Solomon the concentration of test solute is zero inside the cell. Hence the concentration difference Δc_e is just $-c_e$, the negative of the external concentration. Also, if its reflection coefficient is σ the test solute makes a contribution to Δp of $-kT\sigma c_e$.

The ions and other entities inside the cell are not free particles in ideal solution. They nevertheless contribute to the pressure difference Δp. Fortunately, it is not necessary to have a specific expression for their contribution. Instead, for experiments with a single external solute in ideal solution, we can write

$$\Delta p = C - kT\sigma c_e \, , \tag{11.41}$$

where C is a constant which includes effects of particles inside the cell. In the experiments described by Solomon the concentration for each test solute is adjusted to a critical value c_{ec} for which there is (initially) no change in size of the red cell. Referring to Eq. (11.41) the adjustment of c_e to c_{ec} evidently yields a value, say, A for the product c_{ec} such that C - kTA is equal (initially) to the normal value of Δp . (However, as time goes on, some of the solute may leak into the cell, thus changing the "constant" C). When A is known the reflection coefficient σ can be obtained for any solute from knowledge of its critical concentration c_{ec}: specifically, σ is equal to A/c_{ec}. The constant A can be evaluated by determining the critical concentration, call it c_{eo}, for a test solute consisting of large molecules to which the membrane is impermeable. For this solute we have $\sigma = 1$, and hence we see that $A = c_{eo}$. Hence for any other test solute, for which the critical concentration is c_{ec} the reflection coefficient is given by

$$\sigma = \frac{c_{eo}}{c_{ec}} \quad . \tag{11.42}$$

Experimental values of σ have been obtained for red cells relative to a variety of test solutes by making measurements of c_{ec} and c_{eo}. In Fig. 11.4 data obtained in this way for human red cells are used to plot the sieve constant $(1-\sigma)$ for solute molecules of varying radius $\underline{a}$. We see that, as expected, $(1-\sigma)$ decreases (and σ increases) with increasing radius $\underline{a}$. The rate of passage through the membrane is small for molecules with radii greater than 3 A . These results are consistent with the assumption that pores are present of radius no greater than 4.5 A .

More specific comparison can be made with theory for the reflection coefficient, applicable to small pores. However, when the pore is only a little larger than the solute and solvent molecules the theory is complex and still in the process of development. The interested reader is referred to Solomon (1968) (Reference 11.1) and Anderson and Quinn (1974).

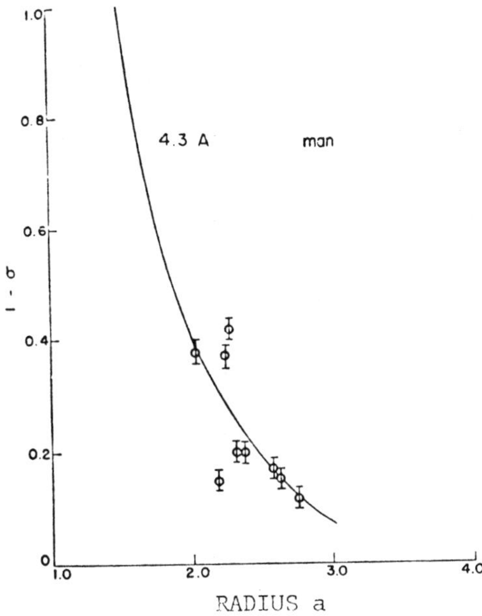

Fig. 11.4 The relationship of the sieve constant $(1 - \sigma)$ to the molecular radius of solute molecules, for membranes of human erythrocytes. From Solomon (1968).

Solomon and Gary-Bobo (1972) have shown that there are striking similarities between transport properties of human red cell membranes and those of lipid bilayers into which certain antibiotic molecules are introducted. Thus the ratios of P_f/P_d for water flow, and the reflection coefficients for several solutes, are nearly the same for the cell membranes as for the bilayers. Assuming the transport occurs through pores, that is, aqueous pathways, the calculated values of the pore radius R in the bilayers are the same as in the red cell, to within about 5% or less.

Especially impressive is a plot of data on permeability ω(see Eq. 11.13) for (hydrophilic non-electrolyte) solute molecules of various sizes. In Fig. 11.5 the circles give the permeability (relative to that for water) of human red cell membrane for molecules with molar volumes V_m varying from about 20 to 80 cm^3/mole. We see that ω decreases sharply as V_m

increases, falling by more than a factor of 100 as V_m increases by a factor of four. (This volume change corresponds to a change in mean molecular radius by only a factor of $4^{1/3}$ or 1.59). The straight line shown on the semilog plot represents similar data for the antibiotic-containing lipid bilayers mentioned earlier. It is remarkable that data for the red cells and bilayers fit so well on the same straight-line plot of log ω vs V_m. Perhaps further research will soon yield insight on the reason for this appealing result.

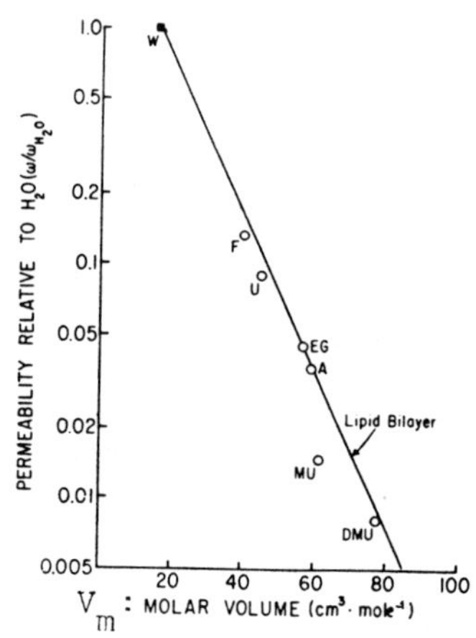

Fig. 11.5

Relative permeability (ratio of permeability to that for water) vs molar volume V_m for various hydrophilic non-electrolyte solutes. Circles are for human red cell membranes. Straight line is a best fit to data for lipid membranes containing certain antibiotics. From Solomon and Gary-Bobo (1972, Reference 11.15). W, water; F, formamide; U, urea; EG, ethylene glycol; A, acetamide; MU, methylurea; DMU, dimethylurea.

11.6.4 Table of Membrane Constants.

Principal constants for describing flow and diffusion through membranes are L_p, σ, ω and P_d, all defined in previous sections. Table 11.2 lists values of these quantities for the human red blood cell and for Visking dialysis tubing (the latter being important for such biomedical applications as the artificial kidney), each for solutions of several different molecules in water. Both $\omega(1)$ and $\omega(2)$ represent permeability; $\omega(1)$ is in units of molecules per dyne-sec, agreeing with the definition of ω in Eq. (11.13). But for $\omega(2)$ the unit is mole per dyne-sec, a unit commonly used in physiology. To calculate $\omega(2)$ we divide $\omega(1)$ by Avogadro's number N_a. To calculate the permeability P_d, defined as in Eq. (11.11), we multiply $\omega(1)$ by kT. Examining the table, we see for each membrane tendencies for the reflection coefficient to increase, and the permeability to decrease, with increasing molecular weight of the solute. These tendencies are as expected from pore theory.

TABLE 11.2

FLOW CHARACTERISTICS OF MEMBRANE-SOLUTE COMBINATIONS

C, human red cell. T, Visking dialysis tubing. Molecular weights in daltons, in ( ) after each solute. L_p, filtration coefficient, in $cm^3\ dyne^{-1}\ sec^{-1}$ when multiplied by 10^{-11}. σ, reflection coefficient, dimensionless.
$\omega(1)$, permeability in molecule $dyne^{-1}\ sec^{-1}$ when multiplied by 10^8.
$\omega(2)$, permeability in mole $dyne^{-1}\ sec^{-1}$ when multiplied by 10^{-15}.
P_d, permeability as defined in Section 11.4; in cm/sec when multiplied by 10^{-3}. All solutes in water. Adapted from Katchalsky and Curran (1965, Reference 10.4), Chap. 10.

Membrane	Solute	L_p	σ	$\omega(1)$	$\omega(2)$	P_d
C	Methanol (32)			740	122	3.11
C	Urea (60)	0.92	0.62	103	17	0.43
T	Urea (60)	3.2	0.013	126	20.8	0.53
T	Glucose (180)		0.123	43.5	7.2	0.18
T	Sucrose (342)		0.163	23.6	3.9	0.099

11.6.5 Fluid Mosaic Model for Biomembranes.

While it has proved possible to develop methods for predicting transport properties of membranes, uncertainties remain about the actual structures of these membranes, In References 7.40, 7.41 and 7.42 emphasis is given to the idea that the phosphilipids of biomembranes are in bilayer form with fluid-like properties, and that proteins play an important role yet to be fully understood. A model including these features is shown in Fig. 11.6. Here globular proteins are shown embedded in a liquid bilayer.

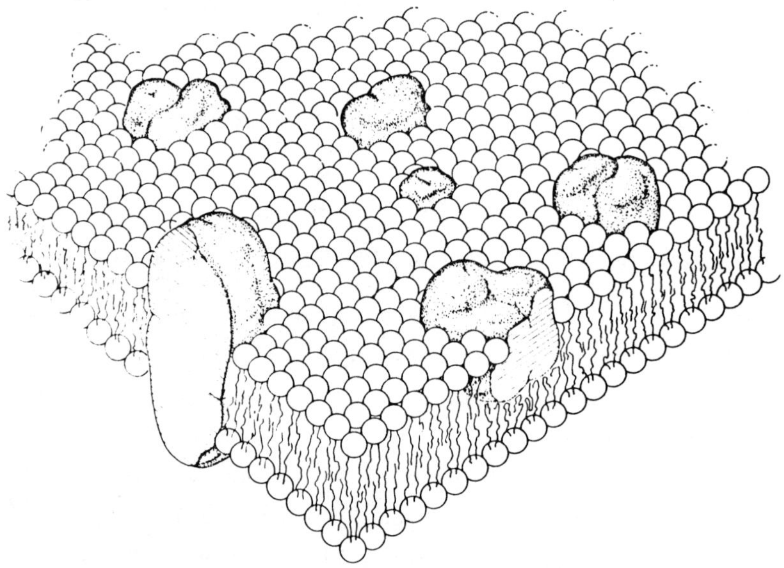

Fig. 11.6 Fluid mosaic model for biomembranes. From Singer and Nicolson (1972, Reference 11.16). Copyright 1972 by the American Association for the Advancement of Science.

In this hypothetical membrane, referred to by Singer and Nicolson (1972) as a fluid mosaic it is assumed that lipid and protein molecules can move by diffusion and flow within the bilayer, though without change in orientation. It is suggested that the effective viscosity of the membrane substance (in respect to protein migration) is about 10^3 to 10^4 times that of water.

11.6.6 Passive Transport in Gels and Cytoplasm.

A convenient method for measuring diffusion coefficients of small molecules in water has been described by Lauffer (1961). In this method the solute is allowed to diffuse in or out of a gel. In a water-based gel, long polymer molecules form a network throughout the water; in this way the water is prevented from taking part in convection which otherwise causes experimental difficulties. For a dilute gel Lauffer states that the effective coefficient for diffusion of a small molecule is D' given by

$$D' = \frac{D(1 - \alpha\phi}{(1 - \phi)} \quad , \tag{11.43}$$

where D is the diffusion coefficient in pure water, ϕ is the volume fraction of gel substance (plus immobilized water) and α is a constant. The quantity $(1 - \alpha\phi)$ is arrived at by considering the polymer molecules as obstacles to diffusion and referring to analogous theory from electrostatics. The quantity $(1 - \phi)$ is simply a correction to the concentration.

In experiments reported by Lauffer the exposed surface of a gel in a cylindrical vessel was bathed with a constant concentration of solute. The distribution of solute after a given time was determined by slicing the gel, determining the concentration in each slice, and using theory similar to that of Section 9.4.3. Equation (11.43) was tested for gels made by dissolving 1.5 gm of agar (a sulfuric acid ester of polygalactose, a long chain carbohydrate much used in bacteriology) in 100 ml of water. For several small molecules as test substances D' was measured in the gel. The fraction ϕ was determined by an independent method to be 0.05 (greater than 0.015 , presumably because of immobilized water; α was taken to be 5/3 , a result from theory for randomly oriented rods. From values of D', ϕ and α the diffusion coefficient D in water was calculated; values agreed very well with results obtained by other means.

In the experiment just described the "effective pore radius" R is much larger than the radius $\underline{a}$ of a solute molecule; the ratio D'/D is then independent of $\underline{a}$. But in more concentrated gels where R is not large compared to $\underline{a}$ results will depend on a/R . An example is discussed by Renkin (1954); he made measurements of flow and diffusion through "Silvania viscose wet gel" (a cellulose product) and found the sieve constant (Section 11.5.3) for various solute molecules to be a function of molecular size.

Questions have been raised about the rate at which diffusion occurs in cytoplasm, that is, the matter contained in biological cells. To obtain information on this matter Lehman and Pollard (1965) carried out measurement on diffusion through a medium produced by crushing bacterial cells. This material, an "amorphous mixture of macromolecules, cell wall debris and water" , had a very high apparent viscosity; from measurements on diluted cell medium with capillary viscometry they suggest that the macroscopic viscosity is 100 poise or more at 25°C. Such a high value of η would lead to a very low value of the diffusion constant for ions and molecules according to Eq. (9.41) (with f given by $6\pi\eta R$) . Diffusion measurements were performed by a method related to those described by Lauffer for gels, but adapted to use of small quantities of material. Letting the measured diffusion coefficient for a given molecule be D* and that for the same molecule in water be D they found D*/D to be about 1/3 for sucrose (molecular weight 342) and about 1/14 for beta galactosidase , an enzyme of molecular weight 250,000-400,000. These values show that hindrance to diffusion is caused by the cellular solids; the "effective pore radius" R is evidently not much larger than the dimension of the enzyme. Nevertheless the diffusion coefficients D^* are fairly high, much higher than would be obtained from $kT/6\pi\eta R$ with η given by the measured macroscopic viscosity. The authors point out important implications of this result for the rate at which biochemical reactions can occur in cells.

PROBLEMS

11.1 Derive the expression for $d\mu^c$ in Eq. (11.9) by carrying out steps indicated in the text.

11.2 For collagen membranes Stenzel, et al. (1971) measured P_f and P_d (the latter for D_2O), obtaining the values 2.8×10^{-11} and 10^{-3}, resp., both in cgs units. Assuming the temperature is 25°C, and that the diffusion coefficient D for D_2O at this temperature is 2.6×10^{-5} cgs, calculate the pore radius R.

11.3 Carry out steps indicated in the text and derive the expression for J in Eq. (11.33) .

11.4 (a) The blood volume V_b of a human being is about 80 ml per kilogram of total weight; for a person weighing 70 kg the volume V_b is then 5.6 liters. In the primary circulatory system the time required for blood to make a "round trip" between heart and foot is about one minute. In a kidney machine part of the blood is shunted from the circulation and may be filtered at the rate of 100 ml/min; for a patient whose volume V_b is 5 liters the concentration c_w of filterable waste products is then reduced by a factor of 0.98 in one minute, and $(0.98)^m$ in m minutes. Calculate the time required to reduce c_w by a factor of 0.5.

(b) Using data given in the text determine the area of Visking membrane required to filter 100 ml/min with a pressure drop of 100 mm Hg. (760 mm Hg corresponds to one atmosphere of pressure).

11.5 Referring to the discussion in Section 11.6.3 show that a flow rate of 0.135 cm^4/osm-sec corresponds to a flow permeability P_f of 5.4×10^{-12} cm^3/dyn-sec . Make use of the fact that a concentration difference of 1 osm/cm^3 is equivalent to a pressure difference of N_a kT .

CHAPTER 12

BIOELASTICITY

12.1 U-TYPE AND S-TYPE FORCES

Elasticity has been defined as "that property of a material which determines the tendency of the stressed material to return to its unstressed geometrical configuration". (M Landowne and R. W. Stacy, 1957). In Chap. 7 situations were discussed where elasticity arises from the tendency of stretched or deformed chemical bonds to return to their normal states. In a biological system another contribution to elasticity arises from thermal agitation as the system (e.g., a complex of flexible macromolecules) seeks a configuration of maximum probability. Thermodynamics can be brought to bear on the topic by referring to the First and Second Laws taken up in Sections 10.2 and 10.3 . For example, the First Law is applied to a fiber under tension in Eq. (10.3) . Combining this with Eq. (10.4) to replace dQ by T dS yields

$$dU = T\,dS + F\,d\ell\ . \tag{12.1}$$

Equation (12.1) allows us to determine the force F for various situations. For example, if the fiber is lengthened (or shortened) with negligible gain or loss of heat, we set T dS = 0 in Eq. (12.1) and obtain $dU = F\,d\ell$, or

$$F = \left(\frac{\partial U}{\partial \ell}\right)_S\ . \tag{12.2}$$

The partial derivative in this equation represents the "ratio of dU to $d\ell$ at constant entropy". On the other hand if the fiber is such that a change in length is accompanied by negligible change in internal energy we have

$$F = -\,T\left(\frac{\partial S}{\partial \ell}\right)_U\ , \tag{12.3}$$

with a corresponding meaning for the partial derivative. Usually both the internal energy and the entropy are altered when a fiber is stretched. In general F is given by the sum of the contributions given in Eq. (12.2) and (12.3), that is, by

$$F = \left[\frac{\partial U}{\partial \ell}\right]_S - T\left[\frac{\partial S}{\partial \ell}\right]_U . \qquad (12.4)$$

If U and S are known as functions of ℓ , Eq. (12.4) provides an expression for calculating F.

When Eq. (12.2) applies the force arises from dependence of U on ℓ , a dependence which is expected if chemical bonds are stretched or bent when ℓ increases. For U represents the total energy of a system and this increases if bonds are deformed. Thus the energy for a diatomic molecule is given as a function of interatomic separation r by Eq. (7.11); this energy corresponds to the internal energy U for the molecule. In a macroscopic system consisting of many atoms the total internal energy is the sum of contributions from interactions between all of the atoms which make up the system. (But usually the interactions between nearest neighbors are the most important ones). In Chap. 7 elastic properties were considered for materials in which restoring forces arise in the way we have just described, that is, from deformation-dependent internal energy as in Eq. (12.2). We sometimes refer to such forces, and the corresponding elasticity, as "U-type" .

A contrasting situation exists for elasticity in elastomers or rubber-like media. Rubber consists of long flexible molecules which are interwoven in a chaotic way; it can be regarded as a solution of chainlike macromolecules (Section 6.4.5.) in a solvent made up of the same molecules. Thermal motion causes each macromolecule to continually change its shape. For nonvulcanized rubber the motion of each molecule is essentially independent of the others. In volcanized rubber crosslinks are formed, that is, bonds between adjacent chain molecules at crossing points or junctions; however the chains remain flexible between the junctions.

In an "ideal" elastomer deformations can occur without any stretching or bending of bonds. This means that a change in configuration can occur, at constant temperature, without any change in the internal energy U .

(It should be realized, however, that the ideal elastomer is a convenient fiction; even rubber is not ideal). For this hypothetical medium a deformation only causes changes in shape (of the macromolecules); bond energies are not affected. The medium nevertheless tends to return to its original state, because of thermal motion, which causes the macromolecules to assume their most probable configurations. Among various configurations those corresponding to equilibrium are the most probable. Hence for an ideal elastomer its elasticity comes about because (1) the configuration of a deformed macromolecule has a relatively low probability and (2) each macromolecule tends toward a configuration of maximum probability. Since probability is closely related to entropy (see Eq. 10.6) we see that the above statements are consistent with the Second Law of Thermodynamics. The process by which the probability (or entropy) of a configuration tends to increase is closely related to the phenomena of Brownian movement and diffusion which we considered in Chap. 9. If a fiber consisting of ideal elastomer is under tension Eq. (12.3) applies; then the tensile force F is proportional to the increase of entropy per unit increase of length. Under these circumstances the force and associated elasticity are sometimes referred to as "S-type" or "entropic".

The difference in U-type and S-type forces may be seen particularly well for the chain model sometimes used for macromolecules (already mentioned in Section 6.4.5). For our present purpose we imagine a chain made up of N links, the end-to-end distance for the chain being ℓ. Suppose that initially the links assume a set of orientations relative to one another, and that the overall length ℓ is then changed under either of the following conditions:

(a) The orientations are held fixed (e.g., by soldering) and the links are stretched. Here the internal energy U increases, Eq. (12.2) applies, and the force F is U-type.

(b) The links are free to re-orient, and thus assume a new configuration when ℓ is increased. Then the entropy S of the chain decreases (assuming the configuration probability decreases) and Eq. (12.3) applies if the change occurs at constant temperature. Here the force is entropic or S-type.

12.2 EXAMPLE: AN ENTROPIC FORCE EXERTED BY GAS MOLECULES

While Eq. (12.1) was derived specifically for a fiber under tension the kind of principle represented there applies more widely. For any system with definite boundaries (which may be internal as well as external) we can imagine a part of the boundary, such as a wall or partition, which is moveable while other parts of the boundary are held fixed; let the displacement of this moveable part be $d\ell$ while the external force on it is F in the direction of the displacement. Then the work done on the system in the displacement is $F\,d\ell$. Referring again to the First Law expressed in Eq. (10.1) we let dW^* be $(-F\,d\ell)$ and let dQ be T dS; the result is Eq. (12.1) just as before, but with a generalized meaning.

For an example we refer to a situation treated in Section 10.3, namely, a gas in two compartments of a vessel, separated by a partition. There, in Eq. (10.9), an expression was derived for the entropy as a function of L_1, the length of the "1" compartment. Let us assume that the gas in both compartments is always at constant temperature T, independent of L_1 ; this condition can be achieved experimentally if the outer wall is a good heat conductor and is maintained at temperature T. The internal energy U of the gas (assumed ideal) depends only on the temperature and, in particular, is independent of L_1 . Hence Eq. (12.3) is applicable, letting ℓ be L_1 , and we obtain from Eq. (10.9) that the force F on the membrane is (Prob. 12.1)

$$F = -T \frac{\partial S}{\partial L_1} = kT(c_2 - c_1) , \tag{12.5}$$

where c_1 and c_2 give the concentration of particles in the "1" and "2" compartments, resp.; it is assumed that the membrane is of unit area. The derivative $\partial S/\partial L_1$ is "at constant U" which in this application means "at constant T". In Eq. (12.5) F is positive when the force is in the direction of increasing L_1 , that is, when it is directed from Compartment 1 to Compartment 2 . When $c_2 > c_1$ we learn from Eq. (12.5) that a positive external force F must be exerted on the membrane to maintain equilibrium; the magnitude of F is proportional to the concentration difference and to the Kelvin temperature. It is easy to show that the same result follows from simple gas laws. (Prob. 12.1).

12.3 Law of Elasticity for an Ideal Elastomer

The stress-strain curves for soft body tissues are often S-shaped ; an example for aortal walls is seen in Fig. 12.8 . Other examples are cited by King and Lawton (1950). This shape can be explained by assuming that the ideal elastomer referred to in Section 12.1 is a relevant model. The theory of stress-strain relations for an elastomer has been discussed by King (1946, Reference 12.3); our treatment here is a simplified version of his. It is supposed that a strip of elastomer contains a large number of flexible molecular chains, each executing continual thermal motion in a "liquid" environment provided by neighboring molecules.

A representative chain is indicated in Fig. 12.1; its two ends are in the planes $x = 0$ and $x = \ell$, respectively. Thus ℓ is the x-projection of the end-to-end distance. For any given value of ℓ the chain assumes random configurations whose mean entropy (related to the probability by Eq. (10.6)) is a function of ℓ . According to Eq. (12.3) the force required to maintain a given length ℓ , for a single chain whose entropy is $s(\ell)$, is proportional to the derivative $\partial s/\partial \ell$ at constant temperature (i.e., at constant internal energy). For a strip of elastomer, which contains many chains extending in all directions, the entropy S for a given strip length ℓ is the sum of contributions from all the chains. The force required to maintain the length ℓ is then proportional to $\partial S/\partial \ell$ at constant temperature .

12.3.1 Folding-Ruler Model

To see what these considerations mean more specifically we proceed first to a calculation of the entropy s for a single chain. We assume the chain consists of N links, each of length b. For simplicity we ignore the 3-dimensional characteristic of the flexible chain and, as Frenkel (1955) has done, consider a one-dimensional "folding-ruler" model. In this model the links are constrained as in the ruler sometimes used by carpenters; here we assume all links lie along the ±x direction. Analysis of the situation is then closely related to that for the one-dimensional random walk problem treated in Section 9.2 . But instead of an unsteady person carrying out a series of steps, we imagine here that our model chain (the ruler) is unfolded one link at a time. Just as the steps were either forward or backward, here the links of the chain may be oriented either along positive or negative x . Let p and q be the probability of the link orientation being forward and backward, respectively; of course $p + q = 1$. Then the probability of n forward links and N-n backward links in a given sequence is $p^n q^{N-n}$. But there are C_n^N sequences of this type, Eq. (8.17), so that the total probability of n forward steps is, as in Eq. (8.18),

$$W_n^N = C_n^N p^n q^{N-n} . \qquad (12.6)$$

The length ℓ of the model chain is

$$\ell = [n - (N - n)]\, b = (2n - N)\, b . \qquad (12.7)$$

If p and q are equal the results for a random walk problem apply. When n is approximately equal to $\frac{1}{2}N$ the probability of n forward links is given by Eq. (8.41); comparing Eqs. (8.39) and (12.7) we find that these are identical if $\delta = \ell/2b$. Writing the probability as $W(\ell)$ we have

$$W(\ell) = \left[\frac{2}{\pi N}\right]^{\frac{1}{2}} \exp \frac{-\ell^2}{2Nb^2} . \qquad (12.8)$$

From Eq. (10.6) the entropy is $k \ln W$, and from Eq. (12.3) the tension force F required to maintain the length at a given value of ℓ is $-T\, \partial s/\ell$; hence we obtain

$$F = \frac{kT\ell}{Nb^2} = \frac{kT\ell}{b\ell_\infty} , \qquad (12.9)$$

where $\ell_\infty = Nb$ represents the maximum length of the chain. Equation (12.9) applies when $\ell << \ell_\infty$. We see that under these conditions the force F required to maintain the chain at a given length ℓ is proportional to ℓ ; that is, the force-length relation is of the Hooke's Law type. Also F is proportional to the Kelvin temperature T ; such an important role for T is reasonable since the force is a result of thermal motions. Finally we see that F is inversely proportional to the extended-chain length ℓ_∞ and to the length b of a link.

As stated above Eq. (12.9) applies when $\ell << \ell_\infty$, that is, when the length ℓ is much less than the fully extended length ℓ_∞. But elastomers are very extensible; they can sometimes be stretched so that the macromolecules are greatly extended and the approximate equations, Eqs. (12.8) and (12.9) do not apply. To treat this situation we return to the more general expression for the probability function in Eq. (12.6). Again replacing W_n^N by the more convenient symbol $W(\ell)$ and letting $p = q = \frac{1}{2}$ we obtain

$$W(\ell) = (\tfrac{1}{2})^N C_n^N . \tag{12.10}$$

As before we let s be k $\ln W$ and let F be $-T\,\partial s/\partial\ell$. Also by virtue of Eq. (12.7) we find that

$$\frac{\partial s}{\partial \ell} = \frac{1}{2b}\frac{\partial s}{\partial n} . \tag{12.11}$$

Hence we obtain (Prob. 12.2) that

$$F = -\frac{kT}{2b}\frac{\partial}{\partial n} \ln C_n^N . \tag{12.12}$$

Using the expression for C_n^N from Eq. (8.17) and assuming n and N-n are both very much greater than unity (so that, e.g., the Stirling approximations in Appendix B can be used) we obtain (Prob. 12.2)

$$-\frac{\partial}{\partial n} \ln C_n^N \cong \ln \frac{n}{N-n} + \frac{1}{2}\left(\frac{1}{n} - \frac{1}{N-n}\right) .$$

$$\cong \ln \frac{n}{N-n} .$$

Equation (12.12) then yields

$$F = \frac{kT}{2b} \ln \frac{n}{N-n} , \tag{12.13}$$

or, in terms of ℓ,

$$F = \frac{kT}{2b} \ln \frac{1+x'}{1-x'} ; \; x' = \frac{\ell}{\ell_\infty} . \tag{12.14}$$

A plot based on Eq. (12.14) appears in Fig. 12.2. The ordinates are proportional to F and the abscissae to ℓ, so the plot has the character

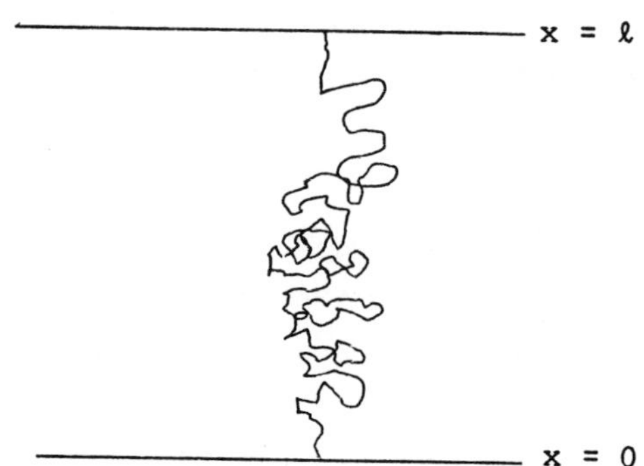

Fig. 12.1 Flexible chain whose ends are at $x = 0$ and $x = \ell$.

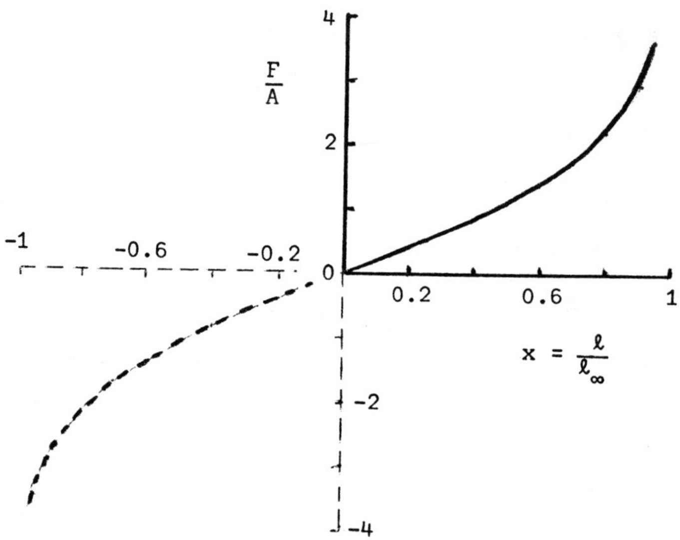

Fig. 12.2 Plot, based on Eq. (12.14), of the force F required to maintain the length ℓ of a flexible chain. $A = kT/2b$; ℓ_∞ is the chain length when fully extended.

of a stress-strain curve. We see that when all positive and negative values of ℓ are considered the curve has an S-shape. But, of course, negative values of ℓ have no physical significance so we restrict our attention to the upper right quadrant of Fig. 12.2 . In this range (all positive ℓ) the F vs ℓ plot is concave upward: the force required to produce unit increase in length becomes greater as ℓ increases, reaching very large values as ℓ approaches its limiting value ℓ_∞ .

At small values of x' it is easy to show (Prob. 12.3) that Eq. (12.14) reduced to Eq. (12.9). Thus Hooke's law applies when F and x' are small and n/N is in the vicinity of ½ . It is for the larger values of $x' = \ell/\ell_\infty$ that the relation is nonlinear as shown in Fig. 12.2 .

12.3.2 Folding Ruler Model: Units along Three Axes

With results for the single "folding ruler" unit as background we now proceed to a model which is finite in three dimensions. This model proves helpful in understanding properties of "soft" biological tissues taken up in Section 12.4. Here we consider a strip of material whose surface consists of two x faces (i.e., faces with normals along the x direction), two y faces and two z faces. (Fig. 12.3). When unstrained its dimension along x is ℓ_o , and those along y and z are each equal to w_o ; its x faces are at $0, \ell_o)$, that is, at $x = 0$ and $x = \ell_o$; its y faces are at $(0, w_o)$ and its z faces at $(0, w_o)$. We suppose the material consists of interpenetrating flexible chains extending along each of the coordinate directions. In the undeformed strip the chains along x end at the faces $x = 0$ and $x = \ell_o$ with density J per unit cross-sectional area; hence there are Jw_o^2 chains ending in each of the faces $x = 0$ and $x = \ell_o$. Also in the undeformed strip there are $J\ell_o w_o$ chains along each of the y and z directions, with endings in the faces $y = 0$, $y = w_o$, etc.

Suppose the strip is stretched along the x direction by exerting an outward force F_x at each of the x faces. The situation is similar to that considered in Section 4.4 with application to a stretched wire or filament. But here we assume that the volume of the strip remains constant in the deformation, since this would be approximately true for soft biological tissue. Hence if the x dimension of the strip increases from ℓ_o to ℓ the dimensions along y and z each decrease from w_o to w so that

$$\ell_o w_o^2 = \ell w^2 . \tag{12.15}$$

In such a deformation the retractive "entropic" force exerted by the chains along x on the x faces is increased, while similar forces exerted by chains along y and z are decreased. These changes in the entropic forces are expected since lengthening the strip, and hence also narrowing it, causes a decrease in entropy of chains along x and an increase for those along y and z . To treat the subject quantitatively we again choose the "folding ruler" model for the flexible chain. An expression for the retractive force F per chain is given in Eq. (12.14); introducing the symbol ϕ_x for this force we write

$$\phi_x = \frac{kT}{2b} \ln \frac{\ell_\infty + \ell}{\ell_\infty - \ell} . \tag{12.16}$$

We now let Φ_x be the total retractive force on either x face because of the chains along x . It is given by a sum of the single-chain contributions ϕ_x , and is just $Jw_o^2\phi_x$. Letting S_x be the total entropy associated with chains along x we realize from Eq. (12.3) that Φ_x is proportional to $\partial S_x/\partial \ell$ at constant temperature. Specifically we obtain

$$\Phi_x = -T\left[\frac{\partial S_x}{\partial \ell}\right]_T = Jw_o^2 \phi_x . \tag{12.17}$$

Similarly for the lateral chains we let the contractile force for a single chain along y or z be ϕ_w given by

$$\phi_w = \frac{kT}{2b} \ln \frac{w_\infty + w}{w_\infty - w} , \tag{12.18}$$

where w_∞ is the length of a fully extended chain. It is assumed that

$$\frac{w_o}{w_\infty} = \frac{\ell_o}{\ell_\infty} . \tag{12.19}$$

We let the total contractile force on a y face or z face be Φ_w, and let the total entropy of the y chains, or of the z chains, be S_w .

Proceeding as for the x chains we then obtain

$$\Phi_w = -T\left[\frac{\partial S_w}{\partial w}\right]_T = J\ell_o w_o \phi_w . \qquad (12.20$$

We wish to obtain an equation relating the length ℓ of the strip to the force F_x required to maintain that length. If there were no chains along y and z, and correspondingly no tendency for the strip to maintain constant volume, F_x would be equal in magnitude to Φ_x, given by Eq. (12.17). But we shall see that the lateral chains, together with the constant-volume condition, lead to an important modification of F_x. The total entropy S of the strip is the sum of all single-chain entropies and is given by $S_x + 2S_w$. From Eq. (12.3) the total force F_x is then given by

$$F_x = -T\left[\frac{\partial S_x}{\partial \ell} + 2\frac{\partial S_w}{\partial \ell}\right]_T . \qquad (12.21)$$

We can relate the derivative $\partial S_w/\partial \ell$ appearing in this equation to $\partial S_w/\partial w$ in Eq. (12.20) if we refer to the constant-volume condition expressed in Eq. (12.15). Differentiating the latter equation with respect to ℓ we find that $dw/d\ell$ is equal to $-w/2\ell$. Hence we obtain

$$\frac{\partial S_w}{\partial \ell} = \frac{\partial S_w}{\partial w}\frac{dw}{d\ell} = -\frac{w}{2\ell}\frac{\partial S_w}{\partial w} . \qquad (12.22)$$

Substituting from this equation into Eq. (12.21) and making use of Eqs. (12.17) and (12.20) we obtain

$$F_x = \Phi_x - \frac{w}{2\ell}\Phi_w , \qquad (12.23)$$

or

$$F_x = Jw_o^2\,\phi_x - \frac{w}{\ell}J\ell_o w_o\,\phi_w . \qquad (12.24)$$

Using the expressions for ϕ_x and ϕ_w from Eqs. (12.16) and (12.18), and using relationships between ℓ, w, etc., from Eqs. (12.15) and (12.19) we obtain (Prob. 12.4) a result which can be written

$$\frac{F_x}{w_o^2} = \frac{JkT}{2b}\left[\ln\frac{1+\gamma\lambda}{1-\gamma\lambda} - \lambda^{-3/2}\ln\frac{1+\gamma\lambda^{-1/2}}{1-\gamma\lambda^{-1/2}}\right] , \qquad (12.25a)$$

where

$$\gamma = \frac{\ell_o}{\ell_\infty} \; ; \quad \lambda = \frac{\ell}{\ell_o} \; . \qquad (12.25b$$

This expression simplifies when the length ℓ of a chain is much smaller than the fully extended length ℓ_∞ . For $\gamma\lambda$ is just ℓ/ℓ_∞ and we have then

$$\gamma\lambda \ll 1 \; ; \qquad (12.26a)$$

also, if λ is of the order of unity or greater (which is true unless the strip is drastically shortened),

$$\gamma\lambda^{-1/2} \ll 1 \; . \qquad (12.26b)$$

Using Eqs. (12.26) in Eqs. (12.25) (Prob. 12.5) we obtain

$$\frac{F_x}{w_o^2} = \frac{\gamma JkT}{b} \; (\lambda - \lambda^{-2}) \; . \qquad (12.27)$$

The form of the function $(\lambda-\lambda^{-2})$ is seen in Fig. 12.4; as λ increases the slope decreases monotonically, approaching unity as a limit. We now introduce the additional assumption that the extension is small; then we can write

$$\ell = \ell_o \, (1 + \varepsilon) \; , \qquad (12.28)$$

where $\varepsilon \ll 1$. Note that ε represents the longitudinal strain, being analogous to the quantity ε_{11} discussed in Section 3.3 . Then λ is just $1 + \varepsilon$ and, when approximations are introduced (Prob. 12.5), the expression $(\lambda-\lambda^{-2})$ reduces to 3ε so that Eq. (12.27) becomes

$$\frac{F_x}{w_o^2} = \frac{3\gamma JkT\varepsilon}{b} \; . \qquad (12.29)$$

The quantity F_x/w_o^2 on the left of this equation is the tensile stress exerted on the strip, and is analogous to the component S_{11} defined in Chap. 2 . Since ε represents strain we see that Hooke's law is obeyed when ε is small. Specifically, we can rewrite Eq. (12.29) as

$$\frac{\text{Stress}}{\text{Strain}} = \frac{3\gamma JkT}{b} \; . \qquad (12.30)$$

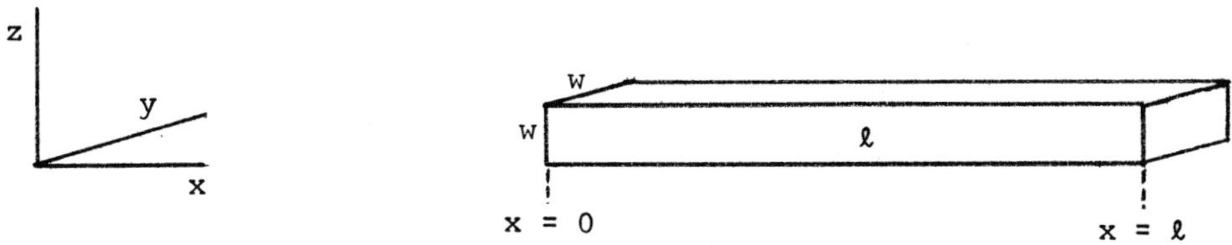

Fig. 12.3 Strip of elastomer.

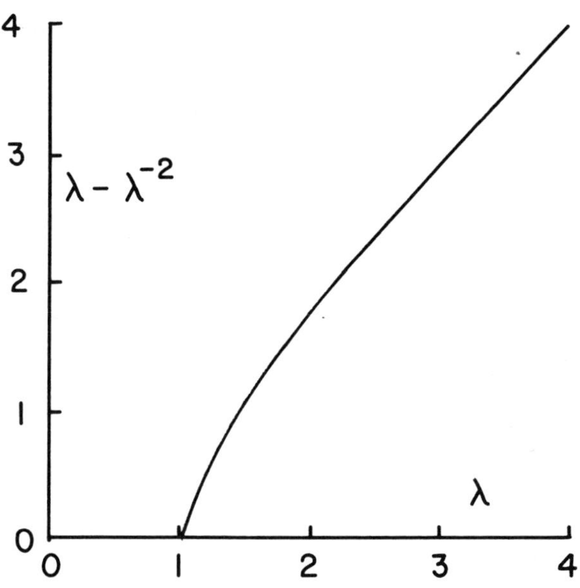

Fig. 12.4 Plot of the function $(\lambda-\lambda^{-2})$ which appears in Eq. (12.27) .

Clearly the right hand side of this equation represents the Young's modulus E of the material, since the situation is like that taken up in Section 4.4 . From results in Probs 4.7 to 4.9 we can now calculate the shear elastic modulus μ . The Poisson ratio ν differs only slightly from 0.5 for the elastomer, since its volume is essentially unchanged when the strip is stretched. Hence μ is about E/3 or

$$\mu = \frac{\gamma JkT}{b} . \tag{12.31}$$

We see that μ is proportional to the Kelvin temperature T, a property characteristic of elastomers. In media such as vulcanized rubber, discussed in Section 12.4.1 , the expression in Eq. (12.31) can sometimes be reduced to ckT, where c is the number of chains per unit volume. It is noteworthy that this latter expression is identical to that for the pressure exerted by an ideal gas, Eq. (8.3) .

When the elastomer strip is stretched to large ratios ℓ/ℓ_o (i.e., to large values of λ) the approximations made in arriving at Eqs. (12.27) and (12.29) are no longer valid. When λ is large the second term in the bracket on the right hand side of Eq. (12.25a) (containing the factor $\lambda^{-3/2}$) is relatively small. Referring back to Eq. (12.24) we see that the expression for F_x then reduces to

$$F_x = Jw_o^{\ 2}\phi_x , \tag{12.32}$$

as if the lateral chains were not present. The dependence of F_x on ℓ is then the same as for single units, as given in Eq. (12.14) .

Plots of the complete function from Eq. (12.25a) are given in Fig. 12.5 . We now recognize the origin of the S-shape . At small values of λ , Eq. (12.27) applies and the slope is monotonically decreasing (see Fig. 12.4) . At large values of λ the lateral chains no longer make a significant contribution to the force. Then Eq. (12.32) applies, representing longitudinal chains only, and the slope is always increasing (as in Fig. 12.2).

12.3.3 Chain Model, with Rotatable Links

The folding-ruler model is, of course, a crude one for representing a flexible macromolecule. A better one is the random chain model, referred to already in Section 6.4.5. Like the folding ruler, the chain consists of N links, each of length b; but in the random chain each link is free to rotate, with all orientations equally probable. The force-versus-length curve for such a chain is found (King, 1946, Reference 12.3) to be similar in a general way to that for a folding ruler. But they differ in that F in Eq. (12.14) is replaced by F^* where

$$F^* = \frac{kT}{b} L^{-1}(x) ; \quad x = \frac{\ell}{\ell_\infty} . \tag{12.33}$$

Here the function $L^{-1}(x)$ is the <u>inverse Langevin function.</u> The Langevin function L(u) is defined by

$$L(u) = \coth u - \frac{1}{u} , \tag{12.34}$$

and the inverse function $L^{-1}(x)$ is then that value of u for which L(u) = x . A plot of $L^{-1}(x)$ is shown in Fig. 12.6 . Comparing this plot with that in Fig. 12.2 we see that the force-length curve for the simple folding ruler is similar to that for the more realistic chain model with its rotatable links. A three-dimensional model can be constructed from chains, analogous to the folding-ruler model. A resulting expression corresponding to Eqs. (12.25) has been derived by King (1946, Reference 12.3) .

12.4 ELASTICITY OF SOFT TISSUES AND RUBBER

By "soft" tissue we mean the material which comprises such units as cell membranes, walls of blood vessels, walls of body organs and skin. As the name implies soft tissues are much more easily deformed than "hard" tissues such as occur in bone and teeth. It has been shown in some soft tissues that their properties are in many ways similar to those of rubber.

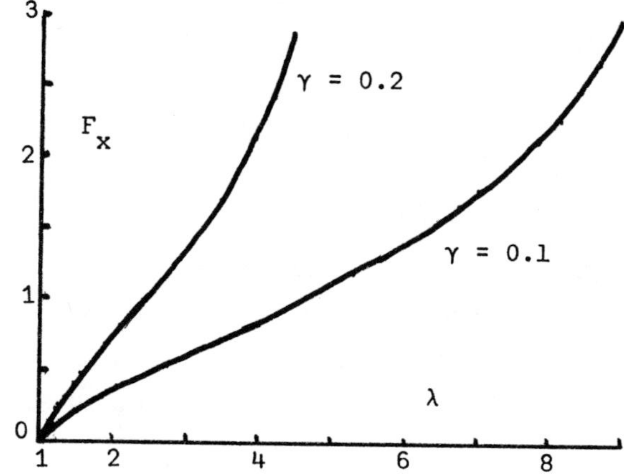

Fig. 12.5 Plots of the force F_x required to elongate an elastomeric strip by the ratio λ . From Eqs. (12.25) . $\gamma = \frac{\ell_o}{\ell_\infty}$.

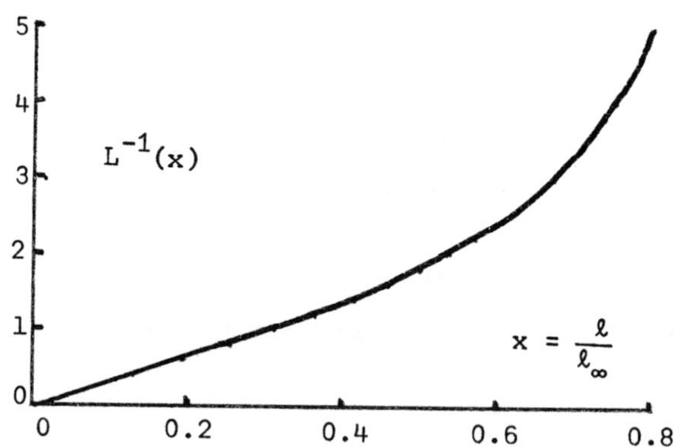

Fig. 12.6 Plot of inverse Langevin function L^{-1} defined in Section 12.3.3 .

Since a great deal of information is available on elastic properties of rubber, we first take up some interesting aspects of this subject.

12.4.1 Rubber

Natural rubber is a hydrocarbon; it is, specifically, a polymer of the isoprene unit whose chemical formula is C_5H_8 . An average molecule of rubber, that is, polyisoprene, contains about 5000 isoprene units arranged end-to-end in a single chain; its molecular weight is about 350,000 and its length, if stretched out straight, would be about 2 μm . However, these long macromolecules are flexible, and thermal motion causes them to assume random configurations. Bulk natural rubber consists of an entangled mesh in which the coiled macromolecules are relatively free to move over and between one another.

Vulcanized rubber is crosslinked by addition of sulphur to natural rubber. A crosslink is a joint between two or more neighboring macromolecules established by interaction of an S atom with double bonds in the polyisoprene neighbors. These junctions are distributed randomly and form a set of points interconnected by flexible molecular segments, or chain segments. For a given value M of the average molecular weight of the free macromolecules, the average molecular weight M_c of the segments is inversely proportional to the number of crosslinks per unit volume. For a highly crosslinked rubber M_c may be in the range 0.1M to 0.01M and may be, for example, as small as 3500 .

Whether vulcanized or not, rubber is very extensible: a strip can be extended to five or ten times its unstretched length without rupture. By contrast bone can only be extended by about 0.5% . The high extensibility of rubber is attributed to the ability of chain segments to change shape in response to stress: applied tensile force causes a segment initially coiled to become more extended. For a flexible unit, increase of its end-to-end distance by a factor of two or more occurs with relatively little distortion of chemical bonds.

When a strip of rubber is subjected to tension it proceeds to lengthen at a rate that depends on properties of the material, the dimensions of the strip and the applied tension. We shall discuss such time-dependent mechanical response in a later chapter. Equilibrium is reached after a period of time which may vary from milliseconds to hours.

When applied tension (or stress) is plotted against the equilibrium elongation λ over a wide range of λ the result for rubber is characteristically an S-shaped curve as shown in Fig. 12.7 . We recognize the similarity of this curve to the plot in Fig. 12.5 , the latter being based on entropic force developed in flexible chains. Tests discussed by Treloar (1958) showed that most of the stress developed in stretched rubber is indeed of entropic origin, though a small amount arises from stretching and bending of bonds.

It is found that when a strip of rubber is stretched it simultaneously becomes thinner to such an extent that its total volume remains nearly constant. For example, Treloar reports measurements on vulcanized rubber in which an increase of length by a factor of two is accompanied by a volume change of only 0.02% or less. In this kind of deformation the rubber is thus nearly incompressible. Referring to the theory in Section 4.4 , and associated Problems 4.7 to 4.9 , we see that this behavior implies a Poisson ratio ν of about 0.5 . Thus in Prob. 4.8 the Poisson ratio is given as approximately

$$\nu = \tfrac{1}{2}(1 - \frac{\mu}{B}) , \qquad (12.35)$$

where μ is the shear coefficient of elasticity (rigidity modulus) and B the bulk modulus; Eq. (12.35) applies when $\mu \ll B$. For rubber B is of the same order as for water, about 2×10^{10} or more in units of dyn/cm^2 . A theoretical expression for μ is given by Treloar as

$$\mu = n_v kT , \qquad (12.36)$$

where n_v is the number of chain segments per unit volume. If m_c is the average mass per chain and ρ is the density of rubber we see that n_v is ρ/m_c . The average "molecular weight" M_c of the chain is $N_a m_c$, where N_a is Avogadro's number. Letting T be 300°K we obtain the simple approximate result

$$\mu = 2.5 \times 10^{10} \frac{\rho}{M_c} , \tag{12.37}$$

in units of dyn/cm^2 when M_c is in daltons (gm/mole) and ρ in gm/cm^3 . Equation (12.37) has been compared with experiment and found to be approximately valid. For highly vulcanized rubber the average length of chain between junctions may have a molecular weight as small as 3500 daltons. Letting ρ be of the order of unity we then obtain from Eq. (12.37) a value for μ of about 7.1×10^6 dyn/cm^2 . Comparing this value with that cited for B we see that μ/B is less than 10^{-3} and hence from Eq. (12.35) the deviation of ν from the value 0.5 is less than 0.1% .

It follows also (from results in Prob. 4.8) that when μ/B is small the Young's modulus E is given approximately by the simple formula

$$E = 3\mu = 3n_v kT . \tag{12.38}$$

Still another striking characteristic of rubber elasticity is its dependence on temperature. The predictions from Eqs. (12.36) and (12.38) that the elastic moduli should be proportional to the Kelvin temperature T has been confirmed experimentally. By contrast the elastic moduli for "hard solids" such as metals and glasses are relatively insensitive to temperature.

12.4.2 Human Arteries

A series of investigations were carried out by Hallock and Benson (1937) on elastic properties of human arteries obtained post mortem. In each experiment a section of the arterial tube about 10 cm long was immersed in a constant temperature bath while the interior, filled with water, was subjected to a pressure p in excess of the external pressure p_o .

For a pressure of p_o the external diameter and wall thickness of the tube were about 12 mm and 1.3 mm , respectively, for the youngest group (20-24 years) and about 16 mm and 0.9 mm, respectively, for the oldest group (71-78 years). The volume V of the tube was measured as a function of the pressure difference $(p-p_o)$. Results are shown in Fig. 12.8 for aorta from individuals of the different age-groups. The plots are from King (1946, Reference 12.8) based on the Hallock-Benson data. Ordinates show the ratio V/V_o where V_o is the volume when $p = p_o$. We see that the curves are about the same for all age groups for pressure differences $(p - p_o)$ up to about 50 mm Hg . At higher pressure differences the curves separate, the relative volume V/V_o for given $p - p_o$ being less for the old-age groups than for the young. Indeed when $p - p_o$ is 200 mm Hg the ratio V/V_o is more than three times as great for the youngest group as for the oldest. The arteries (like the joints) evidently become stiffer with advancing age.

Also the shape of the curve is age-dependent. For an aorta from a young individual the curve has an S-shape , similar to the stress-strain plot for rubber shown in Fig. 12.7 . (Note that stress is plotted vertically in Fig. 12.7 and horizontally in Fig. 12.8). But corresponding curves in Fig. 12.8 for older groups do not show the S-shape; for these the slope (and hence the extensibility) decreases monotonically with increasing pressure difference.

In explaining these results King considered the aortic wall to be a uniformly thick cylindrical elastomeric shell. Actually, as he points out, the wall consists of several layers with elastic material in each layer. A principal constituent is <u>elastin</u>, a protein whose stress-strain characteristics have been shown to be similar to those of rubber. (Ross and Bornstein, 1971).

Experimental results represented by these curves were matched against predictions based on stress-strain theory for elastomers. The specific expression used by King, which we shall not quote here (Eq. 11 of

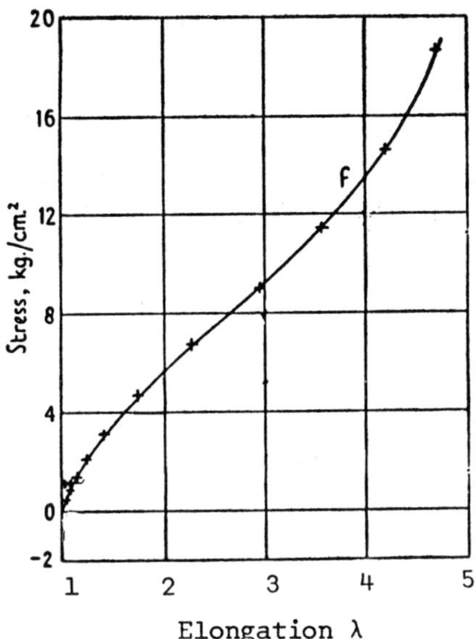

Fig. 12.7 Stress *vs* relative elongation λ for a strip of rubber. From Anthony, Caston and Guth (1942); cited by Treloar (1958, Reference 12.6, Fig. 2.6).

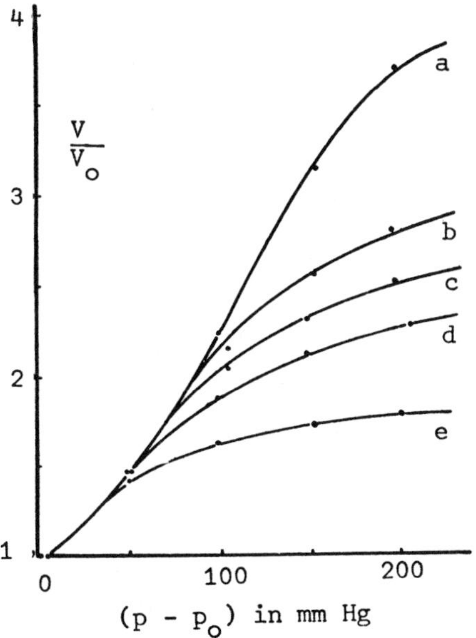

Fig. 12.8 Relative volume V/V_o *vs* pressure difference $(p-p_o)$ for aorta in five age groups. (a) 20-24 yr.; (b) 29-31 yr.; (c) 36-42 yr.; (d) 47-52 yr.; (e) 71-78 yr. From King (1946, Reference 12.8); curves based on data from Hallock and Benson (1937); plotted points from elastomer theory derived by King.

Reference 12.8) is analogous to the "folding ruler" result given in Eqs. 12.25); however, King's expression is based upon the function (for random chains) in Eq. 12.33, and is adapted to the geometry of a cylindrical tube. Predictions are plotted as circles in Fig. 12.8 and agree very well with theory. In fitting theory to experiment two parameters were adjusted, one proportional to the wall thickness w and the other inversely proportional to the mean molecular chain length ℓ . From determinations of w and ℓ for best fit to the different curves in Fig. 12.8 it was found that w and ℓ both decrease with increasing age.

12.4.3 Frog Sarcolemma. See also pp 216-217.

In other studies on tissue elasticity the membranes of single muscle fibers (cells) were investigated, these having been obtained by surgical separation from (semitendinosis) frog muscle. Such a fiber may be about 0.1 mm in diameter and several centimeters in length. The membrane of a muscle cell is called the sarcolemma and consists of several distinct layers with total thickness about 1000 A . When a muscle is stretched both the sarcolemma and the cell substance, called the sarcoplasm, are deformed. To obtain information on mechanical properties of the sarcolemma alone a technique is used to modify the cell so that a portion of its interior is empty of sarcoplasm. In this technique the tube-like cell is squeezed gently at a midway location; in this way the sarcoplasm is injured locally without any apparent damage to the sarcolemma. The injury sets into action a complex process of clotting and coagulation which results in withdrawal of visible cell contents from the site of initial injury. An "empty" region called the retraction zone then exists, of length up to 1 mm and more; while the zone varies in cross-section near its ends, under suitable condition, its central portion is a cylindrical tube. See Fig. 12.9 . The interior of the zone is optically transparent; it is probably a newtonian liquid consisting of ions and molecules in water, whose viscosity is not greatly different from that of water.

These preparations were used by Fields (1970) for making detailed studies on mechanical properties of sarcolemma. In his arrangement it was possible to apply tension longitudinally (i.e., along the axis of the tube); it was also possible to apply internal pressure by forcing water into a retraction zone with a glass micropipet. Strain measurements were made by observing through a microscope the spacing between small particles attached to the sarcolemma. In this way Fields measured both longitudinal and circumferential strain of the sarcolemma in a retraction zone as function of tension and pressure.

Typical results are shown in Fig. 12.10 where circumferential strain e_c is shown plotted against circumferential tension T_c . The slope de_c/dT_c tends to decrease with increasing T_c and e_c ; that is, the sarcolemma becomes less extensible as it becomes more and more stretched. We recall (Section 12.4.2) that this is also the behavior noted for aortic tubes in the range of high extension. Fields also found the <u>circumferential extensibility</u> de_c/dT_c to be affected by the longitudinal strain e_L and tension T_L ; increasing e_L or T_L leads to decreasing de_c/dT_c .

Plots of e_L vs T_L were found to have a similar form to that of e_c vs T_c ; the <u>longitudinal extensibility</u> de_L/dT_L decreases with increasing e_L or T_L , and also with increasing circumferential strain e_c or tension T_c .

In explanation of these results Fields and Faber (1970) assume that the elasticity of sarcolemma arises primarily from collagen macromolecules contained therein. These are considered to be wound around the muscle cell along helical curves. For such a curve the tangent makes a fixed angle θ with the longitudinal axis of the cylindrical cell. The Fields data are fitted by taking θ to be about 55° for the normal fiber and 11° for the retractive zone.

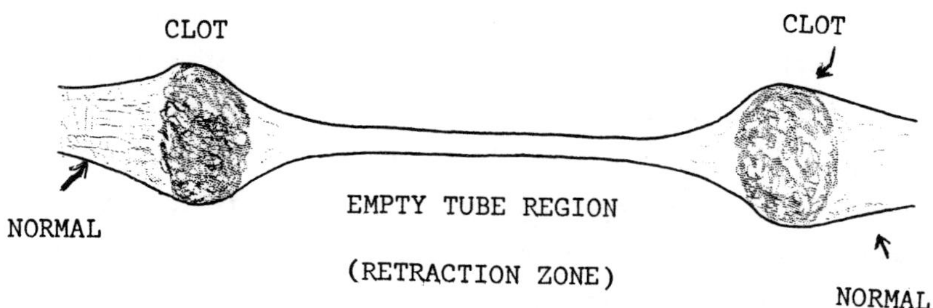

Fig. 12.9 Sarcoplasm-free region in muscle fiber after controlled injury. Used by Fields and Faber (1970) for stress-strain studies on sarcolemma. Section 12.4.3. From Fields, Reference 12.10.

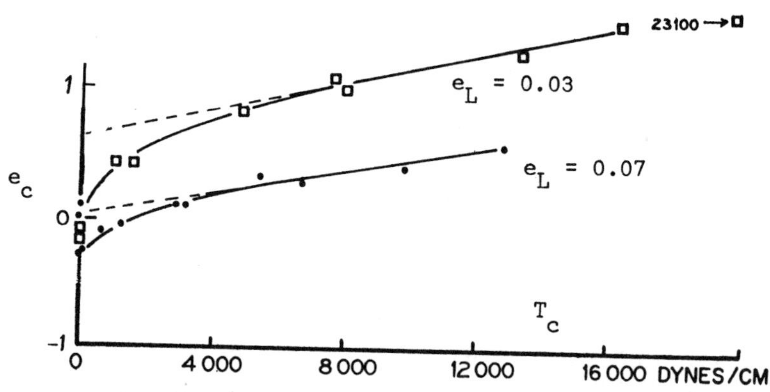

Fig. 12.10 Circumferential strain e_c vs circumferential tension T_c for sarcolemma. From Fields and Faber (1970). Reproduced by permission of the National Research Council of Canada.

It is part of the Fields-Faber hypothesis that the retractive zone represents an unstressed state for the sarcolemma. The data in Fig. 12.10 are for stress and strain relative to that state, where θ is only about 11° and thus the collagen molecules are nearly parallel to the tube axis. In this state the tube is easily deformed (i.e., only a little tension is required to produce a given amount of strain); here collagen molecules are assumed to be somewhat flexible, that is, to contain a small degree of slack. Perhaps the restoring forces in this range of conditions are partly entropic in origin, as for the aortic tubes considered in Section 12.4.2 .

But either circumferential or longitudinal strain causes an increase in length ℓ of the molecule. When ℓ reaches a critical value the molecule is fully extended; further increase of length then requires stretching of bonds. In Fig. 12.10 the straight-line portion of the curve represents a region where bond-stretching is believed to be involved. The least strain e_c , which we call e_{cm} , at which the molecule can be fully extended with bonds still unstretched is obtained by extrapolatin the curve to $T_c = 0$ as shown in Fig. 12.10 . By fairly simple reasoning (which we shall not take space to repeat here) the authors were able to explain a number of experimental findings, including the dependence of e_{cm} on the longitudinal extension e_L . They also were able to estimate the Young's modulus for the collagen fiber, about 2×10^{10} d/cm^2 , in fair agreement with results obtained more directly.

12.5 ELASTICITY OF CRYSTALLINE FIBERS

12.5.1 Keratin

In Section 7.7 we mentioned that the complete stress-strain curve for keratin is a nonlinear one; the "extensibility", that is, the fractional increase in length per unit increase in stress, varies with length. For wool-keratin, experimentally determined curves show three regions:

(1) A Hooke's law region of low extensibility for elongations up to about

2%; (2) A "yield region" of relatively high extensibility for increases of length up to about 30% of the unstretched length; and (3) A post-yield region in which elongations up to about 70% are obtained, where the extensibility is low. Theoretical and experimental results for regions (1) and (3) were discussed in Section 7.7 Data for wool fibers in the yield region are shown in Fig. 12.11; these are from Speakman (1926) as quoted by David, Haukaas, Kalnins and Schor, hereafter called DHKS (1967). These curves show a tendency to be S-shaped; this tendency would be much more evident if the curves were extended at each end. Of course we have considered S-shaped curves previously in this chapter, and explained them with theory for "entropic elasticity", a theory which applies to ideal elastomers. But keratin has a highly-ordered (crystalline) content, as revealed by Xrays, and thus obviously is not constructed from flexible

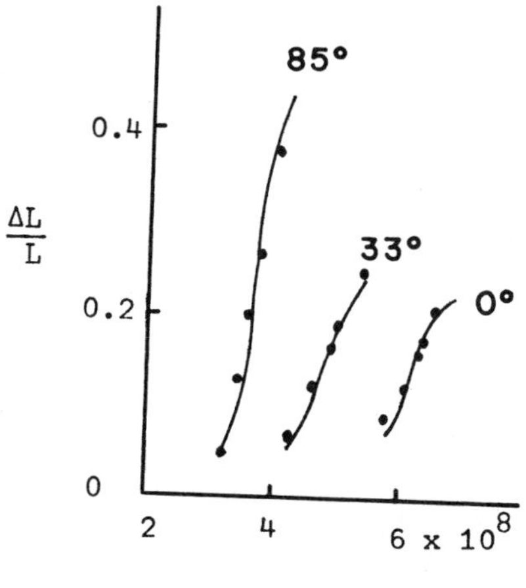

Fig. 12.11 Fractional extension ΔL/L vs stress for Cotswold wool in water, at temperatures given in °C. Experimental data, shown by circles, are from Speakman (1926) as cited by DHKS (1967). Solid curves are from theory by DHKS discussed in Section 12.5.4 .

randomly-coiled macromolecules such as constitute elastomers. Below the yield region shown in Fig. 12.11 the molecular configuration of keratin tends to be α-helical, and above this region the molecules have the more extended β form. (See Section 7.7).

The elasticity of keratin in human hair has been studied experimentally by Bull (1944, 1945, 1954) and Breuer (1972); here, too, the stress-strain curves are nonlinear.

12.5.2 Simple Autone Model

Interesting theories have been advanced for explaining the elastic properties of keratin fibers and other similar fibers. We consider, for example, an approach based on a simple picture sometimes called the autone model. Here the fiber is a linear chain made up of macromolecular units called autones. These units are identical except that each can be either in a short state with length s or a long state with length ℓ. Thermal fluctuations lead to random changes in state of the autone units. When equilibrium exists the fraction of autones in the long state has a definite value p; the quantity p is just the probability that a given autone is in the long state. We assume here that the probability p applies to each autone, regardless of the state of its neighbors. Let the corresponding probability of a short state be q, equal to 1-p. Letting N be the total number of units and n the number of long units, we have for the length L :

$$L = n\ell + (N-n)\, s \, . \qquad (12.39)$$

To determine the average value of n we recall the theory of Chapter 8. The probability that n out of N units are long is given by W_n^N in Eq. (8.18) and $\bar{n}$, the average value of n, is given by Eq. (8.23) as simply pN. We obtain the average length $\bar{L}$ from Eq. (12.39) by letting n have its average value pN; the result can be written

$$\frac{\bar{L}}{N\ell} = p + (1-p)\frac{s}{\ell} = \frac{s}{\ell} + p(1 - \frac{s}{\ell}) \, . \qquad (12.40)$$

The quantity $N\ell$ is the maximum possible length of the chain. In Fig. 12.12 the solid line shows the fraction $\bar{L}/N\ell$ as a function of p, according to Eq. (12.40) when $s/\ell = 0.5$. We see that the average length $\bar{L}$ varies linearly from 50% to 100% of its maximum value $N\ell$ as p varies from 0 to 1.

12.5.3 Monte Carlo Methods

Similar results can be obtained in another way, by using a computer to simulate a random process for choosing which units are long and which are short. In Fig. 12.13 the upper line, labelled p = 0, shows a series of 50 spaces separated by dots; each space represents a unit. This line represents a chain of units all of length s, and its total length is 50s. Below this is a set of three lines, the set being labelled p = 0.2. Each of these lines represents a chain of 50 units, most of which (roughly 80%) are short. Each sequence of short and long units was computer-generated in the following way: for the first unit the computer produced a random number r lying between 0 and 1; the unit was chosen to be long if r fell between 0 and 0.2, and was chosen to be short otherwise. The process was then repeated for each subsequent unit up to the 50th. The difference in length between the three chains for p = 0.2 is a consequence of the random process. The average length is about 20% greater than for p = 0, as expected from Eq. (12.40) for $s/\ell = 0.5$. Similar sets for three chains each are shown for p = 0.4, 0.6 and 0.8; for these the percentage of spacings that are long is about 40%, 60% and 80%, resp. Finally, the bottom chain is for p = 1; all spacings are long and the total length is 50ℓ, just twice that for p = 0.

This procedure of determining sequences of units by a random number generator is sometimes referred to as a "Monte Carlo method"; such methods have proven useful for solving various problems in statistical physics. In Fig. 12.12 the plotted points are chain-lengths determined by the same Monte Carlo method used in preparing Fig. 12.13. We see that these points follow fairly well the curve plotted from Eq. (12.40). The fluctuations reflect the random process involved and would be smaller (relative to the mean value $\bar{L}$) if the number N of units were increased.

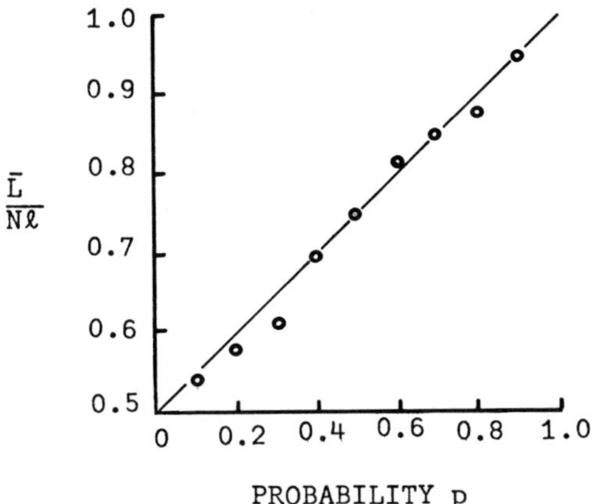

Fig. 12.12 Average length $\bar{L}$ (relative to maximum length $N\ell$) versus probability p for autone model. See Sections 12.5.2 and 12.5.3 .

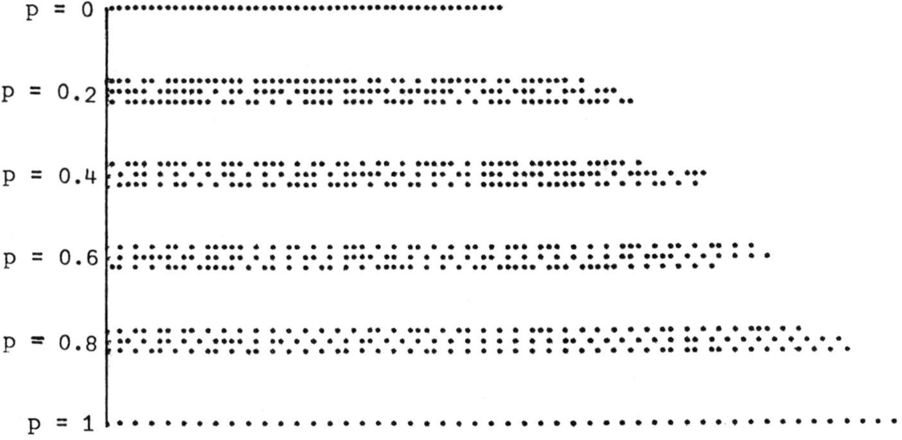

Fig. 12.13 Autone-model chains generated by a Monte Carlo method. See Section 12.5.3 .

12.5.4 Stress vs Strain for an Autone Model

In the previous two sections we took up the problem of predicting the average length $\bar{L}$ of a chain constructed from autones (units which are either of length s or ℓ) as a function of the probability p that a given unit is of length ℓ . We now consider physical factors on which the probability p depends. In keratin it was considered by DHKS (1967) that the autones consist of molecular units whose short state is α-helical and whose long state is the more extended β configuration. (See Sect. 7.7). Let p be the probability that a unit is in the β configuration; then q, which is equal to 1-p, is the corresponding probability that it is in the α configuration. If the chain is not under tension it is assumed that the probabilities p and q are in the ratio

$$\frac{p}{q} = e^{x} , \tag{12.41}$$

where

$$x = -\frac{\varepsilon}{kT} . \tag{12.42}$$

Here the quantity e^{x} can be recognized as a Boltzmann factor such as was noted in connection with Eq. (9.59). The parameter ε is the amount by which the free energy (analogous to potential energy) of state β exceeds that of state α . The autones tend to seek the state of least free energy, namely the α-helix configuration; the ratio p/q of the average number of β units to the average number of α units decreases as ε increases. Since the sum p+q must equal unity we obtain from Eq. (12.41) that p has the value $(1+e^{-x})^{-1}$. Substituting this value into Eq. (12.40) we obtain

$$\frac{\bar{L}}{N\ell} = \frac{s}{\ell} + (1 - \frac{s}{\ell}) \left(\frac{1}{1 + e^{-x}}\right) . \tag{12.43}$$

A plot of this function appears in Fig. 12.14 for a particular value (0.5) of s/ℓ . We see that $\bar{L}$ increases with x , according to an S-shaped curve. The greatest value of the slope $d\bar{L}/dx$ coincides with the inflection point and occurs when x = 0 . Because of the relation between x, the energy difference ε, and the temperature T in Eq. (12.42), we realize that $\bar{L}$ is a decreasing function of ε , and an increasing function of T. The inflection point occurs when ε is zero and marks the condition when $p = q = \frac{1}{2}$; thus when $\varepsilon = 0$ the β units and α units are, on the average, equally numerous.

Proceeding to the goal of finding a force vs elongation (or stress vs strain) curve for the autone-chain, DHKS (1967) introduced the following assumption: When a stretching force F is applied to the chain the quantity x defined in Eq. (12.42) is modified to become

$$x = \frac{F(\ell - s) - \varepsilon}{kT} ; \tag{12.44}$$

thus the energy elevation of the β state relative to the α state is reduced by the amount $F(\ell - s)$. The latter amount is, of course, just the work done in increasing the length of a unit from s to ℓ with an applied force F .

With x given by Eq. (12.44) it is clear that the plot in Fig. (12.14) is similar to a plot of length vs force. This plot is then recognized as having the familiar S-shape, with a low extensibility (small value of $d\bar{L}/dF$) at low and high values of F and maximum extensibility when $x = 0$. Specific plots of relative length $\bar{L}/N\ell$ vs F are obtained by giving specific values to the constants in Eq. (12.44). Curve "3" in Fig. 12.15 is obtained for the following values:

$$\frac{kT}{\ell - s} = 1 \text{ dyn} ; \quad \frac{\varepsilon}{\ell - s} = 3 \text{ dyn} . \tag{12.45}$$

From Eq. (12.44) we then see that the force F in dynes is equal to $x + 3$. Curve 3 is S-shaped with inflection point at $F = 3$ dyn ; this point , corresponding to equal probability for α units and β units, occurs when the numerator on the right hand side of Eq. (12.44) is equal to zero, that is, when

$$F = \frac{\varepsilon}{\ell - s} . \tag{12.46}$$

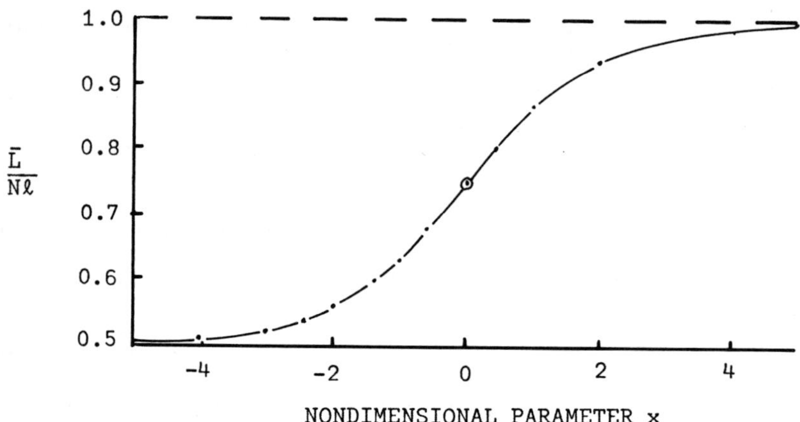

Fig. 12.14 Plot of $\bar{L}/N\ell$ vs x based on Eq. (12.43). The inflection point is marked by a circle.

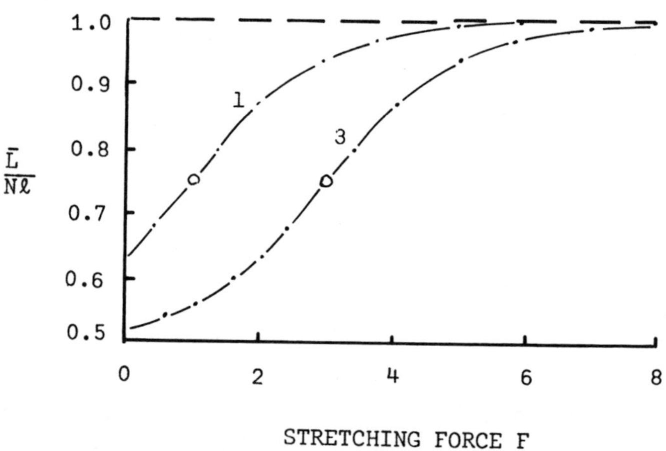

Fig. 12.15 Plot of relative length $\bar{L}/N\ell$ vs stretching force F for an autone model, according to Eqs. (12.43) and (12.44) . For curves 1 and 3, the quantity $\varepsilon/(\ell-s)$ has the values 1 and 3, resp. The inflection points are marked by circles.

From Eq. (12.44) we realize that for any given temperature T the energy difference ε determines the ratio p/q of probabilities for the β units and α units in the absence of any stretching force, that is, the ratio when F = 0 . Curve "1" in Fig. 12.15 is for a smaller value of ε , such that $\varepsilon/(\ell-s)$ is equal to 1. We see that (as expected from Eq. (12.46)) the inflection point now occurs at a stretching force F of only 1 dyn; the curve now hardly exhibits an S-shape, but shows mainly a tendency for the extensibility $d\bar{L}/dF$ to decrease as F increases.

Certainly the force vs elongation curves represented by Eqs. (12.43) and (12.44) (especially as exemplified by Curve 3 in Fig. 12.15) have an S-shape, as was sought in explaining the data for wool-keratin in Fig. 12.11. This suggests that the simple autone model we have described in this and the previous two sections may be appropriate for keratin. However, the curves in Fig. 12.11 are much steeper than those in Fig. 12.15. This increased steepness was accounted for by DHKS (1967) by assuming that a "cooperative" effect occurs. In this assumption it is assumed that the probability of an α or β link depends on the state of its neighbors. Consequences of "cooperation" can be studied by using a Monte Carlo method as described in Section 12.5.3. Suppose the first unit is short; in other words it is an α unit or, simply, an α . Then the probability that the next one is also an α is assumed given by the probability p_α . As successive units are laid down one will, by chance, be a β . Following the β it is assumed that the probability of an α is much reduced and given by σp_α where σ is a "nucleation parameter" and is much less than unity. The physical basis for this assumption is the fact that it is difficult to form a portion of helix except by adding on to an already existing helix. As in crystal growth, the probability that atoms will come together in the desired way is greatly enhanced by having a crystallite "nucleus" present at whose surface the atoms can be assembled.

On the basis of this and related assumptions calculations were carried out partly with Monte Carlo methods (Schor, Haukaas and David, 1968) but mainly with the aid of theory from statistical mechanics (DHKS, 1967) which we do not have the space to develop here; DHKS obtained the solid curves in Fig. 12.11. For these the nucleation parameter σ was assumed to be about 0.05 at room temperature. Since there is satisfactory agreement of their theory with experimental results the α-β model employed by the authors is evidently a very useful one.

A somewhat different two-state model has been proposed by Breuer (1968) to account for keratin stress-strain data; here the two states are assumed to be the α helix and a random coil.

PROBLEMS

12.1 (a) Carry through the indicated steps and derive the expression for F in Eq. (12.5) .

(b) Show that the same expression is obtained by direct application of the perfect gas law, Eq. (1.2), or from simple kinetic theory (Section 8.1) .

12.2 Carry out the indicated steps to derive the expressions for F in Eqs. (12.12) and (12.14) .

12.3 Using series expressions from Appendix B show that the expression for F in Eq. (12.14) reduces to the linear one in Eq. (12.9) when $\ell << \ell_\infty$.

12.4 Carry out the indicated steps and derive the expression for F_x/w_o^2 in Eqs. (12.25) .

12.5 Carry out the indicated steps and obtain the approximate expressions for F_x/w_o^2 in Eqs. (12.27) and (12.29) .

CHAPTER 13

SOUND AND SHEAR WAVES

13.1 TRAVELLING SOUND WAVES

Sound provides one of our principal means of communication. A being, or device, acting as source causes slight fluctuations of the pressure in its own vicinity. Waves of pressure disturbance, or sound waves, propagate outward from the source and thus establish a sound field in the surrounding medium. If reflections occur standing waves may be set up; if not the waves propagate outward until their energy is dissipated, and are called travelling waves.

In treating these, we assume a homogeneous isotropic medium and consider a plane sound wave of frequency f travelling along the x direction. The displacement ξ along x can then be written as

$$\xi = Ae^{-\alpha x} \sin(\omega t - kx) , \qquad (13.1a)$$

where

$$\omega = 2\pi f ; \quad k = \frac{2\pi}{\lambda_s} . \qquad (13.1b)$$

Here λ_s is the wavelength, α is the attenuation coefficient, or absorption coefficient, and A the displacement amplitude at $x = 0$. The velocity u is just $\partial\xi/\partial t$ and is given by

$$u = \partial\xi/\partial t = \omega Ae^{-\alpha x} \cos(\omega t - kx) . \qquad (13.2)$$

Plots of ξ vs x for various values of the time are shown in Fig. 13.1. In each plot zeroes occur at intervals of $\frac{1}{2}\lambda_s$ with alternating peaks and troughs spaced between the zeroes. In Fig. 13.2 all plots are superposed and we see that the envelopes are given by $\pm Ae^{-\alpha x}$. At any given position x the maximum value of ξ with respect to time, that is, the amplitude of ξ, is given by $Ae^{-\alpha x}$. Similar plots would apply to the velocity u, Eq. (13.2), but the velocity amplitude is $\omega Ae^{-\alpha x}$, that is, ω times the displacement amplitude.

It can be shown (Prob. 13.1) that the zeroes move in the positive x direction with velocity c given by

$$c = \frac{\omega}{k} = f\lambda_s . \qquad (13.3)$$

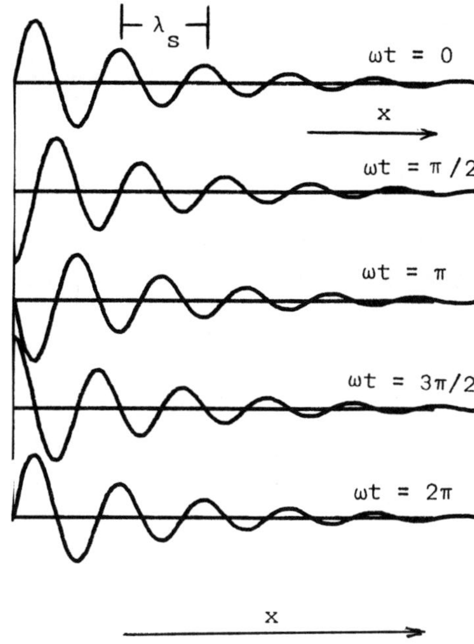

Fig. 13.1 Plots of ξ vs x for an attenuated travelling wave, based on Eqs. (13.1). Consecutive plots are at intervals of one-fourth period.

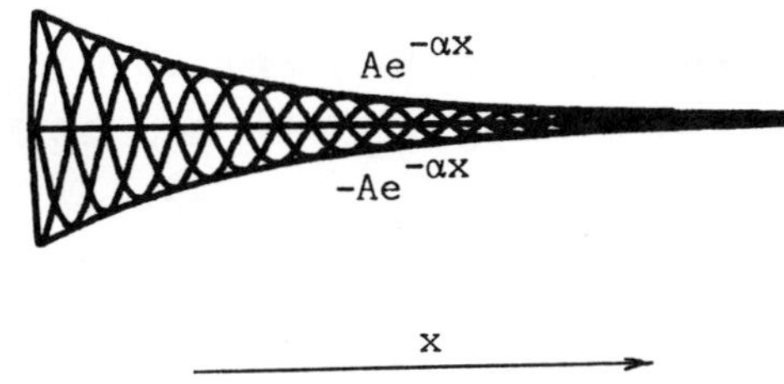

Fig. 13.2 Superposition of ξ vs x plots from Fig. 13.1. The envelopes are $\pm Ae^{-\alpha x}$.

The quantity c is called the velocity of sound (more specifically, the phase velocity) and depends on properties of the medium.

Clearly the attenuation coefficient α gives a measure of the rapidity with which the amplitude decreases as the wave travels. We now consider briefly the units for α . The neper can be used to compare magnitudes. If two amplitudes are A_1 and A_2 the neper ratio N is defined as

$$N = \ln \frac{A_1}{A_2} . \tag{13.4}$$

Another common unit for comparing amplitudes is the decibel or db; the db ratio D for the same amplitudes is defined as

$$D = 20 \log_{10} \frac{A_1}{A_2} . \tag{13.5}$$

It is not hard to show (Prob. 13.2) that

$$D = 8.69\ N . \tag{13.6}$$

We saw that for the wave represented in Eqs. 13.1 the displacement amplitude is $Ae^{-\alpha x}$. Let the amplitudes at x_1 and x_2 , respectively, be A_1 and A_2 . Then the neper ratio between the amplitudes is

$$N = \ln \frac{A_1}{A_2} = \ln \frac{Ae^{-\alpha x_1}}{Ae^{-\alpha x_2}} = \alpha(x_2 - x_1). \tag{13.7}$$

In this application we refer to N as the attenuation in nepers experienced by the wave in travelling from x_1 to x_2 . From Eq. (13.7) we see that N is proportional to the distance $(x_2 - x_1)$; when distances are expressed in centimeters the attenuation coefficient α has the units of nepers/cm. It is also evident that the reciprocal of α in these units gives the distance in centimeters through which the wave travels in experiencing an attenuation of one neper, that is, a decrease in amplitude of e^{-1} or 0.368. (Prob. 13.3)

13.2 WAVE EQUATION FOR SOUND.

We now take up principles from which we justify the expression for ξ given in Eq. (13.1), and, at the same time, obtain information on the constants c and α . In Chap. 5 we discussed the volume force arising from stresses. For a volume element of momentary volume dv the force f_x dv acting on it *via* the stress S_{11} is given by $(\partial S_{11}/\partial x)$ dv . From Newton's Second Law the net force on any body is equal to the product of its mass and the acceleration of its center of mass. If the medium in the element has density ρ and its velocity is u (which can also be written as $\partial\xi/\partial t$ or dx/dt) the element's mass is ρ dv and its acceleration is du/dt. Equating f_x dv to ρ dv du/dt, and cancelling the factor, dv, we obtain

$$\frac{\partial S_{11}}{\partial x} = \rho \frac{du}{dt} . \tag{13.8}$$

From Chap. 3 we know that the density ρ is given by

$$\rho = \rho_o(1 - \frac{\partial\xi}{\partial x}) , \tag{13.9}$$

where ρ_o is the density in the absence of sound. For a plane wave field an expression for ρ can be obtained from the result for ξ in Eqs. (13.1). We can see for the plane wave, and it is also true more generally, that the velocity u is a function of both x and t. Since x refers to the momentary position of the volume element to which Eq. (13.8) applies we realize that x is a function of time. In evaluating the derivative on the right of Eq. (13.8) we therefore follow the rule from calculus for a quantity which depends both explicitly and implicitly on an independent variable t:

$$\frac{du}{dt} = \frac{\partial u}{\partial t} + \frac{\partial u}{\partial x}\frac{dx}{dt} . \tag{13.10}$$

Since dx/dt is just the velocity u, combining Eqs. (13.8) and (13.10) yields

$$\frac{\partial S_{11}}{\partial x} = \rho[\frac{\partial u}{\partial t} + u\frac{\partial u}{\partial x}] . \tag{13.11}$$

In continuum mechanics an equation like Eq. (13.8) or (13.11) which expresses Newton's Second Law for a volume element is sometimes called a dynamical equation.

We now introduce approximations suitable for sound waves of low amplitude. In ordinary sound fields the right hand side of Eq. (13.11) can be approximated simply by $\rho_o \partial^2 \xi/\partial t^2$, where ρ_o is the density of the medium in the absence of sound. (Prob. 13.4). The dynamical equation thus becomes

$$\frac{\partial S_{11}}{\partial x} = \rho_o \frac{\partial^2 \xi}{\partial t^2} . \tag{13.12}$$

In arriving at Eq. (13.8) and its simpler approximate form in Eq. (13.12), no special assumptions were made about the medium except that it is homogeneous and isotropic. While Eq. (13.12) is an important equation, it is only of limited usefulness as it stands, since it involves two dependent variables, S_{11} and ξ. For a given medium an equation involving ξ alone can be obtained by introducing an appropriate constitutive relation for that medium, such as one of those described in Chap. 4. Thus for a linear isotropic elastic solid Eq. (4.2) may be assumed; for a field where the displacements are along x only and are independent of y and z we have

$$S_{11} = \lambda\theta + 2\mu\varepsilon_{11} = (\lambda+2\mu) \frac{\partial \xi}{\partial x} . \tag{13.13}$$

Substituting from Eq. (13.13) into Eq. (13.12) we have an equation in ξ along which can be written

$$\frac{\partial^2 \xi}{\partial x^2} = \frac{1}{c^2} \frac{\partial \xi^2}{\partial t^2} ; \quad c^2 = \frac{\lambda + 2\mu}{\rho} . \tag{13.14}$$

Any function of x and t which satisfies this equation is, of course, consistent with the physical principles on which it is based, namely, Newton's Second Law and the constitutive relation of Eq. (13.13). It is easily shown by substitution that the plane wave function in Eqs. (13.1) is a solution of Eq. (13.14), with $\alpha = 0$ and $k = \omega/c$. Because of the nature of its solutions Eq. (13.14) is reasonably called a wave equation. The velocity c is an important physical quantity, which we have already referred to in Section 13.1 . Other wave equations are obtained when different constitutive relations apply. But it is characteristic of all that the constant c^2 , as in Eq. (13.14), is expressed as the ratio of an elastic modulus (or linear combination of moduli) to the density.

In our development we have assumed the medium to be homogeneous. However any terrestial medium has boundaries, beyond which there is a change in property. A function which is to describe a wave in a given medium must not only satisfy an appropriate differential equation (such as Eq. (13.14)) for that medium, but must also be consistent with suitable statements of the boundary conditions.

In Eq. (13.13) it is assumed that ε_{22} and ε_{33} are zero. This is reasonable if, as in a related situation treated in Section 4.4 (Eqs. (4.16) and (4.17)), the dimensions of the solid medium are sufficiently great in the y and z directions. Put another way, any region of a solid to which Eq. (13.13) applies must be relatively distant from the boundaries which limit the solid laterally. Typically Eqs. (13.13) and (13.14) apply to a plane wave travelling down a bar whose width is much larger than the wavelength λ_s. Where Eq. (13.13) applies the stresses S_{22} and S_{33} are nonzero and in fact are each given by $\lambda \partial\xi/\partial x$ as in Eq. (4.17).

For a wave travelling down a thin rod or filament whose diameter is much smaller than the wavelength λ_s a different situation arises. Here the lateral boundaries are very important since all points in a thin filament are near its surface. It is typical to assume that S_{22} and S_{33} are zero at the surface and also, in fact, at all points within the filament, as in Eq. (4.18). It is then found that S_{11} is given by $E\ \partial\xi/\partial x$, where E is the Young's modulus given in terms of λ and μ in Eqs. (4.21). Using this value for S_{11} in Eq. (13.12) we obtain a wave equation of the same form as in Eq. (13.14), but with the velocity c now given by

$$c^2 = \frac{E}{\rho} . \tag{13.15}$$

In a thin filament the lateral strain quantities ε_{22} and ε_{33} are not zero but, as in Eq. (4.20), are each given by $-\nu\ \partial\xi/\partial x$; here ν is Poisson's ratio for which an expression is given in Eqs. (4.21).

For waves travelling down a bar of intermediate width the situation is more difficult and will not be considered here.

In a liquid where Eq. (4.8) holds the stress S_{11} is equal to $B\ \partial\xi/\partial x$, where B is the bulk modulus of elasticity. Substituting this value for S_{11} into Eq. (13.12) yields again a wave equation of the same form as in Eq. (13.14), but with the sound velocity given by

$$c^2 = \frac{B}{\rho} . \tag{13.16}$$

We now proceed to a more general situation, that of a linear viscoelastic medium for which the constitutive relation is given by Eq. (4.22). For a plane wave along the x direction we have for S_{11} :

$$S_{11} = (\lambda+2\mu)\frac{\partial\xi}{\partial x} + (\lambda' + 2\eta)\frac{\partial^2\xi}{\partial x\partial t} . \tag{13.17}$$

Substituting this expression for S_{11} into Eq. (13.12) yields

$$(\lambda + 2\mu)\frac{\partial^2\xi}{\partial x^2} + (\lambda' + 2\eta)\frac{\partial^3\xi}{\partial x^2\partial t} = \rho_o\frac{\partial^2\xi}{\partial t^2} . \tag{13.18}$$

This equation differs from the wave equation in Eq. (13.14) in that here an extra term appears, including the factor $(\lambda' + 2\eta)$. Equation (13.18) is nevertheless a wave equation; it can be verified by substitution (Prob. 13.5) that the plane wave expression for ξ in Eqs. (13.1) is a solution of Eq. (13.18) when α and k are given by

$$k^2 = \tfrac{1}{2}\beta[1 + (1 + \varepsilon^2)^{\frac{1}{2}}]$$

$$\alpha = \frac{\varepsilon\beta}{2k} , \tag{13.19}$$

where

$$\varepsilon = \frac{\omega(\lambda' + 2\eta)}{\lambda + 2\mu} \quad ; \quad \beta = \frac{k_o^2}{1 + \varepsilon^2} , \tag{13.20}$$

and $k_o = \omega/c_o$, where

$$c_o = \left(\frac{\lambda + 2\mu}{\rho_o}\right)^{\frac{1}{2}} . \tag{13.21}$$

When $\varepsilon << 1$ we find that $\alpha << k$, and simplified expressions result for k and α :

$$k = k_o \left(1 - \frac{3\varepsilon^2}{8}\right) \qquad \alpha = \frac{\varepsilon k_o}{2} \tag{13.22}$$

Since the velocity of sound c can be written as ω/k we obtain

$$c = \frac{c_o}{1 - \frac{3\varepsilon^2}{8}} = \frac{c_o}{1 - \frac{3\alpha^2}{2k_o^2}} . \tag{13.23}$$

We see that c_o is the velocity of sound which would apply if α were zero. When $\alpha << k$ there is little difference between c and c_o .

13.3 SOUND PROPAGATION IN WATER

In Eqs. (13.19) to (13.23) the constants α and c are expressed in terms of the angular frequency ω, the density ρ_o and the elastic and viscous coefficients λ, μ, λ' and η. These equations and other similar ones form the basis for acoustical methods in which information on elastic and viscous constants is obtained from measurements on sound velocity and absorption. In this section we consider some properties of water as determined in this way. Since biological tissue tends to have a high water content it is not surprising that some of the characteristics of tissues are similar to those of water.

Data on the velocity of sound c in water have already been given in Table 7.1. It was from these data and measured values of the density ρ that calculations were made of the elastic modulus B (using Eq. (13.16) listed in the same Table.)

For the absorption coefficient α it is useful to refer to the theoretical results given in Eqs. (13.19) to (13.22). In these we let $\mu = 0$ (since we are dealing with a liquid) and assume $\alpha << k$, an approximation which holds in water except at very high frequencies. We then obtain (Prob. 13.

$$\alpha = \frac{2\omega^2(\eta + \frac{3}{4} B')}{3 \rho_o c_o^3} , \tag{13.24}$$

where $c_o^2 = B/\rho_o$ and $\omega = 2\pi f$; the quantity B' is defined as the coefficient of bulk viscosity and (by analogy to the modulus of bulk elasticity B) is related to λ' and η by

$$B' = \lambda' + \frac{2\eta}{3} . \tag{13.25}$$

(Cf. Eq. (4.13). When B' is zero we represent the absorption by α_o , where

$$\alpha_o = \frac{2\omega^2\eta}{3\rho_o c_o^3} . \tag{13.26}$$

Absorption resulting from shear viscosity alone was treated theoretically by Stokes in the mid-nineteenth century, and is a component of "classical" absorption. Heat conductivity also contributes to classical absorption but its effect is small in water and similar fluids. When Eq. (13.26) applies the absorption coefficient is proportional to ω^2 , that is, to the frequency squared.

On the right hand side of Eq. (13.26) the quantities η, ρ_o and c_o (the latter defined as $\sqrt{B/\rho_o}$) are independent of frequency and can be measured in nonacoustic experiments. But the situation is different with the bulk viscosity coefficient B' in Eq. (13.24); this quantity is usually determined from measurements of the attenuation coefficient α . When α has been measured and Eq. (13.24) is taken as the working equation, B' is defined operationally as "that quantity which, when entered into Eq. (13.24), yields the observed value of α". It is often found that B' is a function of frequency.

For water measurements of α have been made over a range of the frequency f and it is found that at any given temperature α is proportional to f^2, as expected from Eq. (13.24 if B' is constant. Thus α/f^2 is independent of frequency; howaver, α/f^2 is found to decrease with increasing temperature. Results are shown in Table 13.1, based on data from Herzfeld and Litovitz (1959). This table can be used to calculate α for water at a given temperature and frequency f, by noting the value of α/f^2 for the given temperature and multiplying this by f^2. For example, at 20°C and 10^6 Hz (i.e., 1 MHz) we obtain

$$\alpha = 25.3 \times 10^{-17} \times (10^6)^2 = 25.3 \times 10^{-5} \text{ nepers/cm.}$$

Thus at 20°C the amplitude of a plane traveling sound wave of frequency 1 MHz decreases by a factor of e^{-1} or 0.368 when it propagates in water through a distance α^{-1} or 3950 cm .

Information on the bulk viscosity coefficient B' can be obtained by use of the absorption data in Table 13.1, together with Eq. (13.24). It is convenient to refer to the ratio α/α_o, where α_o is given by Eq. (13.26). It is readily shown (Prob. 13.7) that the viscosity ratio B'/η can be written

$$\frac{B'}{\eta} = \frac{4}{3}\left(\frac{\alpha}{\alpha_o} - 1\right) = \frac{4}{3}\left(\frac{\alpha - \alpha_o}{\alpha_o}\right) . \qquad (13.27)$$

From data on η, ρ_o and c_o as given, for example, in Table 7.1 values have been calculated for α_o. Rather than tabulate α_o the ratios α/α_o for various temperatures are listed in Table 13.1. Also shown there are the ratios B'/η calculated from Eq. (13.27). Experimentally determined values of η are listed as well; these are multiplied by corresponding values of B'/η to obtain results for B' in the last row of Table 13.1 . We see that for water the bulk viscosity coefficient B' is roughly equal to three times the shear viscosity coefficient, in the temperature range 0-40°C .

What is the significance of the coefficient B'? In approaching this question we recall the procedure followed in determining B'. Equation (4.22) was assumed to be valid as a constitutive equation and, after a number of steps, Eqs. (13.24) to (13.27) were arrived at. Experimental data on α were then referred to and the quantity B' was so selected as to obtain agreement between experimental α and the theoretical value given in Eq. (13.24). If experimental α had been just equal to α_o we should have set B' equal to zero. Now the coefficient α_o represents "classical absorption", arising from conversion of sonic energy into heat by viscous shearing. Hence when B' is nonzero we know that an extra mechanism exists for sound absorption, over and above the classical viscous shearing process. For most fluids B' is found to be nonzero, and various mechanisms are invoked to account for the excess absorption. Commonly a relaxation process is believed to be involved, that is, one in which a system tends to return to an equilibrium state from which it has been disturbed.

TABLE 13.1

ATTENUATION COEFFICIENTS IN WATER

Values listed for α/f^2 are in cgs units (nepers cm^{-1} sec^2) when multiplied by 10^{-17}. Those for η and B' are in poise when multiplied by 10^{-2}. Absorption data are from Herzfeld and Litovitz (1959).

$T(^oC)$	0	5	10	15	20	30	40
α/f^2	56.9	44.1	36.1	29.6	25.3	19.1	14.6
α/α_o	3.33	3.29	3.18	3.09	3.10	3.10	3.01
B'/η	3.11	3.05	2.91	2.79	2.80	2.80	2.70
η	1.793	1.518	1.309	1.144	1.008	0.800	0.653
B'	5.58	4.63	3.81	3.19	2.82	2.24	1.76

For water Hall (1948) considered relaxation of its intermolecular structure; he assumed that any given water molecule is related to its neighbors in either of two ways: (1) the molecule is part of a small crystal of ice I, whose rather open structure was discussed in Section 7.2, or (2) it is packed with other molecules in a relatively dense liquid structure. Let f_1 be the fraction of all molecules in the ice structure and let f_2 be the corresponding fraction in the liquid structure. The ratio f_1/f_2 has an equilibrium value for any given choice of the state variables, say, the pressure and temperature. If the pressure is perturbed a new equilibrium ratio applies; in a pressure increase the structure with least volume is favored and in a pressure decrease the more open structure is favored. During a sound cycle the pressure changes continuously with time. At any given time the molecules rearrange themselves to approach the equilibrium state applicable at the momentary value of the pressure.

We shall not treat Hall's theory in detail, but only consider a few equations which represent principal ideas. For simplicity in this treatment we assume temporarily that $\eta = 0$; then the only sound absorption comes from the relaxation process under consideration. This absorption can be attributed in a formal way to a bulk viscosity coefficient B', if desired. When $\eta = 0$ the stress tensor is isotropic; the diagonal stresses S_{11} , etc., are all equal to $-p$. For equilibrium conditions the constitutive equations reduce to Eq. (4.10), with $\lambda = B$. We consider a two-structure intermolecular model of water. Suppose the pressure of a small quantity of water is changed quickly from an initial value to a new constant value p at which the ratio f_1/f_2 has a new value and the corresponding new equilibrium value of the dilatation is $\theta_e = -p/B$. We assume the system will not reach equilibrium immediately, but will approach it at a rate which depends on the difference $(\theta_e - \theta)$ where θ is the dilatation at any given time. In particular we assume, following Hall, that the rate of increase of θ is proportional to $(\theta_e - \theta)$, the difference between θ and its equilibrium value θ_e . A convenient way to express this idea in equation form is to write

$$\dot{\theta} = \frac{\theta_e - \theta}{\tau} , \qquad (13.28)$$

where τ is a constant. It is not hard to show that a solution of Eq. (13 is

$$\theta = Ae^{-t/\tau} + \theta_e . \qquad (13.29)$$

This solution represents response of the system to a stepwise change in pressure, as described just previous to Eq. (13.28). According to Eq. (13.29) the difference between θ and θ_e decreases by a factor of e^{-1} whenever the time increases by τ . Thus the time required for θ to appro θ_e is characterized by the relaxation time τ .

Now consider a sound field in which the pressure varies sinusoidally in t with angular frequency ω. If ω is very low, or τ is very short, the rat f_1/f_2 will be approximately equal to its equilibrium value at all times a the dilatation will always be related to the pressure by Eq. (4.10) with $\lambda = B$; for this low-frequency or short-τ situation we represent B by the symbol B_o and obtain

$$\theta = -\frac{p}{B_o} . \quad \text{(Small } \omega\tau\text{)}. \qquad (13.30)$$

On the other hand, at very high frequency or long τ the pressure varies so rapidly that the ratio f_1/f_2 is not able to adjust and, instead, remains essentially constant. We can expect that under these circumstance an equation similar to Eq. (13.30) holds but with a different value for the elastic constant; letting the high-frequency long-τ value of the constant be B_∞ we write

$$\theta = -\frac{p}{B_\infty} . \quad \text{(Large } \omega\tau\text{)} . \qquad (13.31)$$

The time rate of change of θ under the same conditions is

$$\dot{\theta} = -\frac{\dot{p}}{B_\infty} . \quad \text{(Large } \omega\tau\text{)}. \qquad (13.32)$$

At an intermediate value of the frequency we assume that the total rate of change of θ is given by the sum of the right hand sides of Eqs. (13.28) and (13.32):

$$\dot{\theta} = \frac{\theta_e - \theta}{\tau} - \frac{\dot{p}}{B_\infty} . \tag{13.33}$$

Now θ_e is just equal to $-p/B_o$ so Eq. (13.33) can be rewritten

$$\dot{\theta} + \frac{\theta}{\tau} = - \frac{p}{\tau B_o} - \frac{\dot{p}}{B_\infty} . \tag{13.34}$$

This equation is seen to have certain reasonable features:

(i) When τ is very small (meaning that the ratio f_1/f_2 reaches its equilibrium value very quickly) the dominant terms in Eq. (13.34) are those with τ in the denominator, and the equation reduces to Eq. (13.30), as it should.

(ii) When τ is very large, so that the ratio f_1/f_2 hardly changes at all during a sonic cycle, the other terms in Eq. (13.34) are dominant and the equation reduces to Eq. (13.32).

We shall analyze Eq. (13.34) more completely in Section 13.4 using complex-number methods.

13.4 COMPLEX NUMBER REPRESENTATIONS AND THEIR APPLICATIONS

13.4.1 Introduction

In sound fields we are much concerned with quantities which vary sinusoidally with time; see, for example, the expressions in Eqs. (13.1) and (13.2). In dealing with these the notation of complex numbers is convenient. Properties of this number system are described in Appendix E. For any complex number, say, u we let the symbol Re[u] represent the real part of u. From Appendix E we then realize that, for example, Re $[e^{j\omega t}]$ is just cos ωt and if A is a real number

$$\mathrm{Re}[Ae^{j\omega t}] = A \cos \omega t . \tag{13.35}$$

It is not hard to show that Eq. (13.1a) can be rewritten as

$$\xi = R_e\,[-jAe^{-\alpha x}\, e^{j(\omega t - kx)}]\ . \qquad (13.36)$$

It is common to omit the symbol Re[ ] and rewrite Eq. (13.36) simply as

$$\xi = -jAe^{-\alpha x}\, e^{j(\omega t - kx)}\ , \qquad (13.37)$$

with the understanding that the physical quantity represented by ξ is given by the real part of the expression to which ξ is equated. Using the rules of complex operations (Appendix E) we obtain the derivative of the right hand side of Eq. (13.37) with respect to time. From the properties of complex numbers it follows that the real part of this derivative is just the physical quantity $\partial\xi/\partial t$, that is, the velocity u. Thus we write

$$u = \mathrm{Re}[\omega A\, e^{-\alpha x} e^{j(\omega t - kx}] \qquad (13.38)$$

or, again following the convention of omitting the symbol Re [ ],

$$u = \omega Ae^{-\alpha x}\, e^{j(\omega t - kx)}\ . \qquad (13.39)$$

We see that the real expression represented by Eq. (13.38) or (implicitly) by Eq. (13.39) agrees with the expression in Eq. (13.2), obtained by direct differentiation of the real expression for ξ . Use of the complex number representation is often advantageous in that operations are easier to carry out and expressions are simpler. Thus differentiation of $e^{j\omega t}$ with respect to time is accomplished by a simple algebraic operation, namely, multiplication of $e^{j\omega t}$ by the factor $j\omega$. By contrast, differentiation of (for example) sin ωt with respect to time involves a change of function, namely, to cos ωt . Disadvantages of operations with complex numbers are that a few properties and techniques need to be learned in becoming facile with them and complex-number expressions at first seem "less physical". With a little experience, however, these difficulties are overcome. Problem 13.8 provides an exercise. Other applications of complex-number methods are taken up in the remainder of this chapter.

13.4.2 Wave Equations and their Solutions

As one application of complex numbers we reconsider the wave equation in Eq. (13.14). Let us assume ξ varies sinusoidally in time with angular frequency ω. This assumption is expressed by letting ξ be given by $Ae^{j\omega t}$ where A is independent of time but may be complex; it is understood that the physical quantity (the displacement) is given by the real part of this expression. Using rules for differentiation (Appendix E) we easily find that $\partial^2\xi/\partial t^2$ is given by $-\omega^2\xi$. The wave equation in Eq. (13.14) then becomes

$$\frac{\partial^2\xi}{\partial x^2} + k^2\xi = 0 , \tag{13.40}$$

where $k = \omega/c$ is known as the <u>propagation constant</u> and, as noted in Eq. (13.1b), can also be written as $2\pi/\lambda_s$, where λ_s is the sonic wavelength. It is easily shown that a solution of Eq. (13.40) is

$$\xi = Ae^{j(\omega t - kx)} . \tag{13.41}$$

If A is real, the physical displacement, given by the real part of the expression for ξ in Eq. (13.41), is easily seen to be

$$\xi = A \cos (\omega t - kx) . \tag{13.42}$$

If A is complex the cos ( ) function will differ from that in Eq. (13.42) by a phase factor. (Prob. 13.9) It is easily verified that the real expression in Eq. (13.42) is a solution of the wave equation in Eq. (13.40). This result is as expected from Appendix E, and shows how solutions can be obtained by using complex numbers. Either of Eqs. (13.41) or (13.42) represents a plane wave traveling in the positive x direction with velocity ω/k . (Cf Prob. 13.1). Comparing the expressions in Eqs. (13.42) and (13.1) we find two differences: (i) the attenuation coefficient α is zero in Eq. (13.42) and (ii) a cosine function appears here, instead of a sine function. The latter difference signifies only a shift in the time scale and for many purposes is unimportant.

A similar approach can be taken to Eq. (13.18). If ξ is represented by $Ae^{j\omega t}$ this equation can be reduced to

$$\frac{\partial^2\xi}{\partial x^2} + K^2\xi = 0 \ , \tag{13.43}$$

which has the form of Eq. (13.40); but K is a complex number and is given (Prob. 13.10) by

$$K^2 = \frac{k_o^2}{1 + j\varepsilon} \ , \tag{13.44}$$

where $k_o = \omega/c_o$, the quantities ε and c_o having been defined in Eqs. (13.20) and (13.21). Because of the analogy between Eqs. (13.40) and (13.43) the quantity K is often called the "complex propagation constant". If we write K in the form $(k-j\alpha)$ the expression

$$\xi = A\ e^{j(\omega t\ -\ Kx)} \tag{13.45}$$

which satisfies Eq. (13.43), is equivalent to the expression in Eq. (13.37) (except for the factor (-j), which only alters the phase). In order to satisfy Eq. (13.44) we find that k and α must be given by the same expressions as in Eqs. (13.19). (Prob. 13.11).

13.4.3 Hall's Theory for Structural Relaxation.

As still another application of complex-number methods we consider Eq. (13.34); as explained in Section 13.3, this equation represents effects of a relaxation process in water, according to Hall (1948).

In a sound field of angular frequency ω we ignore the static contribution to the pressure p and express p in the form

$$p = p_o e^{j\omega t} , \tag{13.46}$$

where p_o is a real constant; it is understood that the physical quantity, that is, the pressure, is given by the real part of the right hand side of Eq. (13.46). A solution of Eq. (13.34) can be found by assuming that θ is also of the form $Ae^{j\omega t}$ where A is a constant (which may be complex). Then by use of operations for complex numbers we obtain (Prob. 13.12) that

$$\theta = - \frac{\beta_o + j\omega\tau\beta_\infty}{1 + j\omega\tau} p_o e^{j\omega t} , \tag{13.47}$$

where β_o and β_∞ are compressibilities given by B_o^{-1} and B_∞^{-1}, resp. Equation (13.47) can be rewritten (Prob. 13.12) as

$$\theta = - [\beta_\infty + \frac{\beta_o - \beta_\infty}{1 + j\omega\tau}] \ p . \tag{13.48}$$

We see from either of Eqs. (13.47) or (13.48) that when $\omega\tau << 1$ the result reduces to Eq. (13.30), and when $\omega\tau >> 1$ it becomes Eq. (13.31); these results are as expected. At intermediate values of $\omega\tau$ the bracket [ ] in Eq. (13.48) is a complex number; let this number be Z which can alternatively be written as $Z_o e^{j\zeta}$, where Z_o and ζ are real. Then we obtain

$$\theta = - Zp = - Z_o p_o e^{j(\omega t + \zeta)} . \tag{13.49}$$

A consequence of Z being complex is evidently that a nonzero phase angle exists between the dilatation θ and the negative pressure -p. This phase angle results from the relaxation process; Z is real and $\zeta = 0$ if $\tau = 0$.

It is instructive to refer back to Eq. (13.17) and notice that viscosity has a similar effect on the phase between S_{11} and θ . For if we assume ξ has the form $\xi_o e^{j\omega t}$ and replace $\partial\xi/\partial x$ by θ we obtain from Eq. (13.17) that

$$S_{11} = [\lambda + 2\mu + j\omega(\lambda' + 2\eta)]\ \theta\ . \tag{13.50}$$

Thus the ratio between S_{11} and θ is complex if either of the viscosity coefficients λ' or η is nonzero; a phase difference then exists between S_{11} and θ .

We shall see that relaxation causes a travelling wave to be attenuated. For a plane wave propagating in the x direction an approximate dynamic equation is given in Eq. (13.12). In this equation we replace S_{11} by $-p$ since we are imagining our fluid is inviscid, that is, free of viscosity. The result is

$$-\frac{\partial p}{\partial x} = \rho_o \frac{\partial^2 \xi}{\partial t^2}\ . \tag{13.51}$$

Assuming ξ is of the form $Ae^{j\omega t}$ the right hand side of Eq. (13.51) is $-\omega^2 \rho_o \xi$. From Eq. (13.49) we can replace p by $-\theta/Z$. Since θ is just $\partial\xi/\partial x$ in this situation we then obtain from Eq. (13.51):

$$\frac{\partial^2 \xi}{\partial x^2} + \rho_o \omega^2 Z\, \xi = 0 \tag{13.52}$$

This equation has the same form as Eq. (13.43), and we make the association

$$K^2 = \rho_o \omega^2 Z\ . \tag{13.53}$$

We now let K be $(k-j\alpha_r)$ and remember that Z represents the quantity in the bracket [ ] on the right hand side of Eq. (13.48); we obtain

$$\frac{(k - j\alpha_r)^2}{\rho_o \omega^2} = \beta_\infty + \frac{\beta_o - \beta_\infty}{1 + j\omega\tau}\ . \tag{13.54}$$

From the definition for equality (Appendix E) we know that Eq. (13.54) holds if and only if the real and imaginary parts on the two sides of the equation are separately equal. Applying this condition we obtain (Prob. 13.13) the rather lengthy expressions

$$k^2 = \alpha_r^{\ 2} + \rho_o\omega^2 \left[\beta_\infty + \frac{\beta_o - \beta_\infty}{1 + \omega^2\tau^2}\right] ;$$

$$\alpha_r = \tfrac{1}{2}\,\rho_o(\beta_o - \beta_\infty)\,\frac{\omega^3\tau}{k(1 + \omega^2\tau^2)} . \qquad (13.55)$$

Since $k^2 = \omega^2/c^2$ the first of these equations can be solved for c ; The expression simplifies when, as is often true, $\alpha_r^{\ 2}$ is small enough to be neglected. A plot of c vs $\omega\tau$ for this equation is shown in Fig. 13.3 . At low frequencies c is equal to $(\rho_o\beta_o)^{-\frac{1}{2}}$; this is reasonable since for a fluid c is given generally by $(B/\rho_o)^{\frac{1}{2}}$ (see Eq. (13.16) and at low frequencies the bulk modulus B_o is the reciprocal of β_o . As $\omega\tau$ increases the velocity of sound increases, approaching $(\rho_o\beta_\infty)^{-\frac{1}{2}}$ at large values of $\omega\tau$. The frequency dependence of c is associated with the fact that the effective compressibility decreases as the frequency increases, from β_o at the low frequency limit to β_∞ at the upper limit. If $(\beta_o-\beta_\infty)$ is sufficiently small compared to β_∞ the change of c with frequency (often called the "dispersion") is negligible and c is always approximately equal to its low-frequency value c_o .

Under the latter circumstances the expression for α_r in Eqs. (13.55) simplifies and has the form shown in Figure 13.3 . We see that the coefficient α_r increases monotonically with increasing frequency, reaching a limiting value at large values of $\omega\tau$. When $\omega\tau = 1$ the coefficient is just one-half its limiting value. The theoretical curve in a figure to be discussed later, Fig. 13.4 , is based on relaxation theory similar to that represented by Eqs. (13.55); the plot shown there is of $\alpha_r\lambda$, the absorption per wavelength, against the logarithm of frequency. A symmetrical curve appears in such a plot with the maximum of $\alpha_r\lambda$ occurring at a frequency such that $\omega\tau = 1$.

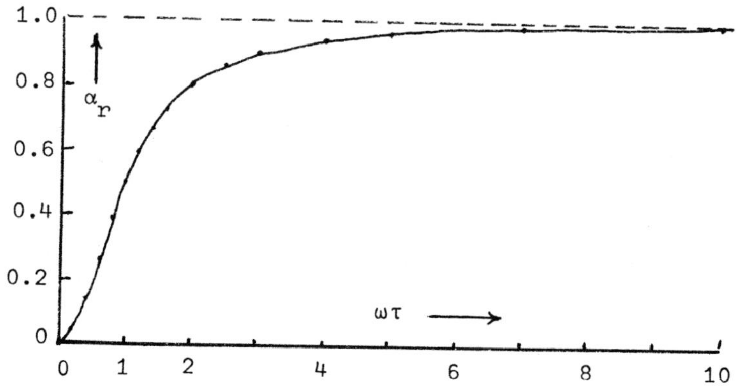

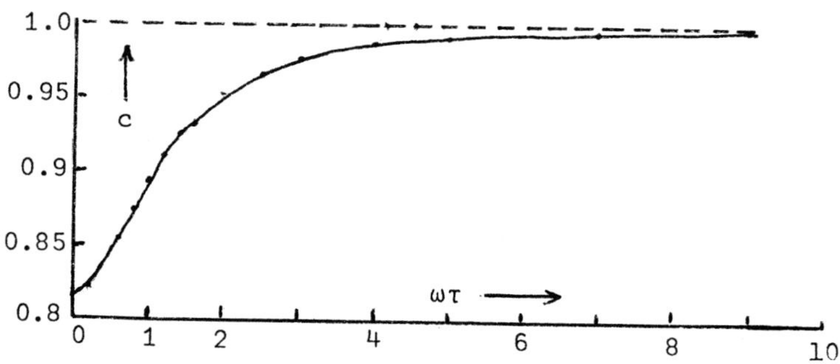

Fig. 13.3 Plots of c and α_r versus a nondimensional quantity $\omega\tau$ proportional to the frequency; based on Eqs. (13.55) when $\alpha_r \ll k$ and $(\beta_o - \beta_\infty) \ll \beta_\infty$. For small $\omega\tau$ the velocity c reduces to c_o given by $(\rho_o\beta_o)^{-\frac{1}{2}}$; for large $\omega\tau$ it approaches c_∞, i.e., $(\rho_o\beta_\infty)^{-\frac{1}{2}}$. For large $\omega\tau$ the coefficient α_r approaches α_∞, given by $\rho_o(\beta_o-\beta_\infty)c_\infty/2\tau$. For these plots we have chosen $c_\infty = \alpha_\infty = 1$ and $c_o = 0.82$.

In water τ is found to be of the order of 10^{-12} sec. Hence $\omega\tau << 1$ at all frequencies in the presently available range for acoustic experimentation, up to somewhat less than 10^9 MHz . It is also found that $\alpha_r << k$ in this range. Equations (13.55) then simplify to yield

$$k = \frac{\omega}{c_o}$$
$$\alpha_r = \tfrac{1}{2}\rho_o c_o \tau(\beta_o - \beta_\infty)\omega^2 . \qquad (13.56)$$

We see that when the above equations apply the coefficient α_r increases with the frequency-squared; the same was found for absorption arising from viscosity when $\alpha << k$; see Eq. (13.24)

In section 13.3 we referred to B' as a quantity introduced into Eq. (13.24) (when $\alpha << k$) to account for absorption in excess of the classical value. A specific expression for determining B'/η is given in Eq. (13.27), valid when $\alpha << k$. To determine B' for a medium in which "Hall-type"

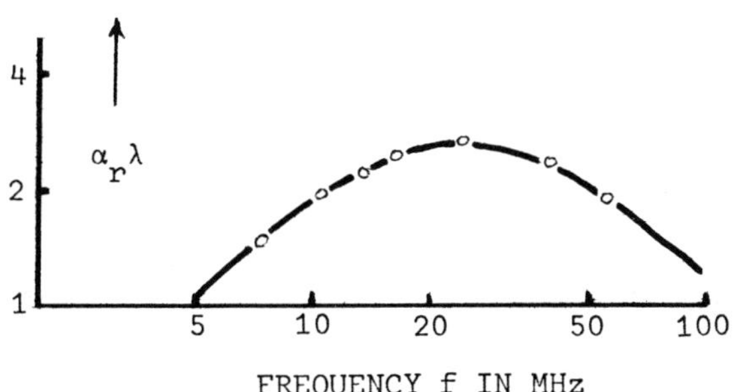

Fig. 13.4 Log-log plot of $\alpha_r\lambda$, the absorption per wavelength (in arbitrary units), versus frequency for an aqueous solution of glycine. Circles show experimental values; the solid curve is from theory for a single relaxation process similar to that represented by Eqs. (13.55). See Section 13.5.3. From Eigen and Hammes (1963, Reference 13.6).

relaxation occurs we use Eq. (13.27), replacing $(\alpha - \alpha_o)$ by α_r , the latter being given (for $\alpha << k$ and $\omega\tau << 1$) by Eqs. (13.56) . We then obtain (Prob. 13.14)

$$B' = \frac{\tau(\beta_o - \beta_\infty)}{\beta_o^2} . \tag{13.57}$$

In comparing his theory with experiment for water Hall found the choice of 18×10^{-12} cm^2/dyn for β_∞ to give good agreement. At 20° we have from Table 7.1 that β_o (given there as β_{ad}) is about 46×10^{-12} cm^2/dyn , or about 2.6 β_∞ .

13.5 ABSORPTION OF ULTRASOUND IN BIOLOGICAL MEDIA

13.5.1 Introduction

Results on the sound velocity c_s in various materials have been cited in Chap. 7. Here we give corresponding attention to measurements of the absorption coefficient α . Many investigators have made experimental determinations of α in solutions and suspensions of biological interest as well as in actual tissues. We shall take up only a few examples here; more complete information can be found in a comprehensive review article by Dunn, Edmonds and Fry (1969). This article is an important source for the present section.

Part of the absorption in any fluid arises from shear viscosity, as expected from the classical theory treated in earlier sections; see especially Eq. (13.26). However, in no instance of biological interest is the absorption coefficient given completely by the latter equation. Excess of the absorption over the classical value is usually attributed to relaxation processes, such as the one treated in the Hall theory. (Section 13.4.3). In general, relaxation involves return of a system to equilibrium after it has been disturbed. A number of mechanisms for such processes have been identified in various media, but in biological materials they are usually not known. The question of relaxation mechanisms can be bypassed, if desired, by simply attributing the excess absorption to bulk viscosity, the coefficient B' being determined for any given situation by use of Eq. (13.27).

In electrolytes such as NaCl, where the ions are singly charged the absorption differs little from that of pure water at frequencies less than 10 MHz. For electrolytes involving multivalent ions absorption coefficients are greater because of special relaxation phenomena which we shall not treat here. This topic is taken up fully by Stuehr and Yeager (1965).

13.5.2 Amino Acids in Solution.

Amino acids are fundamental to biology, since every protein macromolecule is formed of one or more chains in each of which the 20-25 different amino acid units are linked together in a prescribed manner. The simplest amino acid is glycine; its chemical formula can be written NH_2CH_2COOH; its molecular weight is 75.1. For a solution of glycine in water it was found (Barrett, 1966) that the absorption coefficient α is only a little greater than that for water. But the behavior is very strongly dependent on the concentration of ions in the surrounding medium, especially the hydrogen ion concentration. The latter is given by specifying the pH, defined by

$$pH = -\log_{10}[H^+] \,, \tag{13.58}$$

where $[H^+]$ is the hydrogen ion concentraion in mol/ℓ . In pure water the pH is 7, corresponding to a concentration $[H^+]$ of 10^{-7} mol/ ℓ . For glycine in an NaOH solution with pH 10 the absorption was found by Eigen and Hammes (1963) to vary with frequency as shown in Fig. 13.4. In this figure the logarithm of the absorption per wavelength, that is, $\log \alpha\lambda$, is plotted as ordinate and log f is plotted as abscissa. It can be shown that when α results from a single relaxation process (for example, as described by Eqs. (13.55) the resulting plot has a simple bell-shape as in the solid curve plotted in Fig. 13.4 . The peak occurs at a frequency called the <u>relaxation frequency</u> f_r . The quantity $1/2\pi f_r$ gives the <u>relaxation time</u> τ for the process under consideration. Since the data fit very well to the theoretical curve there is good evidence for a single relaxation process. Specifically, Eigen and Hammes believe that the process involves transfer of a proton, that is an H^+ ion, between the glycine molecule and a neighboring water molecule. In equilibrium a certain fraction of existing protons are associated with glycine, and the rest with water molecules. In a small perturbation of the pressure, as takes place in a sound field, the equilibrium fraction will change. The relaxation time τ is a measure of the time required to achieve equilibrium after a small disturbance. For the glycine solution in question the relaxation frequency was a little greater than 20 MHz and the relaxation time was therefore about 0.008μ sec

13.5.3 Macromolecules in Solution; Effects of "Organization"

Among solutions of proteins the most investigated are undoubtedly solutions of hemoglobin, Hb; basic properties of Hb are discussed in Section 7.3.3. In the frequency range 1 to 10 MHz in a human Hb suspension of concentration 0.165 gm/ml the absorption per wavelength $\alpha\lambda$ is roughly constant, and equal to 0.0030 nepers per wavelength. (Carstensen and Schwan, 1959). At 1 MHz the wavelength λ is approximately 0.15 cm so that α is 0.020 nepers/cm or about 80 times α for water at $20^{\circ}C$. The absorption is approximately proportional to the Hb concentration up to 0.165 gm/ml and is not very dependent on temperature in the range $15\text{-}35^{\circ}C$.

Through the efforts of a number of investigators measurements of α have been made on Hb over a wide frequency range, from 0.035 to 400 MHz. Results for bovine Hb are shown in Fig. 13.5 where the ordinate is α/α_w ; the Hb concentration is 0.15 gm/ml and the temperature is 25° C. We see that at 10^4 Hz the ratio of α to the absorption coefficient in water is about 1000, but that with increasing frequency this ratio decreases and approaches a value less than two. It is assumed that the high absorption ratio existing especially at the lower frequencies is a result of multiple relaxation processes; however, a specific theory has not yet been arrived at.

In the range where the absorption coefficient α for a suspension of hemoglobin in water is related linearly to the concentration [Hb] we can write

$$\alpha - \alpha_w = A\,[Hb]\ ; \tag{13.59}$$

here A is a constant which depends on the frequency and α_w is the absorption coefficient for water. At a frequency of 1 MHz it is found that A has the value

$$A = 0.12\ \frac{\text{nepers/cm}}{\text{gm/ml}}\ , \quad \text{(Hemoglobin, 1 MHz)} \tag{13.60}$$

in the density range 0 to 0.15 gm/ml; A is the <u>specific absorption coefficient</u>, in this instance for hemoglobin in water at 1 MHz.

Kremkau (1972) and Kremkau and Carstensen (1972) made comparisons (at a frequency of 10 MHz) of the specific absorption for low-concentration Hb solutions with that for other solutes in water. For example, experiments

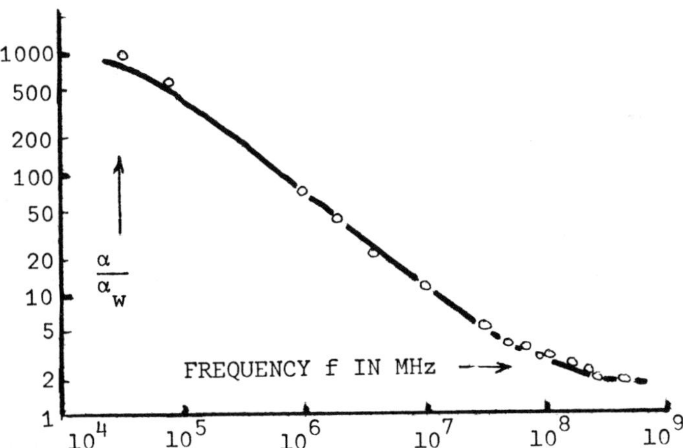

Fig. 13.5 Absorption coefficient α in an aqueous solution of bovine hemoglobin at concentration 0.15 g/ml and 25°C . Ordinate is α/α_w , where α_w is the absorption coefficient in water. Circles show measured values. See Section 13.5.3 . From Dunn, Edmonds and Fry (1969, Reference 13.3).

were done in which Hb was "digested" by an enzyme pronase, so that the molecule whose initial molecular weight is 68,000 was separated into fragments most of which were of molecular weight less than 10,000; the specific absorption was then found to be only a little greater than one-half that for Hb solutions of low concentration. For some reason fragmenting the Hb molecule decreases the absorption coefficient.

The same tendency was exhibited even more when a hemoglobin solution was replaced by a solution of amino acids in which the proportions were about the same as for Hb. The specific absorption coefficient for the solution of separated amino acids was only one-half to one-fourth as great as for an Hb solution.

Kremkau includes the above results under the generalization that the specific absorption caused by macromolecules increases with increasing level of organization. Other examples are cited, for various media. For polysaccharides it had been shown by Hawley and Dunn (1969) that the specific absorption increases as the molecular weight (and thus chain length) increases.

For Hb solutions in which the concentration [Hb] is increased up to 0.6 gm/ml the ratio of $\alpha-\alpha_w$ to the concentration increases up to four times the value at low concentrations. This means, of course, that the linear law in Eq. (13.59) does not apply at high concentrations; but an "effective specific absorption coefficient" can nevertheless be defined, namely, as the ratio of $(\alpha-\alpha_w)$ to the concentration.

13.5.4 Tissues

Intact tissues such as muscle, liver and brain contain a fairly high ratio of "solid" content (mostly protein) to water content. For these the effective specific absorption coefficient is comparable to that for concentrated Hb solutions, suggesting that the "concentration-effect" or "effect of organization" applies also to these complex media. At 1 MHz the absorption coefficient is roughly 0.5 db/cm in fat, 1 db/cm in liver and brain and 2 db/cm in muscle; these values may be converted to nepers/cm by use of Eq. (13.6). Results showing an increase of effective specific absorption with "organization" are attributed by Kremkau and Carstensen to relaxation phenomena, which are not yet identified, but which occur when macromolecules are in high concentration.

Bone is a special tissue in which α is very high, as high as 1 neper/cm at 1 MHz. In lung tissue α is even higher, about 4 nepers/cm at 1 MHz. The high loss in lung is undoubtedly associated with its high air content (density about 0.4 g/ml; see Bauld and Schwan (1974) and Dunn (1974), all cited in Reference 13.12.

13.6 SHEAR WAVES

13.6.1 Introduction

Another kind of field is represented by the situation where the motions are, as before, along x but now vary only in the y direction. We refer to this as a field of <u>simple shear</u>. In such a field S_{12} is the only stress component which acts along x ; the volume force f_x is now $\partial S_{12}/\partial y$. Equation (13.12) is replaced by

$$\frac{\partial S_{12}}{\partial y} = \rho_o \frac{\partial^2 \xi}{\partial t^2} \qquad (13.61)$$

We consider a field in which ξ is a function of y and t. Assuming the time dependence is sinusoidal with angular frequency ω and using complex-number methods a suitable expression for ξ is

$$\xi = \phi(y)\, e^{j\omega t} , \qquad (13.62)$$

where $\phi(y)$ is a function of y . The strain $\partial\xi/\partial y$ is then given by

$$\frac{\partial \xi}{\partial y} = \frac{\partial \phi}{\partial y} e^{j\omega t} ; \qquad (13.63$$

where $\partial\phi/\partial y$ is time-independent.

13.6.2 Isotropic Solid

When the constitutive equation in Eq. (4.2) applies we have

$$S_{12} = \mu \frac{\partial \xi}{\partial y} = \mu \frac{\partial \phi}{\partial y} e^{j\omega t} , \qquad (13.64)$$

in which we have used Eq. (13.63). Since μ is real and positive we see that the stress S_{12} is in phase with the strain $\frac{\partial \xi}{\partial y}$ for a linear isotropic solid. Substituting from Eq. (13.64) into Eq. (13.61) we obtain

$$\frac{\partial^2 \xi}{\partial y^2} = \frac{1}{c^2} \frac{\partial^2 \xi}{\partial t^2} ; \quad c^2 = \frac{\mu}{\rho} . \qquad (13.65)$$

This "wave" equation has the same form as Eq. (13.14) and leads to Eq. (13.40) since ξ varies sinusoidally with time. Hence the solution for a travelling wave in the positive y direction can be written just as in Eqs. (13.41) and (13.42), but with x replaced by y. In the isotropic elastic solid a possible displacement field is a travelling plane wave in which the velocity c is $\sqrt{\mu/\rho}$ and the attenuation is zero.

13.6.3 Newtonian Fluid

When the constitutive relation in Eq. (4.24) applies S_{12} is given by $\eta \partial \dot{\xi}/\partial y$ where the dot, as before, represents differentiation with respect to time. Applying Eq. (13.62) the velocity $\dot{\xi}$ can be written as $j\omega\xi$ and we obtain

$$S_{12} = j\omega\eta \frac{\partial \xi}{\partial y} , \qquad (13.66)$$

as the relation between stress and strain for a Newtonian fluid. Since j can be written as $e^{j\pi/2}$ we see that Eq. (13.66) can be written in the form

$$S_{12} = \omega\eta \frac{\partial \phi}{\partial y} e^{j(\omega t + \pi/2)} . \qquad (13.67)$$

Comparing the phase of S_{12} in Eq. (13.67) with that for $\frac{\partial \xi}{\partial y}$ in Eq. (13.63) we see that for a Newtonian fluid in a field of simple shear the stress and strain differ in phase by $\pi/2$ or 90°. Substituting from Eq. (13.66) into Eq. (13.61) we obtain an equation which can be written

$$\frac{\partial^2 \xi}{\partial y^2} + \kappa^2 \xi = 0 , \tag{13.68}$$

in the form of Eq. (13.40), with κ^2 given by

$$\kappa^2 = \frac{\omega\rho_o}{j\eta} = (1-j)^2 \frac{\omega\rho_o}{2\eta} . \tag{13.69}$$

From Eq. (13.69) an expression for κ can be written as

$$\kappa = \beta - j\beta ; \quad \beta = [\frac{\omega\rho_o}{2\eta}]^{\frac{1}{2}} . \tag{13.70}$$

For a solution we can use the form of Eq. (13.45) but with K replaced by κ and x by y ; this becomes

$$\xi = Ae^{-\beta y} e^{j(\omega t - \beta y)} . \tag{13.71}$$

This represents a highly damped wave; the phase velocity is ω/β and the attenuation coefficient is β. The reciprocal quantity β^{-1} is of some significance; it is given by

$$\beta^{-1} = [\frac{2\eta}{\omega\rho_o}]^{\frac{1}{2}} \tag{13.72}$$

and has the dimensions of length. It is the distance in which the shear wave amplitude is reduced by a factor of e^{-1}, or by 1 neper, and is often called the <u>boundary layer thickness.</u> In water-like fluids and at ultrasonic frequencies it is very small, of the order of microns. For example, in water at room temperature and at a frequency of 1 MHz, β^{-1} is approximately 0.56 μm .

13.6.4 Shear Impedance; The BEL Equation

A biological medium is typically neither an elastic solid nor a Newtonian fluid. One consequence of this fact is that the stress S_{12} is neither exactly in phase with the strain $\partial\xi/\partial y$ (as for a solid) nor exactly 90°

out of phase with it (as for a fluid). It has in fact been found that stress-strain relationships vary considerably from one material to another. Experimental data for shear waves set up in various media provide useful means of characterizing the media.

A surprising finding from shear wave studies is that media ordinarily thought of as liquids show solid-like behavior at high frequencies. Lamb, Barlow and co-workers have made extensive studies in a wide variety of fluids, using a number of different experimental techniques. For brevity we consider their basic setup to consist of an arrangement for setting a plane boundary into oscillation along the x direction in its own plane, the yz plane. This oscillating plane acts a source and generates shear waves which travel in the positive y direction, into a medium of interest. Provision is made for simultaneously measuring the velocity $\dot{\xi}$ of the source plane and the stress S_{12} it exerts on the adjoining medium. For oscillations of angular frequency ω it is possible in this way to determine the impedance Z, defined as the ratio of stress to velocity; that is,

$$Z = \frac{S_{12}}{-\dot{\xi}} = \frac{S_{12}}{-j\omega\xi} . \tag{13.73}$$

By the convention used here Z is positive if the source plane exerts force (on the medium) in the direction of $\dot{\xi}$. For an elastic solid it follows from previous results (Prob. 13.15) that the impedance is Z_S , given by

$$Z_S = -\frac{\mu\partial\xi/\partial y}{j\omega\xi} = (\rho\mu)^{\frac{1}{2}} ; \tag{13.74}$$

here the impedance is real (meaning that the stress and velocity are in phase) and independent of frequency. For a Newtonian liquid it can be shown (Prob. 13.16) that

$$Z_N = -\frac{\eta\partial\dot{\xi}/\partial y}{j\omega\xi} = (1 + j)(\pi f\rho_o\eta)^{\frac{1}{2}} ; \tag{13.75}$$

here the real and imaginary parts of the impedance are equal (meaning that the stress and velocity differ in phase by 45°) and both parts are proportional to $\sqrt{f}$.

A real medium is neither pure liquid nor pure solid, but its impedance is often given approximately by either Eq. (13.74) or Eq. (13.75) in some range of conditions. Many liquids behave in Newtonian fashion at low frequencies, but acquire solid characteristics at high frequencies. For a number of viscous nonpolymerized liquids, Barlow, Erginsav and Lamb (1967) measured the impedance over a wide range of frequency and found excellent fit of their data to an empirical formula where Z, here called Z_L , is given by

$$\frac{1}{Z_L} = \frac{1}{Z_S} + \frac{1}{Z_N} \ . \tag{13.76}$$

In Eq. (13.76), known as the BEL equation, Z_N is given by Eq. (13.75) and Z_S by Eq. (13.74), except that in the latter equation μ is replaced by μ_∞ , a constant which can be interpreted as the effective value of μ at very high frequencies. (See Section 13.6.5). At low frequency the term $1/Z_N$ dominates on the right hand side of Eq. (13.76) and Z_L is approximately equal to the impedance Z_N for a Newtonian liquid of viscosity η . Hence the parameter η in the BEL equation is the ordinary ("static") value of the shear viscosity coefficient, and is determined from impedance data at low frequency; alternatively, η can be determined by other methods.

On the other hand, at high frequencies Z_L becomes approximately equal to Z_S ; the parameter μ_∞ is determined from the impedance data at high frequency, when the "liquid" behaves like a solid. By this unique method it is found that μ_∞ is of the order of 10^{10} dyn/cm^2 for many liquids, and therefore is in the same range as μ for representative solids.

Initially the BEL equation was arrived at on an empirical basis, that is, as a convenient way of fitting experimental results. Later it was found possible to find a satisfactory physical explanation for the equation, in terms of relaxation phenomena involved in viscous flow. (Phillips, Barlow and Lamb, 1972). We shall not attempt to discuss the relaxation theory here, but instead refer readers to the original paper. Implications of the BEL equation are discussed somewhat further in the next section, in terms of other parameters.

13.6.5 Complex Moduli

In the BEL equation, Eq. (13.76), emphasis is given to the impedance Z (or Z_L) as a parameter for characterizing the mechanical behavior of a given medium. Another parameter much in use is the shear rigidity modulus μ^*; for the kind of situation under consideration this is defined as

$$\mu^* = \frac{S_{12}}{\partial\xi/\partial y} , \tag{13.77}$$

that is, as the stress/strain ratio. For an elastic solid Eq. (13.64) applies and μ^* becomes the usual modulus μ . For a Newtonian liquid we see from Eq. (13.66) that μ^* becomes the pure imaginary quantity $j\omega\eta$. In general neither Eq. (13.64) nor Eq. (13.66) applies but μ^* can always be written in either of the forms:

$$\mu^* = \mu' + j\mu'' \tag{13.77a}$$

or

$$\mu^* = |\mu^*|\ e^{j\delta} . \tag{13.77b}$$

Here $|\mu^*|$ is the magnitude of μ^* and hence gives the magnitude of the stress/strain ratio; the angle δ gives the phase angle between stress and strain. Relations between μ', μ'', $|\mu^*|$ and δ are taken up in Prob. 13.17 ; it is also found that a simple relation holds between μ^* and Z , namely

$$Z^2 = \rho\mu^* . \tag{13.78}$$

The loss tangent is a quantity of some significance; it is defined as tan δ , given by

$$\tan\delta = \mu''/\mu' . \tag{13.79}$$

It is also found convenient to refer to the shear compliance J^* defined as the reciprocal of μ^*, that is, by

$$J^* = \frac{1}{\mu^*} . \tag{13.80}$$

Expressions for the real and imaginary parts of J^* are taken up in Prob. 13.18.

An alternative approach to characterizing a medium is to regard Eq. (13.66) as basic, but allow the viscosity coefficient to be complex. Thus we define η^* as

$$\eta^* = \frac{S_{12}}{\partial\dot{\xi}/\partial y} , \tag{13.81}$$

that is, as the complex ratio of stress to rate-of-strain. For a Newtonian liquid Eq. (13.66) applies and η^* becomes the real coefficient η . For an elastic solid Eq. (13.64) applies and, since $\dot{\xi}$ can be written as $j\omega\xi$, the coefficient η^* becomes the pure imaginary quantity $(\mu/j\omega)$. In general we see that

$$\eta^* = \frac{\mu^*}{j\omega} . \tag{13.82}$$

Letting η' and $(-\eta'')$ be, respectively, the real and imaginary parts of η^* we see that

$$\eta^* = \eta' - j\eta'' ,$$

$$\eta' = \frac{\mu''}{\omega} ; \qquad \eta'' = \frac{\mu'}{\omega} . \tag{13.83}$$

Using these relationships Eq. (13.77a) is sometimes rewritten as

$$\mu^* = \mu' + j\omega\eta' . \tag{13.84}$$

In the literature on mechanical behavior of various kinds of media there is much reference to the quantities μ', μ'', η', η'', etc. In general these quantities are not really constant, but are functions of the frequency. (In contrast, the quantities η and μ_∞ in the BEL equation, Eq. (13.76), are frequency independent). For the BEL model explicit expressions for μ', μ'' and η' have been derived (Lamb, 1967) based on relations developed above. These expressions are somewhat lengthy and will not be quoted here; instead plots of the functions are shown in Fig. 13.6 . We see that η' is equal to the ordinary value η at low frequencies but falls monotonically to zero as the frequency rises. On the other hand μ' is zero at low frequencies but rises to the value μ_∞ at high frequencies. The quantities (μ'/μ_∞) and (η'/η) are equal, with value about 0.2, when the angular frequency ω is equal to (μ_∞/η) . Also shown on the graph is the quantity μ''; this is zero at both high and low frequencies, but passes through a maximum of about 0.25 μ_∞ at a frequency somewhat greater than (μ_∞/η) . However, we are not likely to be concerned with behavior at frequencies near or above (μ_∞/η), unless the viscosity coefficient η is very high. For suppose the highest frequency f available is 80 MHz; the corresponding value of ω is $2\pi f$ or about 5×10^8 sec^{-1} which we set equal to μ_∞/η . For a typical value of 10^{10} dyn/cm^2 for μ_∞ we then find that η must be at least 20 P, in order to correspond to that part of Fig. 13.6 where the μ' and η' curves cross.

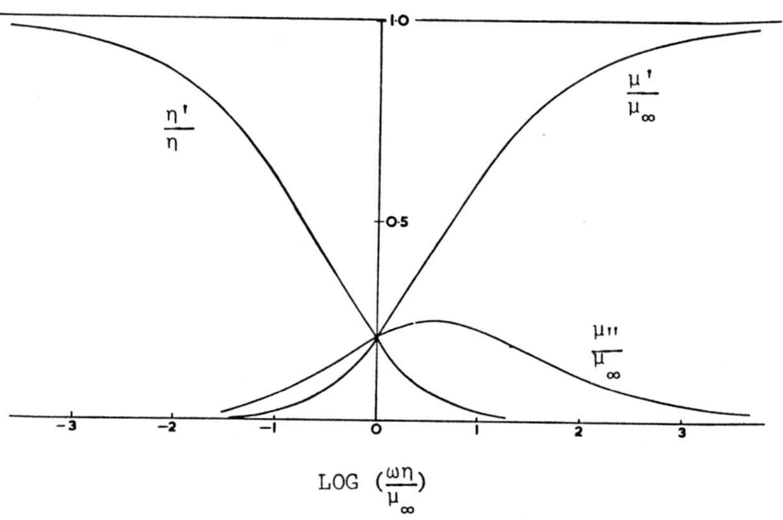

Fig. 13.6 Plots of elastic and viscous coefficients against a nondimensional parameter proportional to frequency, based on BEL theory, discussed in Sections 13.6.4 and 13.6.5 . From Lamb (1967-8, Reference 13.15) .

13.7 VISCOELASTICITY IN POLYMER SOLUTIONS

We consider here a network of flexible macromolecules somewhat as in Section 6.4.5 . Each macromolecule is a chain made up of Z monomer units, each of "effective" length $\underline{a}$, several times the actual monomer length. There are c chains per unit volume immersed in solvent and each is continually rearranging its configuration by random motion. These Brownian motions are opposed by viscous forces, the frictional constant for the monomer unit (Cf "f" in Section 6.2) being ζ_o .

When a dilute solution of these chain molecules is subject to shearing stress each molecule will tend to be slightly perturbed from its most probable shape. Brownian motion will tend to restore the original shape and viscoelasticity comes about through interaction of these two tendencies. In theory developed by Rouse (1953) the response of a solution to oscillatory shear is analyzed. We shall not attempt here to describe the steps involved in this theory, but shall only cite a simplified form of the results. The reader is referred to Rouse's original article for details and to a review by Ferry (1961) for a critical discussion of it.

Rouse arrived at the following expressions for μ' and η' in an oscillating field of shear of angular frequency ω :

$$\mu' = ckT \sum_n \frac{\omega^2 \tau_n^2}{1+\omega^2 \tau_n^2}$$

$$\eta' = ckT \sum_n \frac{\tau_i}{1+\omega^2 \tau_n^2} \quad . \tag{13.85}$$

Here k is the Boltzman constant, T is the Kelvin temperature and τ_n is a relaxation time, given by

$$\tau_n = \frac{a^2 Z^2 \zeta_o}{6\pi^2 n^2 kT} \quad ; \tag{13.86}$$

in the summation n takes on the values n = 1, 2, etc. We consider only the dominant contributions to μ' and η' , namely, those for n = 1; these can be written

$$\mu' = \mu_\infty \frac{\omega^2 \tau_1^2}{1 + \omega^2 \tau_1^2} \tag{13.87a}$$

$$\eta' = \frac{\eta}{1 + \omega^2 \tau_1^2} ,$$

where

$$\mu_\infty = ckT \; ; \quad \eta = ckT\,\tau_1 , \tag{13.87b}$$

and the relaxation time is

$$\tau_1 = \frac{a^2 Z^2 \zeta_o}{6\pi^2 kT} . \tag{13.87c}$$

Plots of μ' and η' against frequency are shown in Fig. 13.7. We see a similarity in the plots to those for the BEL equation in Fig. 13.6. But there are also differences: changes in μ' and η' for the Rouse terms occur over a much smaller range of frequency that for the BEL theory.

The expression for μ_∞ given in Eqs. (13.87) is identical in form to that for the rigidity modulus of a rubber-like solid; see Eq. (12.36). Also the expression for η in Eqs. (13.87) is related to one arrived at by Debye for flexible coils, as discussed in Section 6.4.5. It is interesting that the frequency dependence for μ' and η' in Eqs. (13.87) are just those expected for a Maxwell model. (Prob. 13.19).

The predictions of the Rouse theory have been compared with experimental data on μ' and η' in dilute suspensions of synthetic polymers and good agreement is found (Ferry, 1961).

13.8 VISCOELASTICITY OF BLOOD

Pulsations and unsteady flow are occurring normally in the animal circulatory system. So it is natural that experiments should be done on mechanical behavior of blood under oscillatory conditions. Thurston (1972) carried out investigations in which blood was caused to flow to and fro in circular tubes at a frequency of 10 Hz. Theory for his method is outlined in his 1972 paper and earlier ones. We confine our attention here to some of his results.

Measurements lead to data on the complex shear viscosity coefficient η^* discussed in Section 13.6.5 . See Fig. 13.8 . For blood of 50% hematocrit the real part η' of the viscosity is found to be 0.065 P when the velocity gradient is very low, about 1-2 sec^{-1} or less. It was found that η' decreases somewhat with increasing velocity gradient, approaching about 0.05 P for a gradient of 100 sec^{-1} ; this latter value is in fair agreement with reported values of η in steady flow measurements. (See Section 7.4).

If blood were a Newtonian fluid the viscosity coefficient would be real and η'' would be zero. Indeed at the higher velocity gradients η'' was found to be very small, only about 0.002 P at gradients of 100 sec^{-1} . However Thurston found η'' to rise with decreasing gradient, reaching a value of more than 0.02 P in the low gradient limit. According to Eqs. (13.83) this corresponds to the elastic shear modulus having a real part $\mu' = \omega\eta''$, equal to 1.25 dyn/cm^2 (at 10 Hz).

These findings appear to be consistent with those reported in Section 7.4 , especially with the finding of MGCSBW that blood has solid-like characteristics when it is nearly stationary, but becomes Newtonian under flow conditions such as those existing in most of the circulatory system.

At very low hematocrits (no more than a few per cent) the blood behaves in a Newtonian manner, the intrinsic viscosity being 1.7, somewhat less than the value expected if the red cells were rigid spheres. As the hematocrit increases beyond about 20% both η' and η'' increase rapidly, becoming equal and taking on values greater than 0.2 P as the hematocrit approaches 100%.

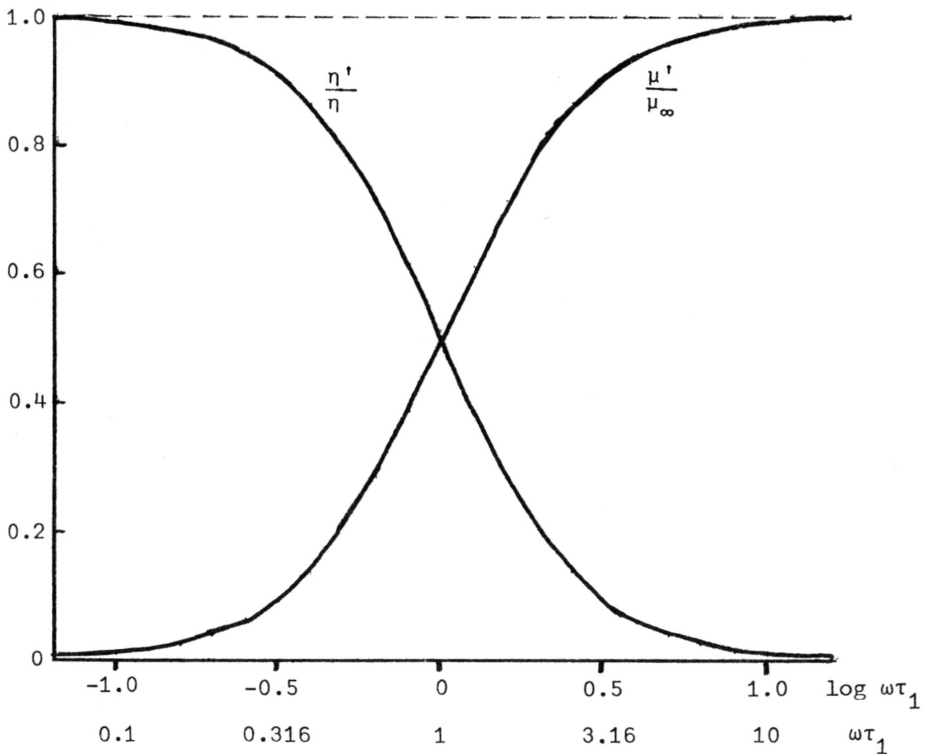

Fig. 13.7 Plots of elastic and viscous parameters against $\omega\tau_1$ and log $\omega\tau_1$, based on Rouse's theory for chain molecules, Eqs. (13.87) . The dimensionless ratios μ'/μ_∞ , η'/η and $\omega\tau_1$ are defined in Section 13.7 .

Fig. 13.8 Plots of the real part η' and imaginary part η'' of the viscosity at a frequency of 10 Hz for human blood. The abscissa G_o is the amplitude of the velocity gradient. Shown for reference is the viscosity coefficient η_{st} measured under steady-flow conditions. See Section 13.8 . From Thurston (1972, Reference 13.18).

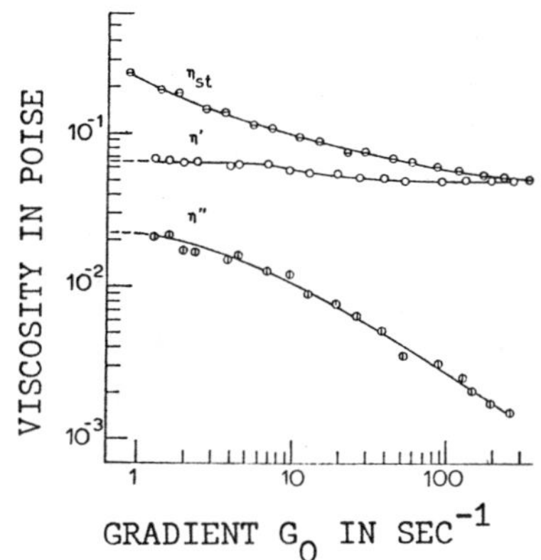

13.9 VISCOELASTICITY IN OTHER BIOMATERIALS

13.9.1 Nuclear Gels

Chromatin is the chief constituent of chromosomes (in eukaryotes, i.e., cells with true nuclei) and is a particularly interesting constituent of the cell because it contains most of the cellular genes. Nuclear chromatin is made up of the nucleic acids DNA and RNA, and the proteins histone and R.P. ("residual protein"). It has been shown by Dounce (1971) that when isolated nuclei are suspended in suitable media stable gels are formed; viscoelastic measurements on these gels lead to useful information on the chromatin and thus on chromosome structure.

A basic question exists on how DNA is distributed along a chromosome. It has been proposed that the DNA extends from one end of a chromosome to the other without interruption, even for chromosomes that are several centimeters or more in length. Dounce offers an alternative hypothesis: the DNA exists in molecules or tracts whose molecular weight is perhaps no more than 10^7 daltons (corresponding to a length of 5 μm), and that in the chromosome these are separated by tracts of R.P. See Fig. 13.9. The fact that gels are formed is regarded as evidence for proteins which protrude from a chromosome and form linkages between molecular chains. That these protruding proteins are not histone is concluded from chemical evidence that histones are very firmly bound to the DNA, as suggested in Fig. 13.9. Further proof comes from mechanical measurements on nuclear gels.

Measurements of the elastic shear modulus μ were made on nuclear gels containing histone and on several media from which the histone was removed. Significantly it was found that even the histone-free medium showed gel-like properties; for a concentration of 1 mg DNA per ml the modulus μ was found to be about 50 dyn/cm^2. While this is less than the modulus for the histone-containing gel the ratio between the two moduli approached unity as the DNA concentration was decreased to ever smaller values. Dounce feels that "these data dispose of any claim that extractable histone is the protein factor necessary for gel formation". Evidence was thus obtained from these mechanical experiments in support of the model in Fig. 13.9.

13.9.2 Viscoelasticity of Intervertebral Discs

It is suggested by Fitzgerald and Freeland (1971) that "An adequate description of the viscoelastic behaviour of living matter might ... lead to significant correlations between cellular and subcellular structures of tissues, and perhaps even to a quantitative measurement of the mechanical differences between healthy and diseased tissue." Some possibilities are seen in measurements by the authors on the discs (from adult dogs) which provide cushioning and flexibility between bones in the spinal column. Fig. 13.10.

These authors subjected discs to shearing at frequencies in the audio range, from 25 to 2500 Hz and made determinations of the complex shear compliance J^* defined in Eq. (13.80) as the reciprocal of the complex shear elastic modulus μ^* . It is convenient to write J^* in the form

$$J^* = J' - jJ'' \; ; \tag{13.88}$$

if desired J' and J'' can be expressed in terms of μ' and μ'' .(Prob. 13.18

Typical results for J' and J'' are shown in Fig. 13.11 where J' and J'' are plotted against frequency, on logarithmic scales. There is a general tendency for J' to decrease with increasing frequency; the authors attribute this to retardation phenomena. We have discussed retardation in connection with the Kelvin-Voigt model (Sections 6.6.1 and 6.7.1); for a medium to which the model applies the stress is related to the strain as in Eq. (6.70); here G_K is equivalent to the quantity $\partial\xi/\partial y$. In oscillations with angular frequency ω we consider ξ as proportional to the complex function $e^{j\omega t}$. Then, conforming to notation in Section 13.6.5 , Eq. (6.70) can be rewritten as

$$S_{12} = (\mu' + j\omega\eta') \frac{\partial\xi}{\partial y} = \mu^* \frac{\partial\xi}{\partial y} \, . \tag{13.89}$$

Here μ^* is the complex shear elastic modulus , its reciprocal (see Eqs. (13.80)) is given by

$$J^* = \frac{1}{\mu' + j\omega\eta'} , \tag{13.90}$$

and the quantities J' and J'' are given (Prob. 13.20) by

$$J' = \frac{\mu'}{(\mu')^2 + \omega^2(\eta')^2} = \frac{(\mu')^{-1}}{1 + \omega^2\tau^2}$$
$$J'' = \frac{\omega\eta}{(\mu')^2 + \omega^2(\eta')^2} = \frac{\omega\tau(\mu')^{-1}}{1 + \omega^2\tau^2} , \tag{13.91}$$

where $\tau = \eta'/\mu'$ is the retardation time.

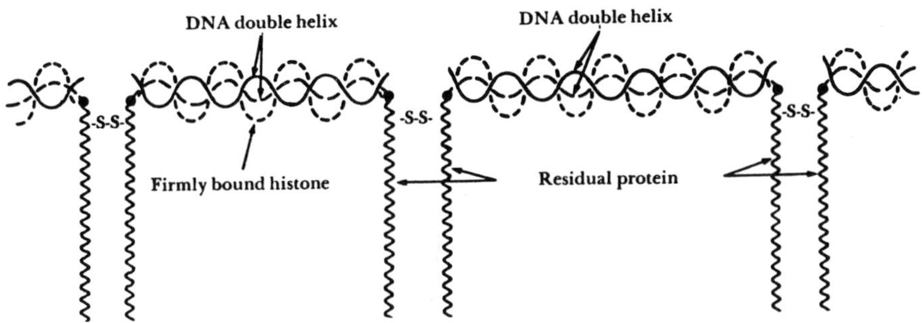

Fig. 13.9 Schematic diagram of model proposed for structure of chromosomal fibers . From Dounce (1971, Reference 13.19) .

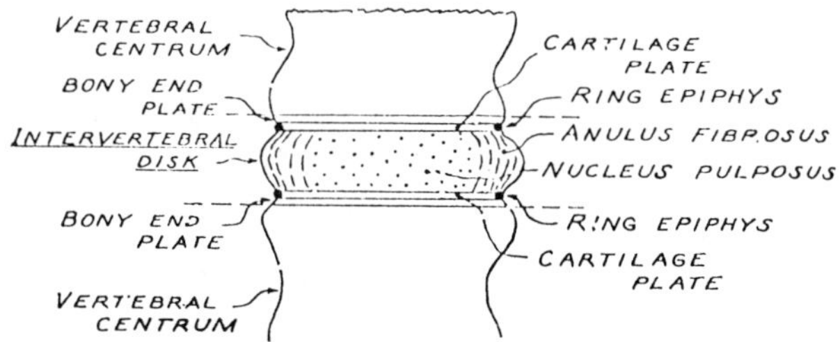

Fig. 13.10 Diagram of intervertebral disk . From Fitzgerald and Freeland (1971, Reference 13.20) .

Assuming μ' and η' are frequency-independent, it is clear from Eqs. (13.91) that for the Kelvin-Voigt "retardation" medium the elastic compliance J' is a decreasing function of frequency. Also J" is seen to increase with frequency at low frequencies, pass through a maximum when $\omega = \tau^{-1}$, and decrease with frequency for ω greater than τ^{-1}. The broad trends in Fig. 13.11 may be roughly as expected for a Kelvin-Voigt medium, assuming the retardation frequency τ^{-1} is beyond the experimental range. But the peaks in J" at about 300 and 2000 Hz, resp., and the corresponding changes in J' are too sharp to be

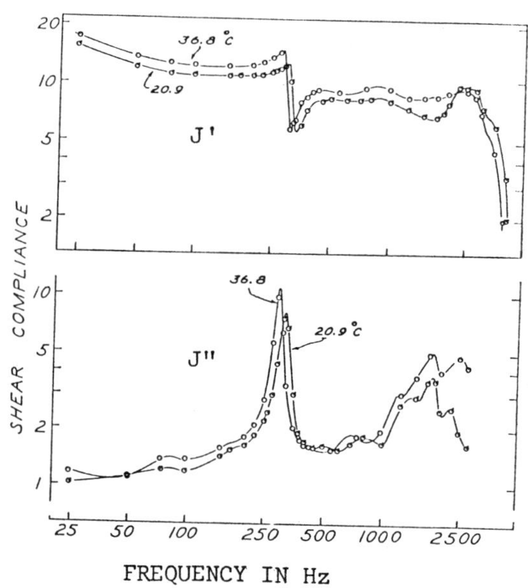

Fig. 13.11 Log-log plot of real part J' and imaginary part (the negative of J") of shear compliance versus frequency for (dog) in-vertebral disks at two temperatures. Units of J' and J" are cm^2/dyn when multiplied by 10^9. See Section 13.9.2. From Fitzgerald and Freeland (1971, Reference 13.20).

accounted for by retardation. Instead these are ascribed by the authors to resonances of an unusual kind, which will be discussed in a later section.

In other data (not shown here) the authors show results for natural and synthetic rubber materials for comparison. The general tendency for J' to decrease with increasing ω occurs for these, as for the biological material. However the natural rubbers are more compliant than the living disc, while the synthetic ones are less compliant. The comparisons are especially significant since replacements for degenerate disks are now being considered in medical practice.

That mechanical measurements can provide a sensitive index of the physiological state of tissue is shown convincingly by the data in Table 13.2 . These show J' and J" for an intervertebrate disc as a function of time after death. We see that in 5 hours J" has decreased by a factor of two and J" by a factor of three. The authors interpret these data as providing a quantitative description of <u>rigor mortis</u>.

13.9.3 Resonances in Bone: Medical Applications

We have mentioned before that sound velocity measurements lead to information on elastic constants. In medical practice this fact is being made use of in (nondestructive!) means for testing bone fragility. For it has been shown that the tensile strength of a given kind of bone is an increasing function of its Young's Modulus E and, in fact, that the strength can be predicted if E is known. (B. S. Mather, 1967). Methods for determining E for bone in living subjects are discussed by Jurist (1970).

Briefly, these methods depend on setting up resonance vibrations in a bone by placing it in contact with a loudspeaker driver unit. In the adult ulna these resonances occur in the region 150-300 Hz, approximately as expected for the fundamental transverse mode. While the exact nature of the vibration pattern may not be known reliance is placed on the

TABLE 13.2

Variation of compliance components J' and J" with time T after death for dog intervertebral disks. Frequency 100 Hz. T in hours. J' and J" in 10^{-9} cm^2/dyn. From Fitzgerald and Freeland, 1971, Reference 13.20.

T	J'	J"
0	34.0	8.0
3.0	19.7	4.16
3.3	22.9	4.37
4.0	20.6	3.47
5.0	17.2	2.55
20.2	14.9	2.14
28.7	15.2	2.16
49.5	15.2	2.14
71.5	16.9	2.60

principle that the velocity of sound should be proportional to the product of a characteristic length L (the bone length) and the frequency f_r at which resonance occurs. Comparisons are then made by observing values of $f_r L$ for various subjects. It is found, for example, that $f_r L$ increases with age until about age 20, remains constant for about 10 years, then decreases. Also $f_r L$ is significantly less for osteoporotic and diabetic subjects than for normal ones. The method may therefore prove useful in evaluation of skeletal status. Related methods have also been suggested for determining the degree of union of healing fractures. (Campbell and Jurist, 1971).

13.9.4 Resonance Spectra in Bone: Momentum Wave Modes

Resonances of an entirely different character have been discovered by a group of workers, hereafter referred to as PRPR, in studies of human cancellous bone. (Pugh, Rose, Paul and Radin, 1973). Studying longitudinally propagated waves they made determinations of the complex Young's modulus E^*. To define E^* we consider a wave propagated along the "1" direction; E^* is then defined as the ratio of the stress S_{11} to the strain ε_{11} . (Cf. Eq. 4.20). Measurements were made of the components E' and E'' defined by

$$E^* = E' + jE'' . \tag{13.92}$$

The real part E' is sometimes called the elastic component and E'' the viscous component, by analogy to interpretations given corresponding components of μ^* . (See Section 13.6.5).

The results of PRPR are shown in Fig. 13.12 . Pronounced well-defined minima appear in E' at a series of characteristic frequencies; at these same frequencies corresponding sharp peaks of E'' appear. These findings are astonishing, especially since the characteristic frequencies were found to be the same for ten different samples, and were not dependent on sample size. Careful testing with different materials showed that the frequencies are not characteristic of the equipment, but depend only on the material. Even more remarkable is the fact that the characteristic frequencies can be grouped into two sets such that in each set the frequencies are approximately in the ratios 1:4:9:16, etc. In Table 13.3 the observed characteristic frequencies in both sequences are listed. Also listed are calculated values for each set obtained in the following way: The lowest frequency is set equal to the observed value; others are obtained by multiplying the lowest by 4, 9 and 16, respectively.

In seeking an explanation for these findings the authors (PRPR) turned to theory which had been advanced earlier by Fitzgerald (1966). In this surprising theory ideas from quantum physics and wave mechanics are used

TABLE 13.3

Resonant frequencies (Hz) for fresh human cancellous bone. From Pugh, Rose, Paul and Rodin, 1973, Reference 13.24 . "Obs": observed. "Cal": calculated.

I (^{40}Ca)		II (^{31}P)	
Obs	Cal	Obs	Cal
145	145	203	203
570	580	708	812
1350	1305	1900	1827
2310	2320	3230	3248

to solve problems in macroscopic mechanics. We shall not attempt to develop Fitzgerald's theory, but only note a few aspects of it involving ideas from elementary physics.

Any particle moving with momentum p has associated with it a de Broglie wavelength λ given by

$$\lambda = \frac{h}{p} , \tag{13.93}$$

where h is Planck's constant. The kinetic energy of the particle can be written as $p^2/2m$; a frequency f can be associated with the particle by setting its energy equal to that for a quantum, that is, hf. We obtain then

$$hf = \frac{p^2}{2m} = \frac{h^2}{2m\lambda^2} ,$$

or

$$f = \frac{h}{2m\lambda^2} . \tag{13.94}$$

Fitzgerald has developed theory, much too extensive to be reported here, showing that resonances can be expected in solids containing structures of suitable size relative to λ . Thus if a solid contains periodic structures with dimension S resonances are expected for values of λ such that

$$S = \frac{n\lambda}{2} , \tag{13.95}$$

where n is any positive integer. Combining Eqs. (13.94) and (13.95) we obtain the Fitzgerald equation:

$$f = \frac{hn^2}{8mS^2} . \qquad (13.96)$$

(It should be emphasized that in these simple steps we have not "derived" Eq. (13.96), but have only illustrated some of the ideas involved).

We see immediately that Eq. (13.96) is in agreement with some features of the PRPR results. For given values of m and S there are a series of values for the frequency f in the ratio 1:4:9:16, etc. The authors proposed that the structures involved are lamellae seen in the bone, with thickness about 3 μm . The mass m is assumed to be that of an atom in hydroxyapatite $Ca_{10}(PO_4)_6(OH)_2$. For one set of frequencies the atom is assumed to be ^{31}P and for the other set, ^{40}Ca . The atomic mass ratio of ^{31}P relative to ^{40}Ca is 0.773 which agrees reasonably well with the

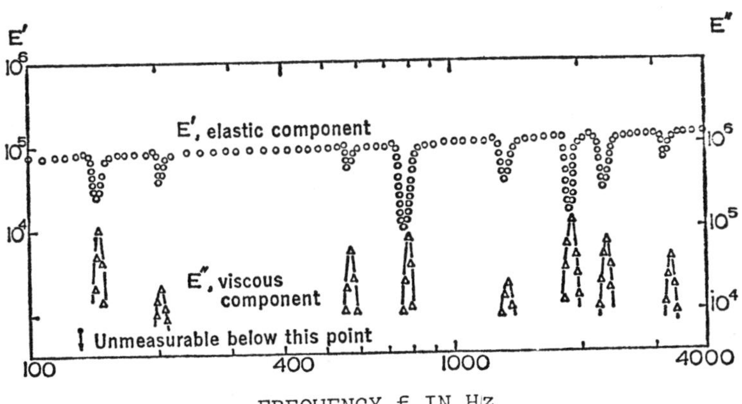

Fig. 13.12 Typical data for real part E' and imaginary part E" of the complex Young's modulus for fresh human cancellous bone. See Section 13.9.4. From Pugh, Rose, Paul and Rodin (1973, Reference 13.24). Copyright 1973 by the American Association for the Advancement of Science.

frequency ratios (0.714, 0.805, 0.711, 0.715) of corresponding members of the two observed series. It is therefore assumed that the set I of characteristic frequencies in Table 13.3 is for the ^{40}Ca atoms and set II is for ^{31}P.

Given the mass m and the value of f for, say, n = 1 in a given set of observed characteristic frequencies the value of S can be calculated. This was done by PRPR, obtaining values of 2.93 μm for the ^{40}Ca series and 2.82 μm for the ^{31}P series, both reasonably close to 3 μm, the observed lamellar thickness.

In summary, the PRPR data provide clear evidence of resonances in cancellous bone, whose frequency relationships are in excellent agreement with the Fitzgerald equation. However, these are by no means the first data to be reported on such resonances. The data of Fitzgerald and Freeland (1971) on J' and J" for intervertebral disks (Fig. 13.11) show evidence of resonances, which the authors find are in agreement with Eq. (13.96) . Other examples, for nonbiological material, are cited by Fitzgerald (1966). But it needs to be pointed out that the Fitzgerald theory has been the subject of controversy, and negative results on the resonances have been reported by other investigators.

In the future we can expect that further experiments will be done in attempts to repeat or extend these results, and also that further development of the theory will be carried out.

PROBLEMS

13.1 (a) Referring to the expression for a travelling wave in Eqs. (13.1), derive an expression for x_n, a value of x at which $\xi = 0$.

(b) Differentiate with respect to time to obtain $\dot{x}_n$, the velocity at which the zero moves. Show that $\dot{x}_n$ is identical to c, given in Eq. (13.3).

13.2 Show that the decibel ratio is 8.7 times the neper ratio; that is, derive Eq. (13.6).

13.3 In brain the acoustic attenuation coefficient α is about 1 db/cm at a frequency of 1 MHz. Calculate the distance in which the amplitude decreases by

(i) A factor of 10

(ii) A factor of 100

(iii) A factor of e.

13.4 Compare the terms on the right hand side of Eq. (13.11) for a sound field in which the displacement ξ is given by Eqs. (13.1). Assuming $\alpha A \ll 1$ and $kA \ll 1$, obtain Eq. (13.12).

13.5 (a) Show that the plane wave expression for ξ in Eq. (13.1) is a solution of Eq. (13.18) when α and k are as given in Eqs. (13.19).

(b) Show that when $\varepsilon \ll 1$ the expressions reduce to those in Eqs. (13.22).

13.6 Carry out steps indicated in the text and derive the expression for α in Eq. (13.24).

13.7 Combine Eqs. (13.24) and (13.26) to obtain the expression for B'/η in Eq. (13.27).

13.8 Show that the complex expression for ξ in Eq. (13.37) is a solution of Eq. (13.18), subject to requirements on k and α . Compare the procedure involved here with that involved in Prob. 13.5, and show that required values of k and α are the same.

13.9 Assume the amplitude A in Eq. (13.41) is complex and given by

$$A = A_o e^{j\phi} ,$$

where A_o and ϕ are real. Obtain the real part of the resulting expression for ξ ; show how this differs from the result in Eq. (13.42) and explain what the difference means physically.

13.10 Verify the expression for K^2 in Eq. (13.44) .

13.11 In Eq. (13.44) set K equal to $(k - j\alpha)$ and show that k and α are given by the same expressions as in Eqs. (13.19).

13.12 (a) Starting from Eqs. (13.34) and (13.46), and assuming that θ has the form $Ae^{j\omega t}$ (where A is independent of time) obtain the expression for θ in Eq. (13.47).

(b) Show that the expression can be rewritten in the form of Eq. (13.48).

13.13 (a) Proceeding from Eq. (13.54) obtain the expressions for α and k in Eqs. (13.55).

(b) Show that when $\omega\tau << 1$ and $\alpha << k$ the expressions reduce to those in Eq. (13.56).

13.14 Carry out indicated steps to derive the expression for B' in Eq. (13.57).

13.15 Show that for an elastic solid the shear impedance Z_S is given by $\sqrt{\rho\mu}$, as in Eq. (13.74).

13.16 Show that for a Newtonian fluid the shear impedance Z_N is given as in Eq. (13.75).

13.17 (a) Show that Eqs. (13.77a) and (13.77b) are equivalent if

$$|\mu^*| = [(\mu')^2 + (\mu'')^2]^{\frac{1}{2}}$$

$$\tan\delta = \mu''/\mu'$$

(b) Also show that Z is equal to $\sqrt{\rho\mu^*}$, in agreement with Eq.(13.78).

13.18 For the complex compliance J* defined in Eq. (13.80), show that it can be written

$$J^* = J' - jJ'' ,$$

where

$$J' = \frac{\mu'}{(\mu')^2+(\mu'')^2}$$

$$J'' = \frac{\mu''}{(\mu')^2+(\mu'')^2}$$

13.19 Referring to Section 6.6.2 on the elastico-viscous model we realize that G_M is the same as $\partial\xi/\partial y$ for this situation. Hence Eqs. (6.79) might be rewritten as

$$\frac{\partial\dot{\xi}}{\partial y} = aS_{12} + b\dot{S}_{12} , \qquad \text{(i)}$$

where a and b are constants.

(a) Show that an analogous equation would relate the velocity $\dot{\xi}$ of extension (analogue to $\partial\dot{\xi}/\partial y$) to the applied force F (analogue to S_{12}) for the Maxwell body pictured in Fig. 6.11.

(b) Considering S_{12} and ξ to vary sinusoidally in time with angular frequency ω (so that the complex representation for each is proportional to $e^{j\omega t}$) put Eq. (i) in the form of Eq. (13.77). Representing the complex rigidity modulus μ^* as in Eq. (13.84) show that μ' and η' have the same frequency dependence as the coefficient derived by Rouse, Eqs. (13.87a).

13.20 Given that the complex compliance J* is given by Eq. (13.90), show that the real part J' and imaginary part -J" are as given in Eq. (13.91).

CHAPTER 14

BIOEFFECTS OF THE PHYSICAL ENVIRONMENT

14.1 INTRODUCTION

We know from ordinary experience that maintenance of normal life requires an environment whose physical parameters lie within certain limits. When these parameters change too much in our surroundings we attempt to protect ourselves by providing a suitable local environment. Thus we are well aware of adverse effects from large changes in temperature or pressure. Hence we heat or air-condition houses to avoid temperatures that are too high or too low. Also, as we ascend in jets or descend in submarines we provide means for maintaining the interior pressure at an ordinary level.

We may also be painfully aware of damage which can arise from tensile, compressional or shearing stresses applied to various parts of the body. Such stresses, whether they arise from skiing falls, automobile accidents, or wrestling matches, can be regarded as another part of our physical environment. Sound is another part of our environment which is mechanical in nature. High level audible sound (from aircraft or other machines, or from rock music) can cause severe damage to the ear and, to a lesser extent, to other parts of the body. Ultrasound, that is, sound whose frequency is above the audible range, is proving increasingly popular in medical diagnostic techniques. In these techniques one or more ultrasonic beams are projected into the body and much useful information is gained by analysis of reflections, or of transmission characteristics. The procedures seem to be harmless under usual conditions but, as the usage becomes widespread and conditions are varied, medical ultrasound becomes another aspect of our environment for which we must show some concern.

In this chapter we take up various topics relating to the physical environment and how it affects biological structures and processes. In treating these topics we shall be concerned primarily with principles, and will emphasize studies with relatively simple systems. In keeping with the theme of this book we will emphasize mechanical properties of the environment, but will include thermal properties.

14.2 ELEVATED TEMPERATURES

14.2.1 Bioeffects of Heating

For most living systems there is a rather narrow range of temperatures which is most favorable. Indices of activity, rate of growth or length of life often show a peak with maximum in the range 0 - 50°C. See Fig. 14.1 . Examples are cited by Johnson, Eyring and Polissar (1954) for such indices as (i) brightness of luminescence from bacteria, (ii) rate of growth of plants and (iii) length of life of water fleas. Biological processes practically cease if the temperature is either much higher, or much lower than the optimum range. The higher animals, including man, normally function at nearly constant body temperature; control mechanisms prevent the temperature from fluctuating more than three or four Centigrade degrees.

But in medicine it is often advantageous to deliberately raise the temperature of selected parts of the body. (Licht, 1965). In physical therapy heat is applied in a controlled manner by means of hotpads, infrared, short-waves, microwaves and ultrasound. It is found that heating is useful for relief of pain, for example, in connection with arthritis or muscular trauma. Healing of wounds may be accelerated by raising the tissue temperature, since this tends to increase blood flow to the area. Ultrasound offers the possibility of applying heat deep within the body, for example, inside a hip joint. (Lehmann, 1965).

Scientific researches on chemical reactions and other processes have contributed much to our understanding of how temperature affects living systems. That chemical reactions are temperature dependent has been explained successfully by reaction-rate theory. (Glasstone, Laidler and Eyring, 1941, Reference 7.7). According to this theory the rate of a reaction is given by an expression of the form

$$\text{Rate} = K\, e^{-E_v/R_g T} , \qquad (14.1)$$

where K and E_v are quantities dependent on the nature of the reaction,

R_g is the gas constant (see Appendix D) and T the Kelvin temperature. Special significance is attached to the "activation energy" E_v which measures the height of an "energy barrier" which must be surmounted in the reaction. Experiments show that E_v is in the neighborhood of 10,000 to 20,000 cal/mole for a typical biochemical reaction (e.g., a digestion reaction) catalyzed by enzymes. (Johnson, et al (1954)). Choosing 15,000 cal/mole for E_v we find, since R_g is 1.99 cal/mole, that the rate increases by a factor of 2.3 when the temperature T increases from 300° to 310°. It is indeed found true that the rate of a typical simple biochemical reaction increases by about a factor (called "Q_{10}") of two to three when the temperature is increased by 10°C .

As noted earlier a life process tends to show a maximum rate at some temperature in the range 0-50°C. This can be explained by assuming two reactions are involved: one, an enzyme-catalyzed biochemical reaction with a normal value for the activation energy; the other involving reversible denaturation of the enzyme, with a much higher activation energy. At temperatures higher than the optimum, the reaction rate is reduced because the enzyme is less effective.

At higher temperatures enzymes and other macromolecules are irreversibly denatured. We saw an example of denaturation in Chapter 7, Fig. 7.4, where a profound change in DNA was observed in the temperature range 80-100°C. Theorists have had considerable success in explaining this alteration of the DNA as a _phase transition_, or as a "melting" process. (Thompson, 1972). Normally, DNA has a helical shape. When the transition occurs hydrogen bonds are broken which are required to maintain the helical structure, and the macromolecule becomes more like a flexible coil. We shall refer to phase transitions again in a later section where effects of hydrodynamic stress are taken up.

Another temperature-dependent process is that of viscous flow. In Section 7.2 we noted that the flow can be thought of as a rate process, governed by theory similar to that for chemical reactions. In Eq. (7.2) an expression for the viscosity coefficient η is given, derived from reaction rate theory. From this it is concluded that η decreases with increasing temperature T, a prediction which is born out in Table 7.1 and 7.3 for water and sucrose solutions. The temperature dependence of viscosity is important in connection with functioning of the circulatory system and, also in connection with transport of ions and other particles across membranes.

14.2.2 Temperature Distribution near an Absorber of Radiation

We now take up a problem which is important when considering effects of radiations of various kinds. Exposure to X-rays, microwaves, ultrasound or other physical agents may cause either desirable or undesireable changes in tissues. Just how these arise depends on the nature of the agent and on exposure conditions. But all radiations produce a certain amount of heat as they pass through matter; if this heat is not removed the temperature of the tissue may rise sufficiently to cause biological change. Specifically, if heat is generated in tissue of density ρ at the rate $\dot{Q}_v$ per unit volume, and is not transported away, the temperature T will increase at the rate

$$\frac{dT}{dt} = \frac{\dot{Q}_v}{\rho c_m}, \tag{14.2}$$

where c_m is the heat capacity per unit mass of tissue. In most situations of biological interest the heat is removed primarily by either convection (e.g., *via* blood flow) or by conduction. The physical principles for these processes are well understood. However the practical problem of predicting the temperature distribution which will exist when a given living system is subject to a given radiation field is difficult in real situations. Some insight into the possibilities is gained by taking up an idealized situation where heat transfer occurs only by conduction. The theory for heat conduction is mathematically similar to that for diffusion, taken up in Chap. 9.

Consider a spherical body of tissue with radius R (which might represent a whole organism, a single organ, or a small portion of tissue) in which heat is generated at the rate $\dot{Q}_v$ per unit volume. We let r measure distance from the center of the sphere. As in other applications of continuum physic we proceed, in principle, by first arriving at statements which apply to a portion of the medium, then obtain a differential equation, and finally seek solutions to the equation. However, we shall leave the derivation of a differential equation as an exercise (Prob. 14.1) and simply quote the result. At points inside the heat-generating sphere (i.e., for $r \leq R$) the applicable equation is

$$\frac{1}{r^2} \frac{\partial}{\partial r} \left(r^2 \frac{\partial T}{\partial r}\right) = -\frac{\dot{Q}_v}{\kappa}, \tag{14.3}$$

where κ is the coefficient of heat conductivity, which we shall assume applies both inside and outside the sphere. Equation (14.3) applies in a temperature field with spherical symmetry. Outside the sphere (for $r > R$) the same equation applies, but with $\dot{Q}_v = 0$. We now seek a suitable function for T(r) which satisfies the applicable differential equation; this function will be the temperature distribution for the situation at hand. Actually we must find two temperature distributions; one for $0 < r < R$ where $\dot{Q}_v$ is constant and nonzero, and the other for $r > R$ where $\dot{Q}_v = 0$. It can be shown by substitution that the following functions are solutions for these two regions:

$$\text{Inside:} \quad T = T_\infty + AR^2 + \tfrac{1}{2} A(R^2 - r^2) \tag{14.4}$$

$$\text{Outside:} \quad T = T_\infty + \frac{AR^3}{r} .$$

Here T_∞ is an arbitrary constant while A is given by

$$A = \frac{\dot{Q}_v}{3\kappa} . \tag{14.5}$$

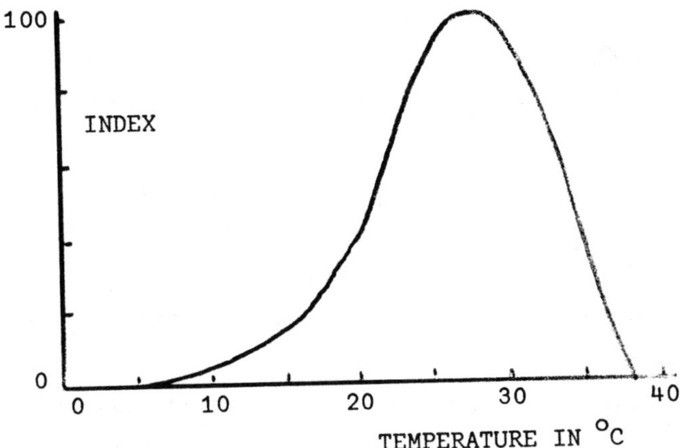

Fig. 14.1 Many indices of biological activity show a peak, such as is shown here, in the temperature range 0 to 50°C . See Section 14.2.1 .

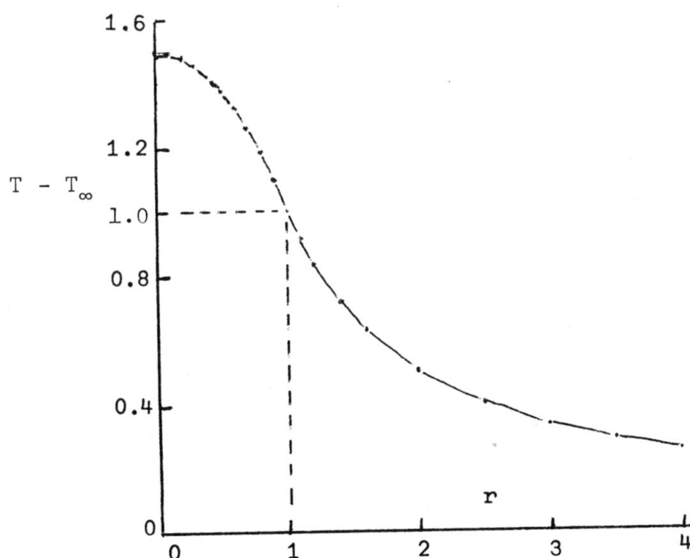

Fig. 14.2 Temperature elevation $T-T_\infty$ versus distance r from the center of a spherical heat source of unit radius. Based on Eqs. (14.4) with R = 1 and A = 1 .

These functions also have the following desirable properties:

(i) At the surface $r = R$ the temperature is continuous; that is, $T(R)$ given by the "inside" solution is equal to $T(R)$ given by the "outside" solution;

(ii) At $r = R$ the heat flow is continuous; that is, $\kappa \partial T/\partial r$ is continuous, or, since κ is the same inside and out, $(\partial T/\partial r)$ has the same value at $r = R$ for both solutions; and

(iii) The temperature approaches a constant value at large distances from the sphere.

The functions in Eqs. (14.4) give the temperature distribution we seek. A plot is shown in Fig. 14.2. We see that the temperature varies parabolically inside the sphere with maximum at the center $r = 0$. Outside the sphere the temperature falls off inversely with the distance r; this may be compared with the fall-off of concentration from a diffusing source, discussed in Section 9.4.4. It is left for an exercise (Prob. 14.2) to show that (as is certainly expected) the total heat flow outward through any surface of radius r is equal to the total rate of heat production within the surface. For $r > R$, that is, outside the heat source, the total heat flow is constant and equal to $4\pi R^3 \dot{Q}_v/3$. At the center and edge of the sphere the temperature has the values

$$T(0) = T_\infty + \frac{\dot{Q}_v}{2\kappa} R^2$$

$$T(R) = T_\infty + \frac{\dot{Q}_v}{3\kappa} R^2 \; . \qquad (14.6)$$

For given Q_v and κ we see that at either $r = 0$ or $r = R$ the temperature ris[e] $(e.g., T(0) - T_\infty)$ is proportional to the square of the heat-source radius [R]. We shall consider a specific numerical example later, in Section 14.9.3 , i[n] discussing heat production by ultrasound.

14.3 REDUCED TEMPERATURES

When the temperature is lower than normal, rates of biochemical reactions and other processes are slowed and, also, phase changes may occur. These alterations may under some circumstances be beneficial to a living system but, as for any change from normal conditions, they are more likely to be harmful. Beneficial effects are seen in medical practice. For example, hypothermia, the lowering of body temperature, is an established technique in open-heart surgery because it leads to reduced oxygen requirements. (Blair, 1964). Cryogenic techniques, in which extreme cold is applied, are increasingly popular in surgery; cataracts are more easily removed if they are first frozen, and tumor tissue can be destroyed cryogenically. (v. Leden and Cahan, 1971). Also methods have been developed for long-term storage of blood and spermatozoa in a frozen state.

Scientific studies (v. Leden and Cahan, 1971; Meryman, 1966 and 1974) of the effects of freezing on biological cells have led to the conclusion that when injury occurs it arises primarily from two causes: (1) changes in concentrations of extracellular solutes and (2) formation of intracellular ice. Of these, the first appears to be more important when cooling takes place slowly, while the latter predominates if the cooling is very rapid. Some aspects of the subject can be understood on the basis of simplified models. We consider both the intracellular and extracellular media to be suspensions of ions, molecules and other "particles" in water. Normally the two media are in osmotic equilibrium. But when an aqueous suspension freezes H_2O is separated from the suspension and forms an ice phase of high purity; the remaining suspension is then of reduced water content or, in other words, has an increased particle concentration c. Now assume, as is usually found true, that freezing occurs first in extracellular space, so that the concentration c is abnormally high in this space. Then the situation is as considered in Sections 2.7 and 11.2 when a cell is immersed in a hypertonic medium; water passes from the inner medium to the outer one, and the cell shrinks. When the cooling takes place slowly enough to permit this process to occur, the cells may not be injured if the shrinking is moderate (even though considerable ice appears in extracellular space) but may be greatly damaged if it is severe. This damage may arise partly from mechanical damage to the cell membrane and intracellular structures, and partly from alterations of biochemical conditions in the shrunken dehydrated cell.

It has been shown by B. J. Luyet and associates (see review by Meryman, Reference 14.8, 1966) that the situation is quite different if the rate of cooling is increased. At an optimum cooling rate there is less time for cell shrinkage and dehydration to occur during the cooling period; once a suitable low temperature is reached these processes occur very slowly, or not at all. Thus damage to cells from these causes is reduced at the increased cooling rate. However, if the temperature falls too rapidly to a low value, ice forms inside the cells and damage occurs because of this.

In connection with the latter phenomenon we note that, in general, the temperature at which ice forms depends on the nature of the medium. It is well known from physical chemistry that the equilibrium freezing point of a suspension of noninteracting particles is lower than that of the pure solvent, by an amount proportional to the particle concentration c . In agreement with this, the equilibrium freezing point for "water" in tissue, with its complement of suspended matter, is less than that for pure water. But freezing occurs by ice formation on pre-existing small ice crystals or "nuclei". If nuclei of suitable size are not present in a suspension it will cool without ice formation, even to temperatures well below the freezing point. The suspension is then not in equilibrium but is in a "supercooled" state; introduction of a nucleus will quickly lead to ice formation. It was calculated by B. Chalmers (see review by Meryman, 1966) that if freezing of water is to occur at $-3^{\circ}C$ an ice nucleus of radius at least 120 A must be present; at $-5^{\circ}C$, $-10^{\circ}C$ and $-20^{\circ}C$ the corresponding radii are about 60 A, 40 A and 20 A resp.

In tissue we assume nuclei exist both inside and outside the cells. That freezing occurs first in extracellular space is probably either because larger nuclei exist there or because conditions are more favorable for growth. Once extracellular freezing begins the cell shrinks through loss of water, if the cooling rate is slow enough to allow this. The concentration of "suspension" inside the cell is then increased and the freezing point depressed so that intracellular ice does not form. But in very rapid cooling, time is not allowed for water to leave the cell, the intracellular freezing point is not lowered and ice forms readily on suitable nuclei.

14.4 ELEVATED PRESSURES

Organisms have been found living in the sea even to depths of 6000 m ; from Eq. (5.16) we find that the hydrostatic pressure is about 600 atm at this depth. That life proceeds under these conditions seems surprising since we are acquainted with harmful effects to both living and nonliving systems arising from stresses much smaller than this. However the stress field in a hydrostatic situation has unique characteristics, discussed in Section 5.2. Here the stress tensor is isotropic; this fact leads to the conclusion that the stress is normal to <u>any plane</u>. Hence at all points on the surface of an immersed biological cell the stress is normal to the surface. Also this stress, that is, the hydrostatic pressure, is essentially the same at all points of an object as small as a biological cell. In short, the hydrostatic pressure field is one in which the stress on a small object is applied uniformly, and always along the inward normal. Because of this it is perhaps reasonable that such a stress field is less damaging to a cell than one in which stresses are applied tangentially to its surface, or along its outward normal, or in an unequal manner. (We consider other stress fields in Sections 14.6 and 14.7).

However, the fact that life can exist at high pressures does not mean that biological processes are independent of pressure. Observations on small aquatic organisms under pressure up to 600 atm have shown a progression of changes including increased activity, paralysis and coma (all of which, under some circumstances, are reversible) as well as death. Observations on dividing cells have shown that the division process can be slowed down or prevented entirely by elevating the hydrostatic pressure. These complex biological responses can sometimes be understood in terms of simpler phenomena, such as chemical reactions. It is pointed out by Johnson, Eyring and Polissar (1954), which we shall call "JEP", that biochemical reactions involving large molecules are significantly pressure-dependent while reactions between small molecules are not. A large molecul can often exist in either of two states, "α" and "β", where the β state has a molar volume greater than the α state by an amount which we call ΔV . Then (by analogy to the discussion of short and long units of keratin, in Section 12.5) the ratio of probabilities for the two states will be given by

$$\frac{P_\alpha}{P_\beta} = C\ e^x \ , \quad x = \frac{p\ \Delta V}{R_g T} \ , \tag{14.7}$$

where C is a constant independent of pressure. The quantity p ΔV gives the work done in changing the volume by the amount ΔV against the pressure p . We see that x is dimensionless and e^x is like the Boltzmann factor considered in Sections 9.5 and 12.5. We see from Eq. (14.7) that an increase of p leads to an increase of P_α/P_β ; hence, as is reasonable, an increase of static pressure leads (on the average) to an increase in the number of molecules in the (small volume) α-state at the expense of those in the β-state. Examples are cited by JEP with respect to denaturation of macromolecules. For some enzymes the denatured state has a smaller volume than the natural one; for them an increase in pressure leads to an increase of the denatured fraction and tends to slow any reaction catalyzed by the enzyme. For other systems, such as tobacco mosaic virus, the natural state has the smaller volume and applied pressure then increases the natural fraction.

When cells are subjected to elevated pressure a considerable change is seen in mechanical properties of the cytoplasm. Marsland (1942) and coworkers carried out experiments with many kinds of cells by observing sedimentation on a very small scale within the cells. (This work is summarized by JEP, 1954). In this method the cells were suspended in a liquid which is _isopycnic_ to them, that is, whose density is approximately equal to that of the mean density of the cells. The suspension is then centrifuged. In this isopycnic suspension the cells do not sediment (or do so only very slowly) but small particles normally distributed through the cells are found to sediment within the cell. In Fig. 14.3 typical results from D. E. S. Brown (1934) are shown for eggs of the sea urchin _Arbacia punctulata_. Distributions of granules are shown after centrifugation for 25 sec at 7200 g . The left hand figure shows a cell for which the experiment was done at atmospheric pressure, while the cell on the right was subject to a pressure of 476 atm during centrifugation. Clearly the sedimentation occurred much more completely for the pressurized cell. Quantitative data were obtained by observing the time required for a granule to move through a certain distance. It was found for a number of kinds of cells (Marsland, 1942) that the time decreased by about 25% for each increase of pressure by 70 atm.

In other experiments it was found by Marsland and associates (cited by JEP, 1954, Reference 14.1) that cell division is prevented in eggs from many animals by applying hydrostatic pressure in the range 200-600 atm. Apparently the application of sufficiently high pressure prevents formation of intracellular structures required for carrying out cell division. An exception was found by Pease and Marsland (1939); they found that eggs of the roundworm _Ascaris_ divide even at pressures as high as 800 atm; _Ascaris_ may thus be similar to organisms that exist and multiply in the ocean depth.

In explaining these results with concepts advanced by JEP we consider intracellular structures to be made up of macromolecular elements. These elements may be arranged either in the form of a normal "β" structure or they may be in a more fluid structureless "α" state. If the β state occupies more volume than the α state, application of pressure will favor the latter. In the sedimentation experiments typified by Fig. 14.3 the normal β structure holds the granules in position while the α state allows them to move. In the experiments with cell division the "β" structure is the normal one required for cell division.

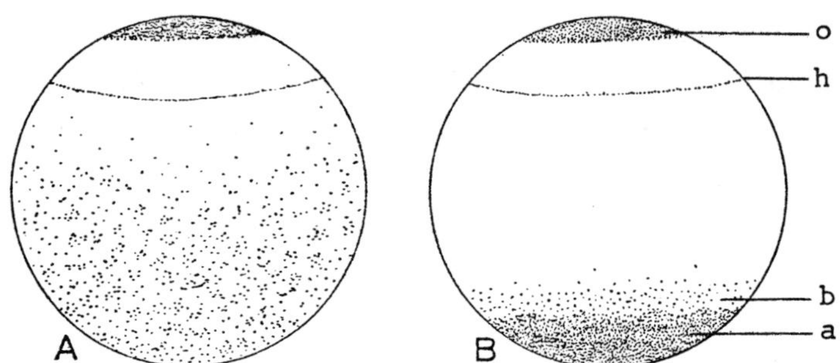

Fig. 14.3 Distribution of pigment granules in eggs of Arbacia punctulata centrifuged for 25 sec at 7200 g . The egg on the left was centrifuged at atmospheric pressure and the one on the right at 476 atm. From Brown (1934). O, oil cap ; h, limit of hyaline zone; b, zone of small scattered granules; a, dense zone.

14.5 DECOMPRESSION

A diver who has adjusted to depths where the pressure is high must be careful in his ascent, in order to avoid injury from decompression sickness or "bends". Also a space pilot must be protected against rapid decrease of pressure in a fast ascent through the earth's atmosphere. In both situations a primary cause for concern is the formation of bubbles in tissues as gas comes out of solution. In decompression studies with animals bubbles are found especially in the vascular system, that is, in the veins and arteries, and in other fluid-filled spaces. They are also found outside the blood vessels, especially in fatty tissues. Bubbles in the blood vessels can retard the circulation. It has also been shown that such intravascular bubbles promote undesirable aggregation of platelets and other blood elements. Many studies have been made on decompression phenomena in animals, including man. See, for example, Fulton (1951).

We can understand some aspects of decompression phenomena by considering principles involved when any gas, such as air, is in contact with a liquid, such as blood. If the gas is under pressure p some of it will dissolve in the liquid. After sufficient time an equilibrium concentration c_e of dissolved gas will be reached, that is, the liquid will be saturated with the gas; the concentration c_e is proportional to p under typical conditions. If then the pressure p is decreased the liquid will, at first, be supersaturated and gas will tend to come out of solution by diffusing toward any liquid-gas or liquid-vapor interfaces that may exist. If small pockets or bubbles of gas or vapor are present, these may grow rapidly to become macroscopic bubbles, such as those seen in boiling water, or those characteristic of decompression illness. These sites from which bubbles grow are sometimes called nuclei. Nuclei of microscopic or submicroscopic size are plentiful in most aqueous solutions and suspensions unless they are removed by special means. Without sufficient nuclei the diffusion of gas outward from a supersaturated liquid (e.g., toward the upper surface of the liquid in a container) may be very slow, and the nonequilibrium state may persist for a relatively long time.

A problem arises in accounting for the existence of nuclei. In a liquid at equilibrium with air at atmospheric pressure we expect a spherical air body to shrink by diffusion, and thus gradually disappear. For consider such a body, or nucleus, of radius R in a liquid for which the interfacial tension is σ . According to Eq. (2.28), a form of Laplace's law, the pressure inside the nucleus will exceed that outside by an amount Δp given by

$$\Delta p = \frac{2\sigma}{R} . \qquad (14.8)$$

For a liquid in which the gas concentration is no greater than that corresponding to saturation for atmospheric pressure, we see that the liquid is undersaturated relative to the nucleus. Hence gas will diffuse out of it into the liquid. The rate at which this occurs, and hence the rate at which the nucleus shrinks, can be determined from theory similar to that discussed in Section 9.6.1. If R is large the shrinking process is slow but for any air nucleus or bubble of submicron size the excess pressure Δp would be of the order of atmospheres according to Eq. (14.8), and we would expect it to disappear very quickly.

In considering this problem Harvey and associates (1944 and 1951, References 14.14, 14.15 and 14.16) suggested, especially for in vitro conditions (i.e., conditions prevailing outside the living body), that nuclei are trapped in cracks or corners on container walls or surfaces of suspended solid particles. In a crack of suitable shape the gas-liquid interface may be flat, or nearly so, even though the nuclear gas volume is small. A nearly flat meniscus implies an indefinitely large value of R ; hence, from Eq. (14.8), Δp is very small and the nucleus is stable.

Another theory for explaining stable gaseous nuclei, advanced by Fox and Herzfeld (1954), is that organic molecules in solution migrate to any air-liquid interface; at the surface of the nucleus they form a sturdy "skin" which prevents diffusion and gives mechanical strength to the nucleus.

In a different category is the gas which exists in the bronchi, tubes and alveoli of lung tissue, and the intercellular gas in plants. Here the gas-filled space is extensive, and losses by diffusion are continually restored from the atmosphere.

Returning to the decompression phenomena described in the first part of this section, a question arises whether the observed bubbles grow from previously existing nuclei, stabilized in crevices or cavities, or by organic skins. From experiments described by Harvey (1951) he concludes that decompression bubbles do not arise from stable nuclei in normal quiescent blood. But there are extensive findings (Fulton, 1951; Harvey, et al., Reference 14.16, 1951) that decompression bubbles are seen copiously in animals which were physically active during or immediately before a compression-decompression cycle. Apparently muscular contraction plays a critical role. While the details of this role are not clear it is possible that tension produced during muscle action produces momentary breaks, cavities, or regions of low pressure, into which gas quickly diffuses. Small nuclei are then formed, possibly stabilized temporarily on vessel walls, which provide sites for growth of bubbles during decompression.

Also interesting are the findings of Gramiak and Shah (1971). Using an ultrasonic technique they observed echoes from blood in human cardiac chambers, certain of which they attribute to gas bubbles of microscopic size. They suggest that these microbubbles grow in zones of low pressure in the complex flow field. Evidence thus exists for small gaseous bodies in blood passing through the normal human heart. How long these bodies persist as the blood proceeds through the normal circulation is as yet unknown, and their significance as nuclei is unknown.

14.6 TENSION

In Section 14.4 we considered biological systems in an environment characterized by an applied hydrostatic pressure; this is an environment where the stress tensor is isotropic, and compressive in sign. In such situations biological cells are able to withstand even rather high values of the stress. The same is not true for situations where stress is applied nonuniformly. If, for example, a cell is squeezed between two horizontal plates, stress is applied to the top and bottom of the cell but not to the lateral surface. As a result the cell membrane is subject to tension and may rupture at rather low values of the stress applied by the horizontal plates.

Little information is available on the conditions required for rupture of biological membranes under tension, except for the red cell; this cell has been studied by several investigators. Rand (1964) used the "sucking pipette" method described in Section 2.7; this provides a way of subjecting the cell to a known tension σ . Rand observed that when a red cell was subjected to a sufficiently high tension rupture occurred (as evidenced by hemolysis , that is, release of Hb into the surrounding solution) but not immediately. The time required for rupture varied from several seconds for relatively high values of applied σ , to 2 hrs for a σ value of only 2 dyn/cm^2. Taking these findings as evidence that the red cell membrane is viscoelastic, Rand assumed that after an initial immediate extension in response to the applied tension the membrane continues to extend as long as the tension exists. This assumption was made more specific by choosing the four-element Burgers model of Eq. (6.86) and Fig. 6.13 to represent the membrane. His results on σ versus time-for-rupture were then found to be consistent with the idea that rupture occurs when the strain reaches a critical value. Because of uncertainties in translating results from the model to those for a real membrane it was not possible to calculate a precise value for the critical strain in the membrane. However, an estimate was arrived at: the breaking strain is that corresponding to an increase of about 8% in the mean cellular radius. This result is reasonably consistent with results obtained by others, using osmotic methods. (Hoffman, et al., 1958; Katchalsky, et al., 1960). The Young's modulus E was found to be of the order of 10^7 d/cm^2 as had also been reported by Katchalsky, et al .

The response of membranes to tension is relevant to material in Section 14.7 , which deals with hydrodynamic shearing. Shearing stress causes distortion of a cell, and thus leads to tension in its bounding membrane.

It is of considerable practical importance to know the breaking strength of bone under tension. For the femur (thigh bone) Bonfield (1971) states that fracture occurs under a tensile stress of about 1.2×10^9 dyn/cm^2 . Mather (1967), Reference 13.21) has developed an empirical formula for predicting the breaking load for the human femur in terms of age and sex of the subject, dimension of the bone and elastic modulus of the bone.

Since the dimensions and moduli can be determined from Xray and acoustic measurements the formula provides a practical method of obtaining information to be used, for example, in designing crash protection equipment.

Since the mechanical strength of membranes, bones and other structural elements of the body depends on their macromolecular components, it would be interesting to know the breaking strength of individual macromolecules under tension. Some information on this subject has come from observing effects of hydrodynamic shear on macromolecules in suspension, a topic which is taken up in Section 14.7.

14.7 HYDRODYNAMIC SHEARING STRESSES

It was anticipated in Section 14.6 that macromolecules and cells are subjected to tension when they are suspended in flowing media. Here we discuss results and some elementary analysis for several examples. One reason for interest in this topic is that hydrodynamic shearing provides an approximate technique for measuring the breaking strength for individual macromolecules. Also shearing methods are in much use for fragmenting cells in the laboratory, or in industry, to obtain enzymes and other intracellular constituents.

Undesired shearing effects may arise from pipetting and other laboratory procedures. Perhaps more important is the fact that artificial aids to the circulation subject cells and macromolecules of the blood to shearing stresses which may be harmful. Surprisingly, even the natural circulation seems to expose blood constituents to shearing stresses that lead to eventual degradation (and replacement).

We first take up effects of hydrodynamic shear on macromolecules, and begin by treating a simple model.

14.7.1 Dumbbells

As anticipated, long molecules can under some conditions be fragmented when suspended in a flowing liquid. Bibliography on the subject is contained in a review article by Casale, Porter and Johnson (1971). To see how this can be we refer to the dumbbell model introduced in Section 6.4.2. Just as in that section we assume the dumbbell, of length 2b, is immersed in a liquid which takes part in steady flow of constant velocity gradient G. See Fig. 6.4. While its center of mass moves with the fluid, the dumbbell is assumed (for simplicity) not to be rotating. The fluid then moves with speed $G\varepsilon$ relative to the body A in the positive x direction, and with the same speed relative to body B in the negative x direction; here ε is $b \sin \theta$ where θ is the inclination of the dumbbell axis relative to the flow direction. Letting the frictional constant for each body be f, as before, we conclude that each body is acted on by a frictional force $fG\varepsilon$; on A the force is along the flow, and on B it is in the opposite direction. The force on each body has a component F directed outward from the center given by $fG\varepsilon \cos \theta$. Hence a tension is set up in the dumbbell which (neglecting friction along the connecting rod) is given by

$$F = fGb \sin \theta \cos \theta = \tfrac{1}{2} fGb \sin 2\theta \ . \qquad (14.9)$$

As stated in Section 9.8.1, the forces on A and B also have components normal to the axis which will tend to cause the dumbbell to rotate; in a more sophisticated calculation we should correct for the effect of this rotation on the relative velocity and frictional forces. Also Brownian movement would need to be considered in an accurate analysis for a dumbbell of molecular dimensions. And, of course, no macromolecule really has a dumbbell shape. But in spite of these deficiencies, the simple result in Eq. (14.9) does very well for illustrating principal features of the situation seen by macromolecules in flow fields. We see that the tension F is proportional to the "molecular" length 2b, and to the velocity gradient G ; F has its greatest value when $\theta = 45^{\circ}$ and is zero when θ is either 0° or 90°. If the bodies A and B are spheres of radius R the constant f is given by $6\pi\eta R$ in a liquid of shear viscosity coefficient η ; hence F is proportional to the product ηG, a quantity with the units of stress.

According to Section 9.8.1 a small "dumbbell molecule" whose density is the same as the liquid will rotate with essentially the same angular velocity ω as the liquid. While Eq. (14.9) does not apply accurately to this situation we can perhaps use it as a guide, and anticipate that the rotating dumbbell will be subject to a tension varying periodically in time with angular frequency G . We expect that if the maximum force F is large enough the "connecting rod" will yield, and the dumbbell molecule will be ruptured.

14.7.2 DNA

Levinthal and Davison (1961) (hereafter referred to as L&D) carried out studies on effects of hydrodynamic shear on DNA, led by a discovery that simple pipetting procedures can cause damage and that, for this reason, previously reported values of molecular weight for DNA were much too low. They forced molecular suspensions of interest through circular tubes of small diameter and, using theory such as that discussed in Section 5.6 calculated the maximum velocity gradient to which the molecules were subject.

In analyzing their data L&D derived an expression for the tension in a long rod-shaped molecule of length L and radius a, analogous to that for the dumbbell in Eq. (14.9). The tension is greatest at the center of the rod and, as for the dumbbell, varies with angle according to the factor $\sin 2\theta$. Their expression for F, the maximum value of the tension, is

$$F = \frac{3\pi\eta G L^2}{16 \ln(L/a)}, \tag{14.10}$$

where η is the coefficient of shear viscosity for the liquid. Again we see that the tension is proportional to the stress product ηG. Also F is an increasing function of L, since $L >> a$ and the logarithmic expressic in the denominator increases only at a slow rate. While the DNA molecule is by no means a straight rigid rod, neither is it a highly coiled molecul It is reasonable that during its tumbling motions in a flow field the molecule will at some time acquire the condition of linear configuration and 45° angle which led to the maximum tension value given in Eq. (14.10). Levinthal and Davidson used DNA of very high molecular weight, up to 13×10^7 daltons, obtained from phage by extraction methods which subjected the DNA to a minimum amount of shear stress. Since the molecular weight of DNA is about 2×10^6 daltons per micron of length the molecules used by L&D were very long, the dimension L varying up to 65 μm. In subjecting the DNA to shearing fields they found a critical value for the velocity gradient G. Below this critical value no change occurred, whereas above it there appeared in suspension molecules about half the original length, namely about 32 μm long. Evidently the tension F produced in the molecule by the shear field was in excess of the molecular breaking strength. Substituting into Eq. (14.1 the appropriate values for L, a, G and η the authors calculated the critical value of F to be 11×10^{-4} dynes.

Considering the structure of DNA the authors assumed the breaks to occur at C-O or C-C bonds; from information on these bonds they calculated 8.9×10^{-4} dynes and 8.1×10^{-4} dynes respectively, for the force required to rupture these bonds. Since the DNA molecule is made up of two strands wound around each other as in a rope, the question had to be considered, whether the critical tension corresponds to a break in one strand at a time, or in both strands simultaneously. Indirect evidence pointed to simultaneous scission of the two strands, leading to a "molecular" calculation of about $16\text{-}18 \times 10^{-4}$ dyn for the tensile strength of DNA.

In view of the approximations made, these molecular calculations are in good agreement with the value 11 x 10^{-4} dyn obtained from shear measurements *via* Eq. (14.10).

Thus L&D not only demonstrated the mechanism for shear degradation of DNA but arrived at a method for calculating the conditions under which the effect occurs. We have already mentioned that the molecules resulting from scission at critical conditions were about half the original length. This was expected from the theory since it predicts maximum tension at the center of the molecule (in a field of uniform G). Further confirmation of theory came from the fact that the "half-molecules" were themselves halved when G was increased by about the amount expected from Eq. (14.10).

14.7.3 Fibrinogen

As stated in Section 7.4, fibrinogen is an important macromolecule in the blood, being indispensable to the clotting process. Charm and Wong (1970) carried out investigations in which they found that shearing of blood plasma causes loss in clotting ability. Their investigations are important for anticipating problems with artificial pumps to aid or replace the heart, since velocity gradients can be relatively high in such arrangements. But especially thought-provoking is the suggestion by the authors that fibrinogen is degraded by shear even as the blood courses through the body in normal circulation.

Their experiments were carried out by subjecting human plasma to shear in two ways: (1) using Couette flow, produced in the space between concentric cylinders (Section 5.8) and (2) using Poiseuille flow by repeated passes of the plasma through cylindrical tubing. The ability to clot was measured by adding *thrombin* (an enzyme which catalyzes the conversion of fibrinogen to fibrin) to the fibrinogen after shearing, and measuring the resulting optical density of the plasma. Carrying out experiments at several values of the velocity gradient G, at each of two temperatures, for times up to 5 days, they found all of their data fit well onto a single plot. See Fig. 14.4. In this plot the abscissa is the nondimensional product $G\theta$ where G is the velocity gradient (or "shear rate") and θ is the "residence time", that is, the length of time during which a fibrinogen macromolecule is subject to the gradient G. In simple shear arrangements G and θ are the same everywhere, but in Poiseuille flow they vary in space, and average values are used. (Prob. 14.3).

From Fig. 14.4 we see that clottability of the plasma was reduced by 50% when $G\theta$ reached a value of about 5×10^8. This reduction in clottability was attributed by Charm and Wong to inactivation of fibrinogen: hence the vertical axis of the plot in Fig. 14.4 is labelled "clottable protein remaining" (CPR). These authors estimated the average value of G in the circulatory system of a 60 kg man to be about 1470 $\sec^{-1}$, the largest gradients being in the pulmonary arteries. The time θ required for the parameter $G\theta$ to take on the value (5×10^8) corresponding to 50% reduction of CPR is then

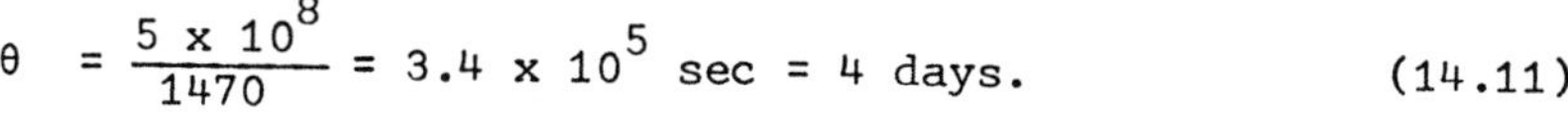

$$\theta = \frac{5 \times 10^8}{1470} = 3.4 \times 10^5 \text{ sec} = 4 \text{ days.} \qquad (14.11)$$

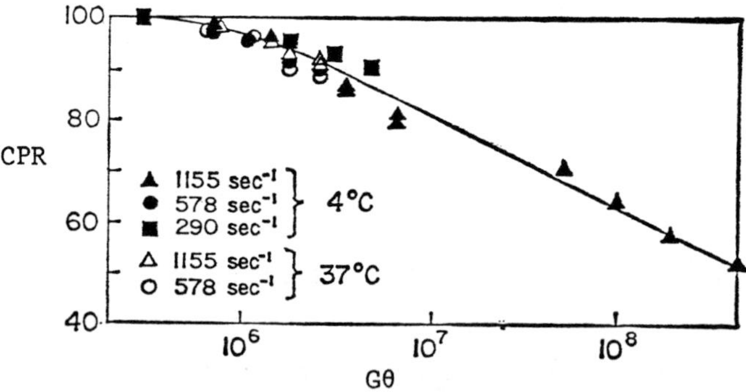

Fig. 14.4 Plot showing effect of hydrodynamic shear on clotting. CPR: % clottable protein remaining. $G\theta$: product of velocity gradient and "residence time". From Charm and Wong (1970, Reference 14.25). Copyright 1970 by the American Association for the Advancement of Science.

Hence the authors conclude that fibrinogen is altered by four days of shearing action in the normal circulation. They remark that this result is consistent with independent chemical evidence that 50% of the fibrinogen is replaced every 4 or 5 days. However, it is not clear just what change the shear has produced in the fibrinogen molecule. Since this molecule is only 700 A long it would not be expected from Eq. (14.10) that halving of its length (like that observed for DNA, Section 14.7.2) would occur at the rather low G values used. (Prob. 14.4).

14.7.4 Droplets: Theory

In fluid flow with velocity gradients, suspended droplets or cells are subjected to nonuniform stress and may be deformed or fragmented. For example, milk is "homogenized" by forcing it through nozzles; here fat globules (about 2 μm in diameter) are subjected to high velocity gradients and thus are dispersed into much smaller units. A systematic study of the deformation of droplets in simple shear fields was carried out by G. I. Taylor (1934). He showed that shear causes a droplet to assume an elongate shape. Because of surface tension such a shape becomes unstable, resulting in fragmentation, if the velocity gradient G exceeds a critical value G_c. For a spherical droplet of liquid with radius _a_ suspended in a liquid with shear viscosity coefficient η, a simplified expression for G_c yields

$$\eta G_c = \frac{\sigma}{2a}, \tag{14.12}$$

where σ is the interfacial tension between the media internal and external to the droplet. (Taylor gives a more general expression which involves the viscosity coefficient for the fluid in the drop, but the value for G_c in that expression never differs by more than 20% from that given in Eq. (14.12)). For an oil droplet of radius 10^{-2} cm in water with viscosity coefficient 10^{-2} P and interfacial tension 10 dyn/cm^2 we obtain 5×10^4 sec^{-1} for G_c and 500 dyn/cm^2 for the product ηG_c.

To see why shear should affect a droplet we refer to material taken up in earlier chapters. In Eq. (4.29) we see the stress components for a simple shearing flow, as they appear in one coordinate system. These give the stresses as seen by any volume element in the fluid, the latter being assumed homogeneous. (When a droplet is present the stress distribution near and in it are altered somewhat, to an extent that depends on the difference in properties between the media inside and outside the droplet. This alteration complicates a complete analysis, but we gain useful insight by simplifying the situation and considering the stresses as given in Eq. (4.29), that is, as if they were unaffected by presence of the droplet.) It is convenient to think of the tensor in Eq. (4.29) as consisting of two parts, just as was done in Section 2.3 (Cf. Eq. 2.10 ff). The isotropic tensor is identical with G_{ij} in Eq. (2.11b), remembering that $\lambda\theta$ is equal to $-p$. This represents a field in which the stress is normal to any plane and is always given by $-p$. Such a field causes no deformation of a spherical droplet but may cause compression or dilatation.

The other tensor is a "deviatric" tensor, like H_{ij} of Eq. (2.11c); its form depends on the coordinate system in which it is viewed. If the flow is along the x_1 direction with velocity given by Gx_2 the tensor will have the form of Eq. (2.11c) in the (1,2,3) system. Letting s_{ij} be the deviatric stress tensor in this system we have

$$s_{ij} = \begin{pmatrix} 0 & \eta G & 0 \\ \eta G & 0 & 0 \\ 0 & 0 & 0 \end{pmatrix} ; \qquad (14.13)$$

thus only s_{12} and s_{21} are nonzero, and these are each equal to ηG, where G is the velocity gradient. Now consider the deviatric tensor in another coordinate system. If axes are rotated through 45° about the "3" direction (Cf. Eq. (2.15)) the tensor now becomes s'_{ij} where

$$s'_{ij} = \begin{pmatrix} \eta G & 0 & 0 \\ 0 & -\eta G & 0 \\ 0 & 0 & 0 \end{pmatrix} . \qquad (14.14)$$

In this coordinate system we see that the deviatric stress tensor is in diagonal form. Similarly we might, if we wished, consider the deviatric tensor in any other coordinate system rotated an angle θ about the "3" axis. Results of a transformation for arbitrary θ are shown in Eq. (2.14). We see that, in general, there are two nonzero stresses acting on each of the 1' and 2' faces, one normal to the face and the other tangential to it.

From results in Eq. (14.13) we know the stresses on planes perpendicular to the "1" and "2" axes, while from those in Eq. (14.14) we know the stresses on planes perpendicular to "1'" and "2'" axes rotated through 45° from the former axes. To see some significance in these results consider a volume element in the shape of an octagonal prism whose section is seen in Fig. 14.5 . Two of its faces are perpendicular to the "1" direction: these are labelled as 0 and 180 in the figure. The stresses on each of these faces is s_{12} (which is equal to s_{21}) whose magnitude is given as ηG in Eq. (14.13). Recalling sign conventions from Chap. 2 we consider the direction of the force exerted <u>by</u> the medium outside the volume element across the 0 and 180 faces <u>on</u> the medium inside the element. These directions are indicated by arrows in Fig. 14.5; the medium to the right of the 0 face exerts an upward force and the medium to the left of the 180 face exerts a downward force. Considering now the 90 and 270 faces, the arrows shown there also indicate the direction of forces exerted by the outer medium on the inner one. As expected, the torque from these is equal and opposite to the torque from forces on the 0 and 180 faces.

From Eq. (14.14) we know that the stresses on faces perpendicular to the 1' and 2' axes are normal to these faces and are of magnitude ηG. Since s_{11}' is positive we know that the outer medium pulls outward on each of the 45 and 225 faces. Similarly, since s_{22}' is negative the forces are inward on the 135 and 315 faces.

Altogether the above results give the stresses on all eight faces of the octagonal volume element. We need not have restricted ourselves to an octagon but could have considered a prismatic volume element with any number of faces, determining the deviatric stress on each face by a result such as that in Eq. (2.14). Letting the number of faces increase without limit we would obtain results for a cylindrical volume element. But we can expect that the results for a cylinder (with its "infinite number of faces") will not be qualitatively different from those for the octagon. Neither would we expect results for a sphere to be different in kind, although we should expect differences in detail.

Considering then the volume element in Fig. 14.5 we see that the stresses will tend to make it elongate in the positive and negative 1' directions and to contract in the 2' directions. This is the effect of pure shearing strain, which is exemplified for a square figure in Fig. 3.8. However a field of simple shear involves not just pure shear but also a superposed rotation with angular velocity G. (See Section 4.6).

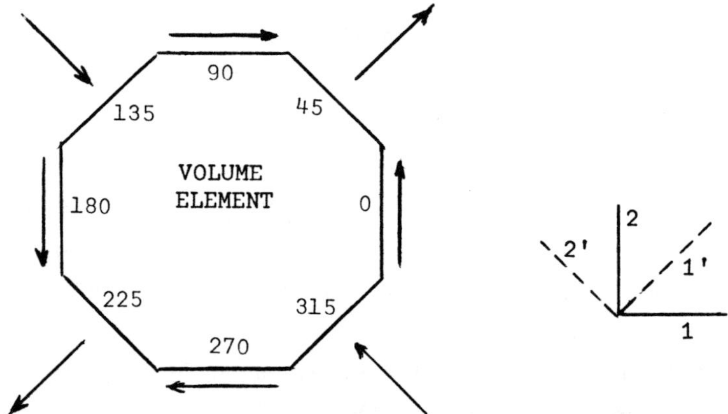

Fig. 14.5 Stresses on the faces of an octagonal volume element in a field of simple shearing flow. Arrows show directions in which the external medium acts on the element.

14.7.5 Droplets: Experiment

We turn now to experimental results on the behavior of droplets in simple shear fields. Many investigations have been carried out on these and related basic phenomena by S. G. Mason and his associates at McGill University. In one experiment droplets in suspension were sheared in Couette flow (Section 5.8) set up between counter-rotating concentric cylinders. In this arrangement a central region of the flow field is at zero velocity; droplets in this region can be observed through a microscope and photographed with motion pictures over a period of time. In this way the deformation of a droplet was studied as a function of time; in Fig. 14.6 a deformation sequence is depicted, sketched by Rumscheidt and Mason (1961) on the basis of their photographs. The numbers below the separate figures are indicative of the time (in arbitrary units) after the shear field is established. At time "0" the droplet is spherical; as time goes on the droplet elongates as expected from the model considered in Section 14.7.4 and Fig. 14.5 . When the length/width ratio for the elongated droplet exceeds a certain critical value (about 3) thinning occurs at the center. With continued shearing the two ends of the droplet may separate, forming separate bodies each of approximately half the original droplet volume. In addition the central thin portion sometimes separates into one or more bodies of much smaller size; breakup of the droplet then yields two large units and several smaller ones. This breakup of an elongated droplet is in

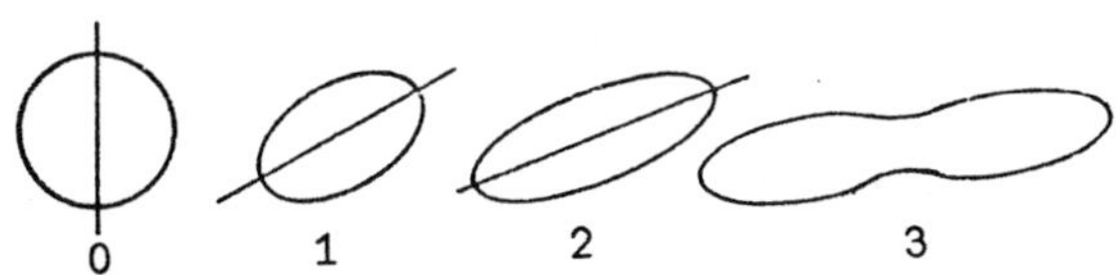

Fig. 14.6 Deformation of a droplet in simple shearing flow. 0: Initial shape of droplet. 1, 2, 3: Appearance of droplet at successive intervals of time after shearing field is established. From Rumscheidt and Mason, 1961, Reference 14.27.

general accordance with theory given by Rayleigh (1879) showing that a thin cylinder of one fluid in another fluid with which there is significant interfacial energy, is unstable and may become fragmented.

14.7.6 Cells

Within the last few years there have been a number of investigations on the effects of shear on biological cells, especially erythrocytes. A review on this subject has been published by Rooney (1973). Schmid-Schonbein and Wells (1969) used a device with transparent walls so that cells could be observed with a microscope while being subjected to simple shear. Deformation was found to occur, of a kind not unlike that observed by Rumscheidt and Mason for droplets. But of course the red cell differs from a droplet in possessing an outer membrane. By watching small particles stuck on the outside of a cell membrane the authors found that simple shear causes the membrane to rotate around the cell contents in a manner resembling the motion of a tread on a military tank. This tank-tread motion was confirmed by Goldsmith and Marlow (1972) who also observed that the concavity of the erythrocyte is still present up to shear stresses (ηG) of 120 dyn/cm^2 .

Nevaril, et al, (1968) found that for shear stresses above 2500 dyn/cm^2 small spherical cell fragments began to separate from the cells. These fragments were of diameter 0.5 to 2.0 μm and have been called <u>microspheres</u> they are probably analogous to the very small spheres which are sometimes formed when droplets fragment. These interesting microspheres have been studied further by Champion, et al, (1971).

A number of groups have found that when the shear stress exceeds a critical value hemolysis occurs, that is, hemoglobin begins to leak from the red cells. Results were summarized by Leverett, et al, (1972) for contemporary experiments. They point out the importance of the time during which cells are subjected to the shearing action. Reported values of the critical shear stress ηG_c vary from a low value of 1500 dyn/cm^2 for a shearing time of 100 sec to a high value of 40,000 dyn/cm^2 when the time is only 10^{-5} sec. Williams (1973) has pursued further the dependence of the critical stress on the shearing time; he has suggested an explanation based on viscoelasticity of the red cell membrane, specifically in terms of the four-element Burgers model proposed by Rand . (Sections 7.9 and 14.6).

It is also found, by Krizan and Williams (1973), that the critical shearing-stress threshold for hemolysis depends critically on the temperature. Some of their findings are shown in Fig. 14.7 , where the experimental points are for human erythrocytes subjected to simple shear for five minutes. The critical stress is about 600 dyn/cm^2 at 25°C but decreases with increasing temperature, dropping essentially to zero at 48°C . The solid curve plotted on the same figure is calculated from theory based on a model from statistical mechanics for phase transitions. This theory, too lengthy for detailed discussion

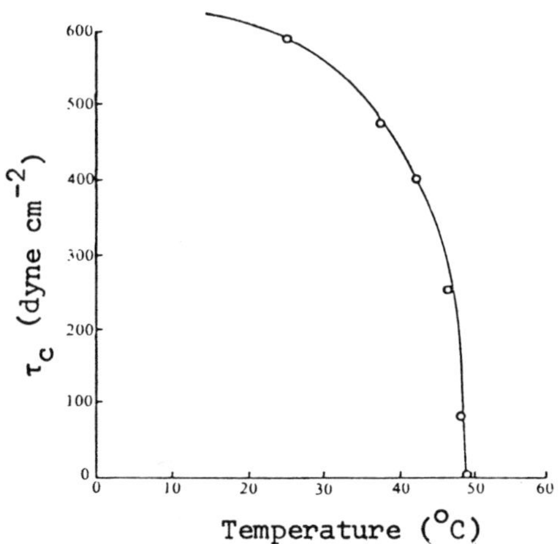

Fig. 14.7 Critical shear stress τ_c versus temperature for red cell membrane. Points are experimental; the solid curve is from theory for phase transition. From Krizan and Williams (1973, Reference 14.36).

here, is based on a two-state model somewhat analogous to the α-β models we have referred to in connection with (i) stress-strain curves in Section 12.5, and (ii) effects of elevated pressures in Section 14.4. The fit between theory and experiment is excellent and gives strong support to the idea that hemolysis is correlated with some kind of phase change in the membrane.

We are led to the question of just how the Hb escapes from the red cell in hemolysis. One possibility is that macroscopic breaks occur in the membrane; that is, gross rupture occurs. This is certainly the situation when the stress is well above the critical value; the damage is then often irreversible. An alternative is that at lower stress levels any "pores" or "pore-equivalents" in the membrane may be enlarged sufficiently to allow Hb molecules to pass. This process may be reversible so that the pores shrink again after the stress is removed. The possibility of reversible pore enlargement has been considered in a lucid manner by Burton (1970).

Other changes produced in canine red cells by shearing have been reported by Nanjappa, et al (1973). These cells have a half-life of 27.5 days in the normal circulation; cells which had been sheared at a stress of 90 dyn/cm^2 (velocity gradient 2110 sec^{-1}) for five minutes, then introduced into the circulation, had a half-life of only 21.4 days, a reduction of 22%. In other tests it was found that cells subject to this stress for 15 minutes were more susceptible to osmotic lysis; furthermore cells subject to the same stress for 60 minutes showed altered permeability to sodium ions and also showed loss of surface enzymatic material into the surrounding medium. We notice that the velocity gradients used in these experiments are not much greater than the values (1470 sec^{-1}) quoted by Charm and Wong(1970) as typical in the normal circulation of an adult human being. Naturally, the question arises(by analogy to the Charm-Wong proposal for fibrinogen) as to whether the normal "aging" of red cells may not be partly a result of shearing action in the circulation process.

14.8 AUDIBLE SOUND

In this section and the next one we consider sound as an aspect of our physical environment which affects biological systems. Sound may be of any frequency but it is convenient to divide the frequency range into three parts: audible sound, extending from 30 Hz to 20,000 Hz, representing the range of normal hearing for human beings; ultrasound, which extends upward from the audible range; and infrasound, which extends downward. Little information is available at present on biological effects of infrasound and we shall consider only the other two: audible sound in this section, and ultrasound in Section 14.9.

Sound affects living systems in a variety of ways. In man the organ most sensitive to audible sound is, of course, the ear. An average person in a quiet environment can hear sounds with pressure amplitudes $|p|$ as low as 2.8×10^{-4} dyn/cm^2 if the frequency lies in the range 1000-4000 Hz. The rms (root-mean-square) pressure corresponding to this is $|p|/\sqrt{2}$ or 2×10^{-4} dyn/cm^2. Estimates suggest that at these very low amplitudes the variation in force on the ear drum during each cycle is not much greater than that expected from "thermal noise", the latter arising from random variations in the bombardment of the ear drum by air molecules. The following analysis is rough, but indicative:

> There are of the order of 10^{20} collisions of air molecules with the tympanic membrane during each half-cycle of sound at a frequency of 1 to 4 kHz. By a "$\sqrt{N}$ law", analogous to that in Eq. (8.42), the rms deviation in the number of collisions from one half-cycle to the next is the square root of 10^{20}, that is, 10^{10}. The fractional rms deviation is $10^{10}/10^{20}$ or 10^{-10}. Since the number of collisions on a membrane during a half-cycle is proportional to the mean pressure on it during that interval, we conclude that randomnesss in molecular collisions leads to a fractional rms deviation in pressure of 10^{-10} between half-cycles. In air at atmospheric pressure this means an absolute rms deviation in pressure of 10^{-10}

atm, or about 10^{-4} dyn/cm^2 . This value serves as an estimate of the rms sound pressure amplitude corresponding to thermal noise in the 1 to 4 kHz frequency range. It is about half the threshold for hearing.

It is convenient and customary to express sound pressure levels in air in decibels (see Eq. (13.5)) relative to a standard reference equal (approximately) to the threshold of hearing at 1000 Hz. Specifically the sound level SL corresponding to an rms pressure p_{rms} is defined as

$$SL = 20 \log_{10} \frac{p_{rms}}{2 \times 10^{-4}} \quad . \tag{14.15}$$

We see, for example, that when p_{rms} is 2 dyn/cm^2 the sound level SL is 80 db.

Determinations have been made of the threshold for hearing over the entire range of audible frequencies. Results for normal young adults are shown in Fig. 14.8. In that figure the curve was arrived at by the American Standard Association as a judicious average of findings by several groups of investigators. (Licklider, 1951). We see that the threshold sound level shows a broad minimum of about 0 to -5 db in the range 1000 to 4000 Hz; the threshold rises at low frequencies to more than 60 db at 30 Hz, and rises at high frequencies to 30 db at 20 kHz.

The human ear has a tremendous dynamic range. For sounds at frequencies between 1000 and 4000 Hz we can hear at levels varying from about 0 db to 120 db, a variation which corresponds to a factor of one million in pressure amplitude. Such a range, which (among other benefits) allows us to enjoy great dramatic effects in music and theater, is difficult to duplicate with the best of electronic and acoustic equipment.

However, unpleasant feelings are experienced when the upper limit is exceeded. Also, damage to hearing results from continued exposure to levels of 80 db or higher. (Kryter, 1970). Results on hearing losses are shown in Fig. 14.9. These were obtained from surveys of large numbers of persons of age between 36 and 45 years. The hearing thresholds at several frequencies were measured for each person, and a determination was made of the typical sound level to which he had been exposed during working hours over a long time period. The population tested included persons working in quiet conditions as well as noisy ones. The ordinates in Fig. 14.9 give the percentage of persons whose hearing was found to be impaired by a standard amount. Specifically the "% impairment" gives the percentage of persons whose threshold was 20-30 db higher than normal, that is, whose hearing loss was 20-30 db relative to persons who worked under quiet conditions. We see that the hearing was impaired for about 5% of those working in noise at an 80 db level, for about 10% of those exposed to 90 db, about 25% of those exposed to 100 db and 65% of those exposed to 110 db. For reference, subway noise is typically at the 90 db level and neighborhoods near large airports are often continuously subject to levels of 100-110 db . "Rock" music reaches the ears of listeners at levels of 110 db (Kryter, Reference 14.40, p. 205) and operators of snowmobiles are typically subject to levels of 105 db or more. (National Research Council of Canada, 1970). It is clear that noise is an aspect of our environment which we must recognize and control.

We do not have space here to discuss the many fascinating researches that have been carried out on the mechanics of hearing. A classic source of information on the subject is by G. v. Bekesy (1960). For a general exposition on the biophysics of hearing see Ackerman (1962), and for accounts of more recent work see Tonndorf (1974).

Neither is there space here to discuss information on the dynamic response of the whole animal body to environmental noise and vibration.

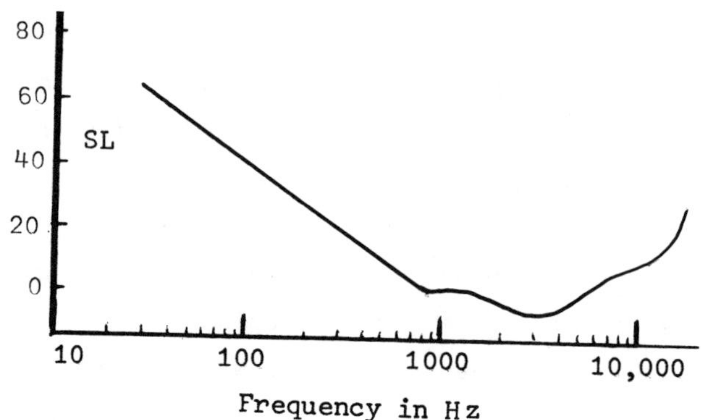

Fig. 14.8 Sound level SL corresponding to the threshold for hearing in a population of young adults. From Licklider, Reference 14.39, 1951.

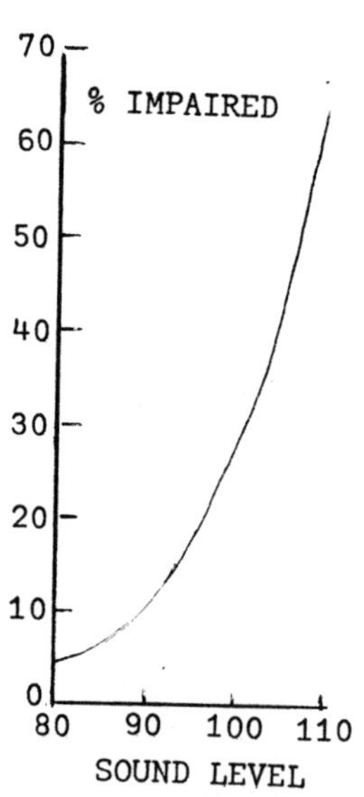

Fig. 14.9 Plot showing loss of hearing as a result of noisy working conditions. Ordinates give percentage of population (aged 36-45 yrs) showing serious hearing impairment (about 20-30 db loss) as a function of the sound level to which they are exposed during working hours. From Cohen et al. (Reference 14.41), cited by Kryter (Reference 14.40).

H. E. v. Gierke (1971) has reviewed this subject and described models which have been developed to represent the body. Among other requirements, a successful model must exhibit resonances in the range 3 to 1000 Hz, thus simulating observed resonances of abdomen, spine and chest.

14.9 ULTRASOUND

14.9.1 Introduction

As stated earlier "ultrasound" is just "sound" whose frequency is too high to be audible. Rather arbitrarily the low-frequency limit is often taken to be 20 kHz. In this section we consider sound as a physical agent for producing alterations in biological structures and processes. While we emphasize ultrasound, it should be realized that many of the phenomena we discuss occur at audible frequencies as well.

It has been known since the late 1920's that ultrasound can produce changes in physical, chemical and biological systems. A comprehensive review has been given by El'piner (1964); in this section we shall take up only a few examples, emphasizing physical principles that seem to be involved. Some of the effects of ultrasound are thermal in origin; that is, they arise because of heat generated by the ultrasound. Other biological changes brought about by ultrasound are consequences of _cavitation_, an activity of gas and vapor bubbles in liquid media. Still others are associated with _radiation pressure_, _acoustic streaming_ and other manifestations of sound which become evident at high amplitudes.

It is characteristic of many sonic effects that they do not occur unless the sound level exceeds a certain minimum value, the threshold level. In specifying the sound level the pressure amplitude is sometimes appropriate, as with audible sound discussed in Section 14.8. But for ultrasonic fields it is common to refer to the _intensity_, a quantity taken up in the next subsection.

14.9.2 The Intensity

The concept of intensity finds its greatest use in connection with travelling plane waves and other fields of somewhat similar character in fluids. For a plane wave travelling in the positive x direction, which direction we consider as being "to the right", we define the intensity at an arbitrary plane $x = x_1$ as follows: The instantaneous intensity is the work done by the medium to the left of this plane on the medium to the right, per unit area per unit time; the quantity of principal interest is the <u>time-averaged</u> intensity at x_1, which we designate as $I(x_1)$. Thus $I(x_1)$ represents the average product of force (component along x) per unit area and velocity, that is, of negative stress $-S_{11}$ and velocity u at x_1. Dropping the subscript from x we have

$$I(x) = < [-S_{11}(x)] [u(x)] > , \qquad (14.16)$$

where the bracket $< >$ represents the average with respect to time. We obtain a specific expression for the intensity I(x) by referring to the results in Sections 13.1 and 13.2. In Eqs. (13.1) and (13.2) we have expressions for the displacement ξ and velocity u; also Eq. (13.17) provides us with an expression for the stress S_{11} in terms of derivatives of ξ. We obtain (Prob. 14.5) for a fluid ($\mu = 0$) that

$$I(x) = \tfrac{1}{2} B\omega k A^2 e^{-2\alpha x} [1 + \frac{\varepsilon\alpha}{k}] , \qquad (14.17)$$

where ε is the nondimensional quantity defined in Eq. (13.20) and B is the bulk elastic modulus defined in Section 4.3. As in Section 13.2 we assume $\varepsilon << 1$ and, as is essentially equivalent, $\alpha << k$; then the quantity $\varepsilon\alpha/k$ can be neglected on the right of Eq. (14.17). Also k is approximately equal to ω/c_o, where c_o is given by Eq. (13.21) (in which $\lambda + 2\mu$ reduces to B). In addition we note (see Eq. (13.2)) that $\omega A e^{-\alpha x}$ is the velocity amplitude at x; designating this quantity as $|u|$ we find (Prob. 14.5) that the intensity I at the arbitrary point x can be written simply as

$$I = \tfrac{1}{2} \rho_o c_o |u|^2 . \qquad (14.18)$$

Since $\varepsilon << 1$ we find that the diagonal stress components S_{11}, S_{22}, and S_{33} are all nearly equal. (This can be shown by forming expressions for S_{22} and S_{33} analogous to that given for S_{11} in Eq. (13.17)). The stress tensor is thus nearly isotropic, having the form of Eq. (5.9). Setting S_{11} equal to $-p$ we find (Prob. 14.6) that in the travelling plane wave the pressure amplitude $|p|$ is given simply by

$$|p| = \rho_o c_o |u| \quad . \tag{14.19}$$

Hence the intensity can be expressed alternatively as

$$I = \tfrac{1}{2} |p| \, |u| = \tfrac{1}{2} \frac{|p|^2}{\rho_o c_o} \, . \tag{14.20}$$

In water and similar media $\rho_o c_o$ is about 1.5×10^5 cgs units at room temperature. (Table 7.1). Remembering that an atmosphere is about 10^6 dyn/cm^2 and that one watt is 10^7 ergs sec^{-1} we find from Eq. (14.20) that a pressure amplitude of one atmosphere (in a plane travelling wave) corresponds to an intensity in water of 0.33 W/cm^2.

14.9.3 Temperature Elevation by Ultrasound

From Eq. (13.2) the velocity amplitude in a travelling plane wave is $\omega A e^{-\alpha x}$ and hence from Eq. (14.18) the intensity $I(x)$ at arbitrary x can be written

$$I(x) = I_o e^{-2\alpha x} \quad , \tag{14.21}$$

where I_o is the intensity at $x = 0$. This equation tells us that the intensity of a wave falls off exponentially with distance, decreasing by a factor of $e^{-1} = 0.368$ whenever the distance increases by the amount $(2\alpha)^{-1}$. When the attenuation is a result of viscous absorption the coefficient α is given by Eq. (13.24), and the decrease of intensity in the travelling wave is accompanied by irreversible production of heat.

We now proceed to derive an expression for $\dot{Q}_v$ (see Section 14.2.2), which we now take to be the heat generated per unit volume in unit time because of sound absorption. Consider a volume element bounded by the planes $x = x_1$ and $x = x_2$, where $x_2 > x_1$, and with unit cross-sectional area. The volume of the element is then just $x_2 - x_1$ and the heat generated in it per unit time is $(x_2 - x_1)\dot{Q}_{va}$ if $\dot{Q}_{va}$ is the average value of $\dot{Q}_v$ in the element. By conservation of energy the rate of heat generation must equal the difference between the rate $I(x_1)$ at which acoustic energy enters the element across the plane $x = x_1$ and the corresponding rate $I(x_2)$ at which acoustic energy leaves the element at $x = x_2$. We obtain

$$(x_2 - x_1)\dot{Q}_{va} = I(x_1) - I(x_2) . \qquad (14.22)$$

Dividing both sides of this equation by $x_2 - x_1$ and considering limits as x_2 approaches x_1 yields

$$\dot{Q}_v = -\frac{dI}{cx} . \qquad (14.23)$$

When the intensity I is given by Eq. (14.21) we obtain the simple and useful expression

$$\dot{Q}_v = 2\alpha I . \qquad (14.24)$$

Thus the heat $\dot{Q}_v$ produced per unit volume per unit time is proportional to both the attenuation coefficient α and the intensity I . In animal tissue such as liver and brain a typical absorption coefficient is 0.1 nepers/cm (0.87 db/cm) . (See Section 13.5.4) . In such tissue a sound wave of frequency 1 MHz and intensity 1 watt/cm^2 produces heat at a rate $\dot{Q}_v$ of 0.2 joules or 0.048 calories per milliliter per second. If heat is not transported away the temperature will rise at a rate dT/dt given by Eq. (14.2). For water, and approximately for tissue, the heat capacity ρc_m per unit volume is 1 calory per milliliter per Centigrade degree rise in temperature. Hence on these assumptions, when a 1 MHz ultrasonic wave of intensity 1 W/cm^2 passes through liver or brain the temperature will rise at the rate

$$\frac{dT}{dt} = 0.048^{o}C/sec = 2.9^{o}C/min . \qquad (14.25)$$

When the circulation is unimpaired the rate of temperature rise will be less than this because heat is conducted away by blood flow. Also ultrasound is often applied in a rather narrow focussed beam; hence when the temperature rises in irradiated tissue heat is readily conducted outward to the cooler surroundings.

To arrive at an estimate of possible temperature elevation in the presence of heat conduction we recall theory in Section 14.2.2 for a spherical source of heat. Suppose the "source" is a portion of tissue with a relatively high absorption coefficient α , and that the heat produced in it results from sound absorption. The situation considered in Section 14.2.2 would be realized if a uniform beam of sound impinged on a spherical "absorber" which was surrounded by tissue of low absorption coefficient. After equilibrium is established in such a situation the temperature distribution is as shown in Fig. 14.2 . According to Eqs. (14.6) the temperature at the center T(0) exceeds that in the surrounding tissue (T_∞) by $\dot{Q}_v R^2/2\kappa$. For water, and approximately for biological tissue, the heat conductivity coefficient κ has the value 0.0060 W/cm$^\circ$C . For the condition considered previously we find that $\dot{Q}_v/\kappa$ has the value 0.2/0.006 or 33.3 cm^{-2} $^\circ$C^{-1} . Hence the temperature rise ΔT at the center of the absorber is given by

$$\Delta T = 16.7\ R^2\ (^\circ C)\ . \qquad (14.26)$$

For an absorber of radius 1 cm the temperature rise ΔT is 16.7 $^\circ$C; but if R is only 1 mm the temperature rise is less by a factor of 100, namely, 0.17 $^\circ$C. It can be shown that a characteristic time required for the steady-state temperature distribution to develop is of the order of R^2/D_{th} , where D_{th} is the thermal diffusivity (defined as $\kappa/\rho c_m$) and is analogous to the (mass) diffusion coefficient discussed in Chapter 9. For water, and approximately for biological tissue, D_{th} has the value 0.00144 cm^2/sec. Hence for R = 1 cm the characteristic time is 700 sec or about 12 minutes; for R = 1 mm the time is about 7 sec.

14.9.4 Bioeffects of Sonic Heating

As was stated in Section 14.2 ultrasound is used in physical therapy for applying heat deep within the body. From Eq. (14.24) we see that heat is generated in tissue at a rate proportional to both the absorption coefficient α and the intensity I. Hence as an ultrasonic beam of given intensity passes through tissue heat is produced preferentially in tissues with high absorption coefficient α . Since for muscle α is higher than for fat (Section 13.5.4) ultrasound will pass through fatty layers to cause selective heating in underlying muscle. For bone the coefficient α is even higher, a face which is probably important in explaining the success of ultrasonic therapy in selectively heating joint structures. For therapy it is desirable to achieve tissue temperatures in the range 40-45 $^{\circ}$C; higher temperatures cause destructive changes. (Lehmann, 1965, Reference 14.3).

In tissues, where the velocity of sound is usually about the same as in water the wavelength in millimeters is given approximately by

$$\lambda \text{ (mm)} = 1.5/f \qquad (f \text{ in MHz}) . \tag{14.27}$$

For example, when f is 1 MHz the wavelength is 1.5 mm. Because the wavelength is small in the megahertz frequency range, it is possible to use lenses and other devices to focus ultrasonic beams in this range, as is done with light beams. This provides another means of causing selective temperature elevation; in a focussed ultrasonic field the intensity I, and hence the rate of heating, is much higher in the focal region than elsewhere. Focused ultrasound has been made use of in medical applications of ultrasound to treat diseases of the central nervous system. Near its focus an ultrasonic beam can cause a small <u>lesion</u> , that is, a small volume in which the tissue has been destroyed, with little harm to surrounding tissue. The usefulness and potential of this capability for medical purposes has been reviewed by Lele (1967).

In careful experiments with 2.7 MHz ultrasound Robinson and Lele (1972) produced lesions (of ellipsoidal shape) in cat brain and measured their dimensions. For a lesion produced with a focal intensity of 1140 W/cm^2 maintained for one second the reported lesion length and diameter were about 6 and 0.8 mm , resp.; these values increased with increasing intensity and irradiation time. They also made calculations (with the aid of computers) based on theory for sonic heat generation and heat conduction; the lesions were found to be regions where the temperature exceeds a threshold temperature of about 55 ^oC. Evidence is thus given for a thermal mechanism in ultrasonic production of lesions. In other work with lesions in brain, under other conditions, evidence has been given for nonthermal mechanisms. (See Section 14.9.8.)

14.9.5 Sonic "Cavitation": Small Volume Oscillations

As was learned in Section 14.5 gas bubbles or "nuclei" often exist in suspensions of biological macromolecules, cells and other particles. These nuclei are important since they can grow into larger bubbles under suitable conditions, for example, when decompression occurs. Nuclei also grow in sound fields when the amplitude is sufficiently high and other conditions are right; here the reasons for growth are not immediately obvious, but are taken up later.

When gas bubbles are present in a sonicated medium they are set into pulsation by the alternating sound pressure. To gain insight on the situation we consider the bubble as a sphere of initial radius R_o in a fluid of density ρ_o . A sound field is set up such that the pressure p in the vicinity of the bubble is given by

$$p = P_o + p_o \sin \omega t \; ; \qquad (14.28)$$

here P_o is the prevailing static pressure in the absence of sound (usually approximately equal to atmospheric pressure); ω is the angular frequency and p_o a constant equal to the amplitude of the "driving pressure". In response to the driving pressure $p_o \sin \omega t$ the bubble pulsates. Let us assume its radius R varies with time according to the relation

$$R = R_o + \xi_b \; ; \qquad \xi_b = \xi_{bo} \sin (\omega t + \alpha) \; ; \qquad (14.29)$$

hrere ξ_{bo} is the amplitude of the displacement ξ_b and α is the phase angle between the displacement ξ_b and the driving pressure $p_o \sin \omega t$. As the bubble pulsates the surrounding liquid is set into motion. Assuming spherical symmetry the displacement ξ in the liquid at any given time is the same at all points at the same distance r from the center of the bubble, and is along the radial direction. If the frequency is low the liquid moves as if it were incompressible. This means that after any given time t the total volume which has been displaced outward through any sphere of constant r is independent of r . Thus the volume displacement $4\pi R^2 \xi_b$ at the bubble surface at a given time is equal to the volume displacement $4\pi r^2 \xi$ at any arbitrary value of r at that same time. When ξ_b is small compared to R_o we can assume $R \simeq R_o$ and obtain

$$\xi = \frac{R_o^{\,2}}{r^2} \xi_b \; ; \qquad (14.30)$$

thus the displacement in the fluid varies inversely with r^2 . It follows that the velocity $\dot{\xi}$ also varies as r^{-2} . The assumption of incompressible liquid flow, on which these "inverse-square laws" are based, is valid for distances r small compared to the wavelength λ for sound waves in the liquid.

An interesting result is obtained by determining the total kinetic energy (KE) of the liquid outside the bubble, assuming Eqs. (14.29) and (14.30) hold. In Prob. 14.7 this is shown to be just

$$KE = \tfrac{1}{2} m \dot{\xi}_b^{\,2} \; , \qquad (14.31)$$

where

$$m = 4\pi R_o^{\,3} \rho_o \; . \qquad (14.32)$$

Thus the kinetic energy of the liquid is the same as that of a hypothetical small body, moving radially with the surface of the bubble, whose mass m is just three times that of the liquid displaced by the bubble.

Because the gas is of low density its kinetic energy is negligible but we must consider changes in its potential energy PE . It is shown in Prob. 14.8 that the latter quantity is given simply by

$$PE = \tfrac{1}{2} k\xi_b^2 + P_o(V_o - V) , \tag{14.33}$$

where V is the volume, V_o is the equilibrium volume, and

$$k = 12\pi\gamma R_o P_o ; \tag{14.34}$$

here γ is the ratio of specific heats for the gas and P_o , as before, is the atmospheric pressure. It is assumed that $\xi_b \ll R_o$ and that changes in state occur adiabatically. Because of the low compressibility of the liquid, changes in its potential energy are negligible. Thinking of the gas in the interior of the bubble and the liquid on its exterior as a single system we can take the total energy of the system to be equal approximately to KE plus PE , where KE is given by Eq. (14.31) and PE by Eq. (14.33). The total energy varies with time as work is done by the external medium in which the pressure, the "ambient" pressure, is the sum of P_o and $p_o \sin \omega t$. (The energy is ultimately supplied by an acoustic generator whose details are not important here.) The work done by the external medium in unit time is just the rate of decrease $(-\dot{V})$ of the bubble volume multiplied by the ambient pressure. Hence we have

$$\frac{d}{dt}(PE + KE) = -\dot{V}\,(P_o + p_o \sin \omega t) . \tag{14.35}$$

Letting $\dot{V}$ be given approximately by $4\pi R_o^2\,\dot{\xi}_b$ and substituting expressions for PE and KE from Eqs. (14.31) and (14.33) the following result is readily obtained (Prob. 14.9)

$$m\ddot{\xi}_b + k\xi_b = -4\pi R_o^2\, p_o \sin \omega t . \tag{14.36}$$

This differential equation is a "forced harmonic oscillator" equation, much referred to in physics; it applies to a body of mass m attached to a spring of stiffness k and acted on by a force which varies sinusoidally with time. In the bubble problem under consideration the "effective mass" is given by m in Eq. (14.32), the "effective stiffness" by k in Eq. (14.34), and the "force" by the right hand side of Eq. (14.36). Another term typically appears on the left hand side of Eq. (14.36), of the form $b\dot{\xi}$, where b is a constant; such a term, representing "friction" or "damping", is important in bubble dynamics but we shall not take space to discuss it here. It is easily verified by substitution that a solution ξ of Eq. (14.36) can be written as $\xi_{bo} \sin \omega t$, where

$$\xi_{bo} = \frac{4\pi R_o^2 P_o}{m\omega^2 - k} \quad . \tag{14.37}$$

A plot of the amplitude ξ_o versus frequency is shown in Fig. 14.10. At low frequencies where $m\omega^2 << k$ the denominator on the right hand side of Eq. (14.37) is approximately $-k$, where k is $12\pi\gamma R_o P_o$; the bubble is then "stiffness controlled" and responds to pressure variations as expected for slow adiabatic changes. (Prob. 14.10). At high frequencies the denominator becomes approximately $m\omega^2$ and the bubble is "mass controlled" ; since $m = 4\pi\rho_o R_o^3$ the displacement amplitude ξ_o is then given

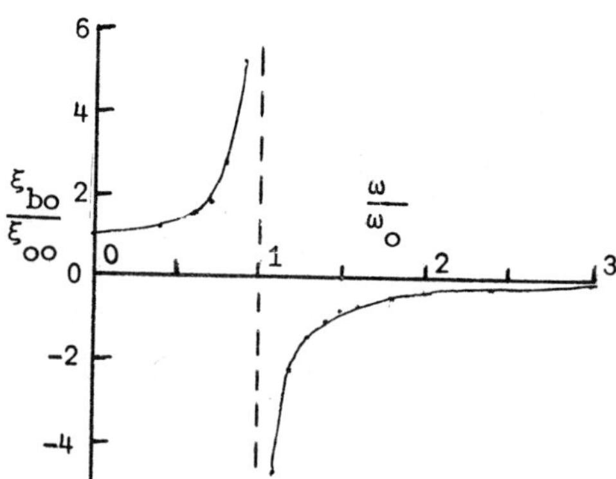

Fig. 14.10 Plot of amplitude versus frequency for a spherical gas bubble, based on Eq. 14.37. Ordinates give the nondimensional quantity ξ_{bo}/ξ_{oo} where ξ_{oo} is the value of the amplitude ξ_o when $\omega = 0$. Abscissae give the nondimensional quantity ω/ω_o , i.e., the ratio of the frequency to the resonant frequency.

by $p_o/\rho_o\omega^2R_o$ and thus varies inversely with the frequency squared. The denominator on the right hand side of Eq. (14.37) is zero and the expression for ξ_{bo} becomes infinite when $m\omega^2$ is equal to k; the angular frequency corresponding to this condition has the resonant value ω_o. If ξ_{oo} is the value of ξ_{bo} when $\omega = o$ we find from Eq. (14.37) that ξ_{bo}/ξ_{oo} is equal to the reciprocal of $(1-\omega^2\omega_o^{-2})$; this result is used in Fig. 14.10. We see that for $\omega << \omega_o$ the ratio ξ_{bo}/ξ_{oo} is approximately constant and equal to unity, but the ratio increases as ω approaches ω_o. For $\omega > \omega_o$ the ratio ξ_{bo}/ξ_{oo} is negative, signifying a reversal in phase by 180°. At high frequencies the amplitude ξ_{bo} becomes vanishingly small compared to ξ_{oo}. We find that the resonant angular frequency is given by

$$\omega_o^2 = \frac{3\gamma P_o}{\rho_o R_o^2}; \tag{14.38}$$

for an air bubble in water this leads to the formula

$$f_o = \frac{3.28}{R_o}, \tag{14.39}$$

where R_o is in millimeters while f_o is the resonant frequency (given by $\omega_o/2\pi$) in kHz. (Prob. 14.10). We see that a bubble of radius 1 mm is resonant at a frequency of 3.28 kHz and a bubble of radius 0.1 mm at a frequency of 32.8 kHz.

Equations (14.38) and (14.39) for the resonant frequency have been tested experimentally and found valid at frequencies as high as 100 kHz. At frequencies of 1 MHz and higher these equations need to be modified by including effects of surface tension. (Flynn, 1964). The fact that Eq. (14.37) yields an infinite value for ξ_{bo} when ω has its resonant value ω_o shows a deficiency in the differential equation, Eq. (14.36). For a relaistic determination of the bubble oscillation amplitude near resonance the processes which give rise to damping should be included.

In deriving results for bubble dynamics the liquid was assumed to flow as if it were incompressible, an assumption which holds approximately (at the relatively low amplitudes we are considering here) when the region of interest is small compared to the wavelength λ for compressional waves. We can now test the applicability of this approximation for bubbles of near-resonant size. The velocity of sound c_s in water is given in Table 7.1; at 20°C and frequency f the wavelength (given by c_s/f) is 1483/f, in meters when f is in Hz. From Eq. (14.39) converted to the same units, the resonant value of R_o is given by 3.28/f, smaller than the wavelength by about a factor of 450. Since the region of interest often extends only to a distance of about 3-4 R_o and seldom beyond 10 R_o we see that the assumption in question is well justified for bubbles whose size is near that for resonance.

14.9.6 Physical Concomitants of Bubble Vibration

When small pockets or bubbles of gas are present in a biological medium they typically have a marked influence on sound propagation, and on bioeffects of sound. This influence comes about in a surprizing variety of ways. We shall have speace here only to list some of these and discuss them very briefly.

1. Absorption and scattering

When sound waves pass through a medium containing a gaseous body the latter is set into pulsation. Because of various dissipative processes this leads to heat generation (which can be accounted for in modifying Eq. (14.36) to include a term $b\dot{\xi}$ on the left hand side). Since the energy comes ultimately from the sound wave its intensity is reduced; bubble vibration thus leads to sound adsorption.

Also, as the bubble pulsates it acts as a secondary source of sound by generating a wave which spreads outward from the bubble. Energy may then be lost from the primary sound field, especially if this is in the form of a well-defined "beam"; this kind of loss is attributed to "scattering".

2. Bubble growth by rectified diffusion

In solution there are typically gas bubbles or "nuclei" of microscopic size stabilized in cracks or crevices at the surfaces of suspended small solid particles or container walls as described in Section 14.5. When a gas nucleus is caused to pulsate by a sound field the condition for equilibrium under diffusion is altered and it may grow by a process initially suggested by Harvey, et al (1944, Reference 14.14) and which has been called "rectified diffusion". We shall not take up this process in detail but only indicate principles involved and give illustrative results.

From Sections 9.4.4 and 9.6.1 we realize that in the absence of sound gas will diffuse outward from a spherical gas bubble into the surrounding liquid at a rate Q given by

$$Q = 4\pi DR_o(c_o - c_\infty) \; ; \tag{14.40}$$

here D is the diffusion coefficient for the gas molecule relative to the liquid, c_o is the concentration of gas in the liquid just outside the bubble surface, and c_∞ is the concentration in the absence of a bubble. We shall assume, for simplicity, that the bubble is relatively large, so that effects of surface tension are negligible and the pressure in the bubble is essentially the atmospheric value P_o. Then c_o is the concentration of gas corresponding to saturation at pressure P_o.

When the bubble vibrates its surface area changes with time; also the concentration of gas in the liquid varies in both space and time. The process is rather complicated, but a number of investigators have shown mathematically as well as experimentally that vibration causes gas to enter the bubble. A simple equation was derived by Hsieh and Plesset (1961), valid for bubbles that are sufficiently large; according to this, the average gas influx rate $- < Q^* >$ resulting from vibration is given by

$$- < Q^* > = 24\pi DR_o c_o \left(\frac{\xi_{bo}}{R_o}\right)^2 , \tag{14.41}$$

where ξ_{bo}, as before, is the displacement amplitude of the bubble surface. The influx (of rate $< -Q^* >$ is superposed on the efflux (of rate Q); when $< Q^* >$ exceeds Q the bubble grows. By equating Q to $< Q^* >$, using Eqs. (14.40) and (14.41), we obtain an expression for the critical amplitude ξ_{boc}, such that growth occurs for any higher amplitude:

$$\left(\frac{\xi_{boc}}{R_o}\right)^2 = \frac{1 - \psi}{6} , \qquad (14.42)$$

where

$$\psi = c_\infty / c_o . \qquad (14.43)$$

Here ψ is the ratio of the concentration of dissolved gas to the saturation value. The more nearly saturated the liquid is, the smaller the amplitude ξ_{boc} required for bubble growth.

For a more detailed discussion, and references to literature on this topic see Flynn (1964).

3. Radiation forces on gas bubbles.

Another kind of behavior appears when a gas bubble is present in a sound field where the pressure amplitude varies in space. The bubble is found to move in the field, in a direction which depends on its size relative to that of a resonant bubble. Consider, for example, a "standing wave" field such as is typically set up when reflectors are present; here the sonic pressure might be given, for example, by

$$p(x,t) = p_x \sin \omega t , \quad p_x = p_o \sin kx . \qquad (14.44)$$

According to theory in Section 14.9.5 a spherical gas bubble placed in such a field will be set into vibration and its radial displacement will be given by Eq. (14.37). From this it is found (Prob. 14.11) that when $\xi_b << R_o$ the volume V at any instant is given by

$$V = V_o - Mp , \quad M = \frac{4\pi R_o}{\rho_o(\omega_o^2 - \omega^2)} , \qquad (14.45)$$

where V_o is a constant equal to the mean volume. The force F on any small object of volume V in a field where the pressure p varies with x is given by

$$F = -V \frac{\partial p}{\partial x} . \qquad (14.46)$$

We are particularly interested in the time-average of the force over one or more cycles, indicated by $\langle F \rangle$. Combining the above three equations we obtain

$$\langle F \rangle = \tfrac{1}{2} M p_x \frac{\partial p_x}{\partial x} = \frac{M}{4} \frac{\partial}{\partial x} (p_x)^2 . \qquad (14.47)$$

We see that when M is positive ($\omega < \omega_o$) the force is in the direction of increasing p_x^2 or, as is equally true, increasing pressure amplitude $|p_x|$; the opposite is true when M is negative ($\omega > \omega_o$). In the standing wave situation represented by Eq. (14.44) the zeroes of the pressure function p_x occur at planes (one-half wavelength apart) where kx is equal to $0, \pi$, 2π , etc. and maxima of p_x lie in between. We conclude that in such a field small bubbles, for which $\omega_o > \omega$, will move to the pressure maxima while bubbles larger than resonant size will move to the pressure zeroes. These predictions have been confirmed experimentally (Crum and Eller, 1969); they have implications for sonication of biological suspensions.

It is also a consequence of radiation forces that two bubbles, each smaller than resonant size, are attracted to each other; because of this they tend to move together, combine and form a single bubble which, of course, is of larger size.

4. Acoustic microstreaming

If a bubble oscillates in an asymmetrical pattern it is found that small scale time-independent eddying motion occurs in the liquid near the bubble. This is especially true if the bubble rests on a solid boundary. (Elder, 1959). This steady motion, circulatory in nature, is called acoustic streaming or, because of its small scale, microstreaming. Theory exists for the motion; we shall not discuss it here except to say that in the approximation usually used the streaming velocity near a pulsating bubble is proportional to ξ_{bo}^{2}. It is found that velocity gradients in the microstreaming can be great enough to damage biological cells; see Section 14.9.7.

5. Radiation forces on suspended particles near a pulsating bubble. Near a small sound source, such as a pulsating bubble, a small particle of volume V in suspension is acted on by a steady force F_x along the x direction given by

$$F_x = VB\frac{\partial\bar{T}}{\partial x}, \quad B = \frac{3(1 - \beta)}{2 + \beta}, \tag{14.48}$$

where $\bar{T}$ is the time-averaged kinetic energy per unit volume in the sound field, and β is ρ_o/ρ where ρ_o is the density of the liquid and ρ that of the particle. (Nyborg, 1967). Near a pulsating bubble the kinetic energy density $\bar{T}$ is relatively high, but drops off rapidly with increasing distance from the bubble. Hence if x is measured radially outward from a pulsating bubble the gradient $\partial\bar{T}/\partial x$ is negative. For a particle whose density ρ is greater than that of the liquid the quantity β is less than unity, and B is positive. Hence F_x is in the negative x direction, and the particle is "attracted" to the bubble. If the particle is less dense than the liquid we see that B is negative and that the force is "repulsive." Since $\bar{T}$ is proportional to the velocity-amplitude squared the force F_x, called a radiation force, is also proportional to the amplitude squared. The existence of radiation forces on particles in suspension near a vibrating bubble has been confirmed by D. L. Miller (unpublished work) in experiments with biological suspensions.

6. Surface waves and microbubble generation

The symmetry which we have assumed in some of the preceding discussion on motion of a vibrating bubble actually holds only in special circumstances. More typically the bubble surface is covered with waves, and has a rippled character. Even at modest ultrasonic levels these surface waves achieve rather high amplitudes, then become distorted and chaotic. A peculiarity the wave distortion, as shown by Storm (1974), is the appearance of evaginations which form very small bubbles, or "microbubbles"; these separate from the major bubble and appear in the surrounding liquid. These microbubbles probably are significant in serving as nuclei from which other bubbles grow in the sound field. (Neppiras and Fill, 1969).

7. Nonlinear behavior of the bubble.

In much of the discussion so far in this section we have assumed that the bubble responds linearly to the driving pressure; that is, the displacement amplitude ξ_{bo} has been assumed proportional to the driving pressure amplitude p_o, as in Eq. (14.37). In a linear response a driving pressure p which varies sinusoidally in time with angular frequency ω leads to a displacement ξ_b which (after the bubble has been oscillating for a sufficient length of time) also varies sinusoidally in time with the same frequency ω. Linear theory holds well when the driving pressure amplitude p_o is sufficiently small, but when p_o becomes of the order of atmospheric pressure (or less, if the bubble is nearly of resonant size) Eq. (14.37) is no longer a good approximation.

Theory applicable to the higher amplitudes was developed first by Noltingk and Neppiras (1950, 1951). These authors derived a nonlinear differential equation for the pulsation of a gas bubble, assuming it to remain spherical in shape. While such nonlinear equations are difficult to solve Noltingk and Neppiras were able, by the use of various approaches, to obtain valuable insight on the cavitation phenomena which occur commonly in sonicated liquids and suspensions. Of special importance is their finding that under typical conditions the bubble will shrink to a relatively small volume during a contraction phase. Assuming the volume change occurs nearly adiabatically (i.e., without much conduction of heat away from the bubble) the gas will reach a high temperature when the volume is small. The authors suggest that temperatures as high as 10,000°C might be achieved momentarily druing contraction. At such temperatures molecules in the gas dissociate and chemical reactions occur as the dissociation products combine with other species.

It is implicit in the nonlinear equations that a bubble subject to a driving pressure p_o sin ωt responds with a displacement ξ_b which does not vary sinusoidially with time. This affects the surrounding fluid since th bubble, as a secondary source, radiates sound whose characteristics depen on the displacement-versus-time profile of the source. In a fluid where cavitation occurs the sonic spectrum characteristically contains not only the driving frequency f (equal to $\omega/2\pi$) but some of its multiples (2f, 3f etc.), submultiples (such as f/2) , and other discrete frequencies, as we as fluctuating noise. Techniques for detecting harmonics, subharmonics an noise have been made use of in tests for the presence of cavitation in biological suspensions and tissues; see Section 14.9.7.

An extensive literature on the physics of cavitation exists, for which we do not have space here. A review of the subject has been given by Flynn (1964), Reference 14.50.

14.9.7 Bioeffects of Sonic Cavitation

Many research papers have been written on changes brought about in biological suspensions subject to ultrasonic fields. These changes are usually mediated by bubble activity or cavitation. Reviews have been given by El'piner (1964, Reference 14.47) and by Flynn (1964, Reference 14.50). Here we cite only a few examples to illustrate principles. Usually the cavitation field is much too complicated to permit detailed analysis of results. It is not unusual for all aspects of cavitation activity described in Section 14.9.6 to be occurring simultaneously. Our chosen examples are special in that attempts were made to simplify conditions and arrive at interpretations.

Goldman and Lepeschkin (1952) carried out interesting experiments with selected water plants. Positioning plant filaments along a standing wave field (such as that represented in Eq. (14.44)) they found that damage to the cells occurred principally near planes one-half wavelength apart corresponding to pressure maxima. Since gas bubbles were visible in the field they attributed the effect to "cavitation".

Considering some aspects of the behavior of gas bubbles we can arrive at a plausible explanation of the Goldman-Lepeschkin results. For as we have noted in connection with radiation forces on gas bubbles (sometimes called "Bjerknes forces"), and as was pointed out by Blake (1949), bubbles smaller than resonant size tend to move to pressure maxima. They then vibrate with enhanced amplitude and are especially effective in causing cellular changes. Also, as mentioned in connection with radiation force, a pair of bubbles which are each smaller than resonant size will tend to merge and form a new bubble. If the new bubble is not too large it will be more nearly of resonant size than the "parent" bubble and will vibrate more vigorously for that reason. But when bubbles are formed that are larger than resonant size these are moved, again <u>via</u> radiation force, to the pressure minima where they are inactive. In view of these expectations for bubble behavior it is reasonable, as found by Goldman and Lepeschkin, that biological effects take place especially at pressure maxima.

There is voluminous evidence that sonic cavitation causes changes in biological cells. One way this comes about is evidently by hydrodynamic shearing stress, as discussed in Section 14.7. Near a bubble there is oscillatory shear associated with the to-and-fro motion of the liquid and the bubble. But evidence is accumulating which shows that time-independent shear associated with acoustic microstreaming is even more important, at least, at the lower amplitudes. Especially convincing is the work of J. A. Rooney (1970); in his experimental arrangement he subjected a small amount of red cell suspension to the action of a single gas bubble set into pulsation at 20 kHz. He assayed cellular damage by measuring the extent to which Hb was released from the red cells. Using a bubble of mean radius 125 μm he found little Hb release at bubble amplitudes ξ_{bo} below about 18 μm , but above this amplitude Hb was found outside the cells in an amount increasing with amplitude as seen in Fig. 14.11 . Using theory for acoustic microstreaming he calculated that at the 18 μm threshold the red cells are subject to velocity gradients G up to 14,000 sec^{-1} as they near the bubble. The viscosity of his suspending liquid was high, about 0.31 P; hence the viscous stress ηG corresponding to threshold is about 4300 dyn/cm^2 . Because the velocity gradient is nonuniform in the microstreaming situation, there is uncertainty in the stress-threshold determination which Rooney estimates to be 1500 dyn/cm^2 or less. Another aspect of the situation is that any given cell is subject to high shear only for an instant, while it is swept near the bubble by the flow. This is important since, as noted in Section 14.7.6, the stress threshold increases as the time of stress application decreases. It is not surprising, then, that the stress threshold for damage to a red cell in a microstreaming field is several times greater than that in a steady shearing field.

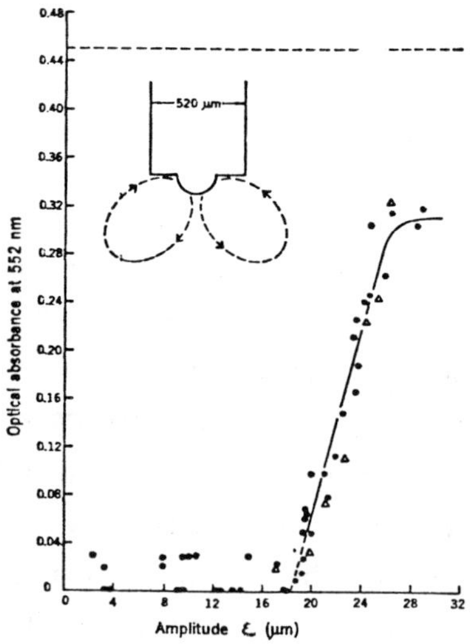

Fig. 14.11 Damage to red cells resulting from pulsation of a single stable bubble. Ordinate is proportional to quantity of Hb released from the cells in five minutes of sonation. Abscissae give radial oscillation amplitude ξ_{bo}. Dashed line represents 100% hemolysis. From Rooney (1970, Reference 14.60). Copyright 1970 by the American Association for the Advancement of Science.

Another way in which sound acts on cells *via* vibrating gas volumes has been reported by Harvey, et al (1928) and recently by Nyborg, Gershoy and Miller (1974). In these experiments they show that the gas contained in spaces between cells in plant tissue causes eddying movements to occur in these adjoining cells. In the most recent work the eddying is identified as acoustic streaming which occurs when cell boundaries are set into nonuniform vibration at sufficiently high amplitudes. The motion is similar to that found when the cell boundaries are vibrated by other means, discussed in Section 14.9.9.

At higher amplitudes sonic cavitation causes a variety of effects in bio-suspensions. The subject is reviewed by El'piner (1964, Reference 14.47). Effects included are rupture of cell membranes (probably caused by shear), degradation of nucleic acids, enzymes and other macromolecules (resulting partly from chemical action) and many others. A number of studies have been carried out by W. T. Coakley and associates, using focused ultrasound in aqueous suspensions. Under typical conditions they find there that cavitation occurs primarily in the form of discrete "events", each event being marked by the sudden appearance of a bubble cloud accompanied by a radiated acoustic signal. In one series of experiments Coakley, Hampton and Dunn (1971) studied effects on amoebae. They made the interesting finding that the number of amoebae destroyed in a given experiment was proportional to the number of cavitation events occurring during that experiment. Specifically, the results were as if each event damaged all cells in a spherical volume of radius 0.8 mm.

In animal tissue there appear to have been no convincing direct observations of sonic cavitation. In optically opaque organs, such as brain or liver, special methods would be required for detecting and characterizing any cavitation that might be occurring. In view of the evidence for gaseous nuclei in blood vessels (Section 14.5), especially after muscular activity, it is possible that sonication will cause these to grow and become active as cavitation sites. We will probably not be well informed on this matter until better techniques are devised for detecting small nuclei, and for studying the growth and activity of bubbles in optically opaque tissue.

14.9.8 Consequences of Nonlinearity

In Sections 14.9.4 and 14.9.7 we considered experimentally observed changes in bio-systems brought about with ultrasound by way of heating and cavitation, resp. There are other observed effects for which these explanations do not suffice. In considering other mechanisms we return to basic principles. It is important to realize that wave equations such as Eqs. (13.14) and (13.18), on which much of acoustics is based, are inexact. Hence solutions of these equations, such as the expression for a plane travelling wave in Eqs. (13.1) are not exact descriptions of real sound fields.

Wave equations are derived by proceeding as we did in Section 13.2 . A fundamental step in the derivation is to invoke a dynamical equation, such as Eq. (13.11). On the right hand side of the latter equation we have terms involving the density ρ and velocity u . Since ρ and u are both dependent variables the term $\partial(\rho u)/\partial t$, in which they appear as a product, is nonlinear. Approximations are made in arriving at the linearized form of the dynamical equation in Eq. (13.12).

Approximations are also introduced in other steps leading to the wave equations in Section 13.2 , for example, in assuming linear constitutive relations. Errors in these approximations are negligible at low amplitudes and hence the linearized wave equations, such as Eqs.(13.14) and (13.18) are valid if the amplitudes are not too high.

The exact equations of acoustics are nonlinear and are difficult to solve. However, much information is obtained by using methods for approximating the solutions. A first approximation, applicable only at low amplitudes, yields the linear wave equations, and solutions such as Eqs. (13.1) . In the latter equations the displacement ξ varies sinusoidally in time with angular frequency ω and amplitude $A\ e^{-\alpha x}$; this motion can be thought of as arising from a sound source (such as a piezoelectric plate) which vibrates with the same angular frequency ω and with displacement amplitude proportional to A . Similarly, in a linear approximation other quantities such as the velocity $\dot{\xi}$ and stress S_{11} vary sinusoidally in time with the same frequency as the source, and with amplitude proportional to the source amplitude.

It is clear that the linear approximation does not, by itself, account for any observed biological changes brought about by sound. For in this approximation each part of the medium simply moves to and fro, and at the end of any number of sound cycles is back at its starting place.

Much more relevant to bioeffects is the second approximation, which holds at moderate amplitudes. This approximation leads to contributions (to the various field quantities, such as displacement, velocity, pressure and temperature) which are independent of time. For example, we find from this approximation that the time-averaged pressure at any given point in a sound field is somewhat altered from its value (typically, atmospheric pressure) in the absence of sound. The magnitude of this excess "static" pressure, sometimes called <u>radiation pressure</u>, is proportional to the square of the source amplitude. Also we learn from the second approximation that when sound is set up in a fluid a steady circulatory flow will usually result, in which the fluid velocity is proportional to the square of the sonic amplitude. This flow is referred to as <u>acoustic streaming</u>.

Radiation pressure, acoustic streaming and other phenomena can all be thought of as resulting from a field of "effective force" $\underline{F}$, set up by the sound. This fictitious force F , obtained rather directly in the second-approximation approach, is time-independent with magnitude proportional to the square of the source amplitude. It acts on each part of the medium as if it were, for example, a gravitational force. Its magnitude and direction vary in space in a manner that depends on the nature of the sound field. In a fluid this force causes time-independent pressure distributions and flow fields to be established. In solid-like media typical of biological tissues we can expect time-independent stress distributions of various kinds to be set up . Since the stresses and movements caused by $\underline{F}$ are proportional to the square of the source amplitude, we expect these quantities, sometimes called"second-order quantities", to be insignificant (relative to linear quantities) at low amplitudes, but to become more and more important as the amplitude increases.

In the next subsection we consider briefly selected bioeffects of ultrasound, some of which can be explained in terms of second-approximation theory.

14.9.9 Other Bioeffects

Complications from heating and cavitation have been avoided in some experiments by use of small-scale arrangements for sonating single cells or small quantities of suspension. In one method longitudinal oscillations are set up in a solid bar, one end of which is tapered down to a small rounded tip. The bar plus its driver is sometimes called a <u>microvibrator</u>. A tissue or cell is positioned on the stage of an optical microscope, and the microvibrator positioned so that its tip contacts a selected part of a cell boundary. This method has been used with plant cells by Nyborg and Dyer (1960) and also by Gershoy and Nyborg (1973). Eddying motions and other movements of intracellular particles were seen which, it was concluded, could be partially accounted for by theory for acoustic streaming and radiation pressure. Movements within the nuclei of marine eggs sonated in this way were seen by Schmitt (1929) and by Wilson, et al. (1966). Wilson and Schnitzler (1963) showed that when a fertilized marine egg is contacted with a microvibrator at the time of cell division the division is delayed, probably because of a "gentle churning" of the cytoplasm which they observed. Dyer (1965) found that microvibration of the boundary of moss protonema during the process of cell division set the incipient crosswall into steady rotation. After the vibration ended he found that the cell division proceeded in a number of instances, but that the daughter cells and their progeny were abnormal. Ravitz and Schnitzler (1970) applied microvibration to the boundary of frog muscle cells and, by use of electron microscopy, showed that alterations had occurred in intracellular structures. While it has not been shown convincingly it is possible that all of these phenomena produced by localized vibration can be accounted for in terms of the effective force field $\underline{F}$ discussed in Section 14.9.8. In fluid-like media the field $\underline{F}$ causes rotations and flow, while in solid-like media deformations leading to damage may be produced, especially in fragile biological structures. A study of strains produced by microvibration in soft plastic solids has been carried out by Frost (1974), from which insight is gained on effects in solid-like tissue.

Using another kind of technique for microsonation Williams, Hughes and Nyborg (1970) devised an arrangement for subjecting biological suspensions to the field of a small wire vibrating at 20 kHz. They found effects on suspended cells, such as red cells, which are attributed to shearing stresses associated with small scale acoustic streaming. From theory for acoustic streaming they obtained estimates for the least value of the shearing stress which would bring about rupture of the red cell membrane. Rooney (1972) extended the work and showed that the shearing stress threshold decreases with increasing temperature; the trend is consistent to that shown by Krizan and Williams (1973, Reference 14.36) in Fig. 14.7, discussed in Section 14.7.6 In other experiments it has been shown that platelets (Williams, 1975) and white blood cells (Crowell, Kusserow and Nyborg, 1975) are altered by microsonation, using a vibrating wire at amplitudes significantly less than those required to reproduce hemolysis.

Methods for microsonation are useful in obtaining information about effects of ultrasound under conditions where temperature elevation and cavitation are not important. Also they sometimes present conditions where physical theory can be applied, and quantitative understanding achieved. A disadvantage is that they are different in scale and nature from the macroscopic arrangements more typically used in ultrasonic experiments, and there is some difficulty in relating conditions from one situation to the other.

In other experiments ultrasound is applied to animal tissues in beams at frequencies of 1 MHz and higher under conditions where it appears that temperature elevation and cavitation are not significant factors. Thus Fry, et al. (1950, 1951) found that pulses of ultrasound produced paralysis of frogs under conditions which were quite reproducible. The temperature elevation produced by the ultrasound was measured by thermocouples and judged insufficient to cause paralysis. The investigators tested for the possibility of cavitation by applying superpressures up to 13 atm during sonation. Since this superpressure was regarded as sufficient to suppress cavitation, and since sonic production of paralysis was not prevented, it was concluded that the paralysis was not caused by cavitation.

Further evidence was obtained by Fry and Dunn (1956) in experiments with day-old mice. These are poikilothermic, that is, they lack the usual control mechanisms; hence they can be cooled without damage. It was found that the ultrasound produced paralysis in these mice, even though they were pre-cooled to nearly $0^{\circ}C$; hence support was obtained for a nonthermal mechanism. Another interesting feature of the experiments on paralysis was revealed by determining the intensity I required for paralysis (actually the intensity required to produce paralysis in 50% of the experiments) for various values of the pulse duration t . The product $(I^{\frac{1}{2}}t)$ was found to be quite accurately constant for all experiments, including values of t varying over about a factor of ten. This shows that the total delivered energy is not constant, for this would be proportional to the quantity (It) .

In other experiments it has been shown by Connolly (1968) that red cells are altered by 1 MHz ultrasound under conditions where the action is believed mechanical. Specifically (see Connolly and Pond, 1967) the cause is thought to be viscous stress brought about by small-scale acoustic streaming. Connolly found the stress required for cell damage to decrease as the temperature is increased, a finding confirmed by the later results of Rooney (1972, Reference 14.73) and of Krizan and Williams (1973, Reference 14.36).

Pond, Woodward and Dyson have reported a striking phenomenon observed in blood vessels of chick embryos. When 1 MHz ultrasound is applied the red cells collect rather quickly into striae one-half wavelength apart. Presumeably radiation pressure causes this phenomenon, but the detailed mechanics are not yet known.

A summary of available data on thresholds for producing certain bio-effects is shown in Fig. 14.12 from Dunn and Fry (1971). Curve A shows the intensity in a pulse of ultrasound required to produce lesions in cat and rat brain as a function of pulse duration. It is based on data from three groups of observers and shows the threshold varying from about 10^4 W/cm^2 for 10^{-3} sec pulses to 100 W/cm^2 for 10 sec pulses. It has been suggested that curve A might be divided into three parts:

(1) A lower part, for intensities of 150 W/cm^2 or less, where the lesions are caused by sonic heating;

(2) An upper part, for intensities greater than 1500 W/cm^2 where the mechanism is cavitation; and

(3) An intermediate region where the mechanism is unknown.

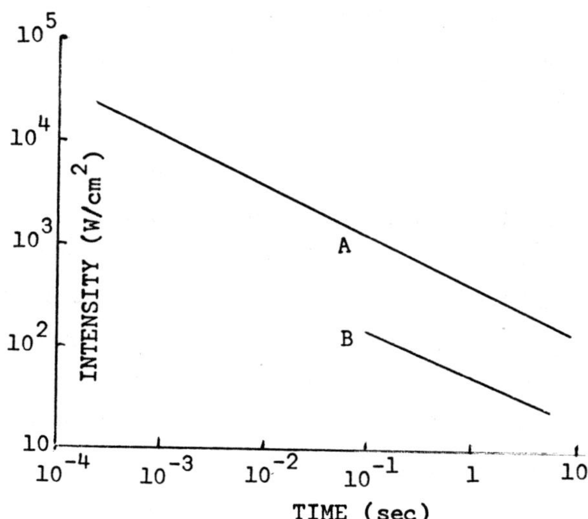

Fig. 14.12 Plots of intensity for pulses of 1 MHz ultrasound versus pulse duration corresponding to threshold conditions. Curve A: Lesions in brain. Curve B: Paralysis. From Dunn and Fry (1971, Reference 14.81).

Curve B shows similar data for paralysis of mice. The thresholds for curve B are less by about a factor of 8 than those for curve A .

In this subsection we have considered only a few of the topics which might have been taken up. Further information on bioeffects of ultrasound may be found in References 14.82 and 14.83 .

PROBLEMS

14.1 Derive the differential equation for steady spherically symmetrical heat flow quoted in Eq. (14.3). In doing so, follow these steps:
(a) Obtain an expression for the total heat generated per unit time in a shell lying between the spherical surfaces $r = r_1$ and $r = r_2$.
(b) Obtain an expression for the heat conducted away from the same shell in unit time. In doing this, assume a law for heat flow analogous to Fick's First Law (Eq. 9.27) for diffusion; specifically, assume that the heat flow outward across unit area of a spherical surface of radius r is given by $- \kappa\, \partial T/\partial r$, evaluated at that surface. (Remember that the shell has two surfaces!).
(c) Equate the above two expressions and obtain Eq. (14.3) by carrying out a limiting process.

14.2 Using Eqs. (14.4), verify the statement made in the text that the heat flow outward through any spherical surface of radius r (concentric with the heat source) is equal to the total rate of heat production within the surface. Show this both for $r < R$ and for $r > R$.

14.3 For Poiseuille flow along the x direction in a circular tube of inner radius a the velocity u is

$$u = A(a^2 - r^2) ,$$

where r is distance from the center of the tube and A is a constant.
(a) Obtain an expression for the velocity gradient $G = \partial u/\partial r$.
(b) Obtain an expression for the time θ required for a particle at a distance r from the axis to travel the length L of the tube.
(c) Show that the magnitude of $G\theta$ is

$$|G\theta| = \frac{2Lr}{a^2 - r^2} .$$

(d) Show that the volume rate of flow Q is $\frac{1}{2} \pi A a^4$.
(e) Show that the average value of $G\theta$ is $8L/3a$. Charm and Wong call θ the residence time and attach considerable significance to the product $G\theta$.

14.4 Using Eq. (14.10) calculate the maximum tension F to which a fibrinogen molecule is subjected in blood if the velocity gradient is 1470 sec^{-1} . (Use values for constants given elsewhere in the book.) Compare F with the breaking strength for a macromolecule, using the discussion in Section 14.7.2 (for DNA) as a guide.

14.5 Proceeding from the definition for the intensity I(x) in Eq. (14.16) and the expressions for ξ , u and S_{11} in Eqs. (13.1), (13.2) and (13.17) derive the expression cited in Eq. (14.17).

14.6 Show, as stated in the discussion following Eq. (14.18), that S_{11} , S_{22} and S_{33} are nearly equal in a travelling plane wave when $\varepsilon \ll 1$; also arrive at Eq. (14.19).

14.7 Show that the kinetic energy KE and the effective mass m for a spherical bubble oscillating in a liquid of density ρ_o are as given in Eqs. (14.31) and (14.32). In carrying out the proof make use of the fact that the kinetic energy KE of the entire mass of liquid is

$$KE = \int_{r=R}^{r=\infty} (\tfrac{1}{2}\rho_o \dot{\xi}^2)(4\pi r^2\, dr) ,$$

where $\dot{\xi}$ is the velocity at any distance r from the center of the bubble. Assume $\dot{\xi}$ is just $d\xi/dt$, where ξ is given by Eqs. (14.29) and (14.30). Also assume that the lower limit R is given sufficiently accurately by R_o (since the displacement ξ_b is always small compared to R_o) .

14.8 Show that the potential energy PE of the gas in a spherical bubble is given by Eqs. (14.33) and (14.34) when the volume is V, corresponding to a radius $R_o + \xi_b$; assume that changes in state occur adiabatically and that $\xi_b \ll R_o$. In doing this make use of the fact that PE is just the work done on the bubble in changing its volume adiabatically from V_o to V and is given by

$$PE = -\int_{V_o}^{V} p_g\, dV ,$$

where p_g is the pressure of the gas, which changes with V. Letting ξ' be an intermediate value of the radial displacement, consider ξ' varying from 0 to ξ during the change in volume, and proceed by these steps:

Prob. 14.8 (continued)

(i) Note from Eqs. (4.11) and (4.14) that for a small adiabatic change in volume dV the corresponding increase in pressure is $-\gamma P_o\, dV/V_o$;
(ii) Show that when the fractional change in volume dV/V_o is small it is approximately equal to $3\xi'/R_o$ and hence (considering the result from (i)) the pressure p_g can be written approximately as

$$p_g = P_o - \frac{3\gamma P_o}{R_o}\,\xi' \,.$$

(iii) Show that dV can be approximated as $4\pi R_o^2\, d\xi'$ and carry out the integration for PE.

14.9 Carry out indicated steps and obtain the differential equations for bubble vibration given in Eq. (14.36).

14.10 In this problem we consider properties of the solution given in Eq. (14.37) for the pulsation of a gas bubble.
(a) Show that at low frequencies the displacement amplitude ξ_{bo} is given by

$$\xi_{bo} = \frac{-R_o p_o}{3\gamma P_o} \,,$$

and explain why this is an expected result for slow adiabatic changes.
(b) Show that at high frequencies the expression becomes

$$\xi_o = \frac{p_o}{\rho_o \omega^2 R_o} \,.$$

(c) Verify Eqs. (14.38) and (14.39) for ω_o and f_o .

14.11 Verify the expression for the volume V given in Eq. (14.45), including the expression for M.

APPENDIX A. TABULATED CONSTANTS FOR VARIOUS MEDIA: UNITS

TABLE A.1 SURFACE TENSION

The <u>surface</u> (or <u>interfacial</u>) tension σ is a parameter characteristic of a pair of media. It is equally well called <u>surface</u> (<u>interfacial</u>) <u>energy</u> and gives the work required to form unit area of interface between two media. For example, consider a spherical droplet of Medium 1 immersed in Medium 2. If the droplet is deformed slightly its surface area will increase, say, by dA, and the work required to produce the deformation will be σ dA. The units of σ are force/length or energy/area.

$$1 \text{ dyn/cm} = 1 \text{ erg/cm}^2 = 10^{-3} \text{ newtons/m} = 10^{-3} \text{ joules/m}^2 .$$

When one of the media is a liquid or solid and the second is a gas at ordinary pressures σ is about the same as if the second medium were vacuum. In the following table Medium 1 is always a liquid and Medium 2 a gas or liquid. We see from the table that σ decreases with increasing temperature (for water and carbon tetrachloride), that water has an unusually high value of σ among nonmetallic liquids, and that organic solutes (such as acetic acid and methyl alcohol) in water tend to decrease σ. Units for the temperature T are degrees Centigrade and those for σ are dyn/cm .

MEDIUM 1 (VARIOUS LIQUIDS)	MEDIUM 2	T	σ
Acetone	air	0	26.21
Benzene	air	10	30.22
Carbon bisulphide	vapor	20	32.33
Carbon tetrachloride	vapor	20	26.95
		100	17.26
Chloroform	air	20	27.14
Ethyl alcohol	air	0	24.05
Glycerol	air	20	63.4
Mercury	air	20	435.5
Methyl alcohol	air	0	24.49
Toluene	vapor	10	27.7

TABLE A.1 (Continued)

MEDIUM 1	MEDIUM 2	T	σ
Water	air	0	75.6
		20	72.75
		30	71.18
		40	69.56
		60	66.18
		100	58.9
(SOLUTIONS WITH WATER AS SOLVENT)			
NaCl			
% by wt. 2.84%	air	20	73.75
% by wt. 10.46%			76.05
Acetic acid			
% by wt. 2.48%	air	30	64.40
% by wt. 10.0%			54.60
Methyl alcohol			
% by vol. 10%	air	20	59.04
% by vol. 25%			46.38
(LIQUID AGAINST LIQUID)			
Water	Benzene	20	35
Water	$C\,Cl_4$	20	45
Water	Mercury	20	375

TABLE A.2 CONVERSION OF UNITS

Quantities in a given line are equal. Thus from the first line under "Units for Stress" we find that 1 atm is equal to 1.013×10^6 dyn/cm^2 to 1.033 kg/cm^2 , and to 1.013×10^5 n/m^2. Abbreviations are listed in Table A.3 .

UNITS FOR STRESS

atm	dyn/cm^2	kg/cm^2	n/m^2
1	1.013×10^6	1.033	1.013×10^5
9.87×10^{-7}	1	1.020×10^{-6}	10^{-1}
9.68×10^{-1}	9.81×10^5	1	9.81×10^4
9.87×10^{-6}	10	1.020×10^{-5}	1

UNITS FOR FORCE

dyn	n	g	kg
1	10^{-5}	1.020×10^{-3}	1.020×10^{-6}
10^5	1	1.020×10^2	1.020×10^{-1}
9.81×10^2	9.81×10^{-3}	1	10^{-3}
9.81×10^5	9.81	10^3	1

UNITS FOR LENGTH

m	cm	mm	μm	A
1	10^2	10^3	10^6	10^{10}
10^{-2}	1	10	10^4	10^8
10^{-3}	10^{-1}	1	10^3	10^7
10^{-6}	10^{-4}	10^{-3}	1	10^4
10^{-10}	10^{-8}	10^{-7}	10^{-4}	1

TABLE A.2 (Continued)

UNITS FOR ENERGY

erg	joule	ev	calorie
1	10^{-7}	6.24×10^{11}	2.390×10^{-8}
10^{7}	1	6.24×10^{18}	2.390×10^{-1}
1.602×10^{-12}	1.602×10^{-19}	1	3.83×10^{-20}
4.18×10^{7}	4.18	2.611×10^{19}	1

UNITS FOR POWER

erg/sec	watt	mw
1	10^{-7}	10^{-4}
10^{7}	1	10^{3}
10^{4}	10^{-3}	1

TABLE A.3 ABBREVIATIONS

cm	centimeter	mW	milliwatt
dyn	dyne	P	poise
g	gram	cP	centipoise
kg	kilogram	s	stoke
atm	atmosphere	cs	centistoke
N	newton	m	meter
A	Angstrom	mm	millimeter
ev	electron volt	μm	micron
W	watt		

TABLE A.4

APPROXIMATE VALUES OF ELASTIC CONSTANTS

For some materials the constants vary significantly with such parameters as impurity content and details of preparation. Values given below are representative but are not necessarily highly accurate. Multiply values of E, μ, λ, and B by 10^{10} for units of N/m^2, or by 10^{11} for units of dyn/cm^2. Multiply values of β by 10^{-11} for units of m^2/N, or by 10^{-12} for units of cm^2/dyn. The ratio ν is dimensionless. All quantities defined in Chap. 4. $\beta = B^{-1}$.

Material	E	μ	λ	ν	B	β
Aluminum, rolled	6.9	2.5	6.1	0.36	7.7	1.30
Copper, rolled	12.5	4.6	13.1	0.37	16.2	0.62
Gold, hard-drawn	8.12	2.85	15.0	0.42	16.9	0.59
Iron, electrolytic	20.6	18.2	11.3	0.29	16.8	0.59
Lead, rolled	1.6	0.54	3.3	0.43	3.7	2.7
Nickel	21.4	8.0	16.4	0.34	21.7	0.46
Platinum	16.7	6.4	9.9	0.30	14.2	0.70
Silver	7.5	2.7	8.6	0.38	10.4	0.96
Tin, rolled	5.5	2.1	4.0	0.34	5.4	1.85
Tungsten, drawn	36.2	13.4	31.3	0.35	40.3	0.25
Pyrex glass	6.2	2.5	2.3	0.24	3.9	2.6
Lucite	0.40	0.143	0.56	0.4	0.66	15.1
Polyethylene	0.076	0.026	0.288	0.46	0.30	33
Polystyrene	0.53	0.12	0.34	0.41	0.42	24
Water, 25°C	0	0	0.225	(0.5)	0.225	44.4

For gelatin at 10°C the value of μ varies from 10^3 to 5×10^5 dyn/cm^2 for concentrations in the range 1% to 20%. For water and gelatin the expected value of ν is 0.5; for gelatin the values of λ, B and β are close to those for water and the expected value of E is 3μ.

APPENDIX B MISCELLANEOUS FORMULAE

SERIES

$$e^x = \sum_{n=o}^{\infty} x^n/n! = 1 + x + \tfrac{1}{2}x^2 + - - - - \quad . \tag{B.1}$$

$$e^x \cong \sum_{n=o}^{M} x^n/n! \quad \text{if } M >> x \ . \tag{B.2}$$

$$\ell n(1 + x) = x - \tfrac{1}{2}x^2 + (1/3)x^3 - (1/4)x^4 + - - \quad . \qquad (\text{valid for } x^2 < 1) \tag{B.3}$$

$$(1 + x)^n = 1 + nx + \frac{n(n-1)}{2!} x^2 + \frac{n(n-1)(n-2)}{3!} x^3 + - - \quad . \qquad (\text{valid for } x^2 < 1) \tag{B.4}$$

$$\frac{1}{1 - x} = 1 + x + x^2 + - - - - \quad , \quad (x^2 < 1) \tag{B.5}$$

$$\frac{1 - x^{N+1}}{1 - x} = 1 + x + x^2 + - - - - - + x^N \ . \tag{B.6}$$

$$f(x) = f(x_o) + (x - x_o)\left(\frac{df}{dx}\right)_{x_o} + \tfrac{1}{2}(x - x_o)^2 \left(\frac{d^2f}{dx^2}\right)_{x_o} + - - \ , \tag{B.7}$$

STIRLING'S APPROXIMATION: N >> 1

$$N! \cong (2\pi)^{\frac{1}{2}} N^{N+\frac{1}{2}} e^{-N} \ . \tag{B.8}$$

$$\ell n\ N! \cong N\ \ell n\ N - N + \tfrac{1}{2}\ \ell n\ 2\pi N \ , \tag{B.9}$$

DEFINITE INTEGRALS

$$\int_o^{\infty} e^{-a^2x^2}\, dx = \sqrt{\pi}/2a \ . \tag{B.10}$$

$$\int_o^{\infty} x\, e^{-x^2}\, dx = \tfrac{1}{2} \ . \tag{B.11}$$

$$\int_o^{\infty} x^2 e^{-x^2}\, dx = \sqrt{\pi}/4 \ . \tag{B.12}$$

APPENDIX C

PROBABILITY AND STATISTICS

In this Appendix we give a number of definitions, theorems, formulae, proofs and tables, useful in connection with Chap. 8 of the text.

C.1 DEFINITION OF PROBABILITY

Suppose an experiment has event A as a possible outcome and that in a large number M of experiments this event occurs M_A times. Then the probability of the event A (in such a set of experiments) is P_A, given by

$$P_A = M_A / M . \tag{C.1}$$

Thus the probability of event A might be called the fractional rate of occurrence of that event.

C.2 THEOREM I: MUTUALLY EXCLUSIVE EVENTS

The probability that some one of a set of mutually exclusive events will occur is the sum of the probabilities for the separate events.

In proving this theorem, suppose that when a particular kind of experiment is done the result is always one of a number of mutually exclusive possibilities or events which we call A, B, C - - - F . If a large number M of these experiments are done event A occurs M_A times, event B occurs M_B times, etc., where

$$M = M_A + M_B + - - - M_F . \tag{C.2}$$

Let p_A be the probability of event A, p_B the probability of event B, etc.; then as in Eq. (C.1) we have

$$p_A = M_A / M ; \quad p_B = M_B / M ; \quad \text{etc.} \tag{C.3}$$

Now consider the probability p(A,B,C) that in M experiments some one (i.e., one and only one) of the set of three events (A, B, C) will occur. We can write

$$p(A,B,C) = M(A,B,C)/M , \tag{C.4}$$

where M(A,B,C) is the number of occurrences of some one of (A,B,C). But we see also, from the definitions of M_A , etc., that

$$M(A,B,C) = M_A + M_B + M_C \; . \tag{C.5}$$

Combining Eqs. (C.3), (C.4) and (C.5) we obtain

$$p(A,B,C) = p_A + p_B + p_C \; . \tag{C.6}$$

Clearly this result can be generalized, and thus leads to the principle enunciated in Theorem I .

Corollary to Theorem I : If A, B, . . . F are a set of mutually exclusive events and if it is certain that some one of these will occur then

$$p_A + p_B + - - - p_F = 1 \; . \tag{C.7}$$

The proof of this Corollary is left as an exercise. (Prob. 8.6).

C.3 THEOREM II: INDEPENDENT EXPERIMENTS

The probability for occurrence of a specified event in each of a set of independent experiments is the product of the separate probabilities.

In proving this theorem consider a set of K independent Experiments identified as No. 1, No. 2, etc., up to No. K. By stating that these are independent we mean that the probability of a given outcome for, say Experiment No. 7, is not at all affected by results of other Experiments. Then for any of the Experiments we can assume the possible outcomes are A, B, etc., with probabilities p_A, p_B, etc., as in the discussion of Theorem I . Suppose the entire set of K Experiments is carried out M times, where M is very large. We use the following terminology:

M_A: Number of sets for which Event A occurs in a specified Experiment, say, No. 1 .

M_{AB}: Number of sets for which Events A and B each occur in specified Experiments, say, A in No. 1 and B. in No. 2.

M_{ABC}: Number of sets for which Events A, B and C each occur in specified Experiments, say, A in No. 1, B in No. 2 and C in No. 3 .

p_A: Probability that Event A will occur in a specified Experiment

p_B: Probability that Event B will occur in a specified Experiment

p_C: Probability that Event C will occur in a specified Experiment

p_{AB}: Probability that both of the two Events A and B will occur in specified Experiments, e.g. A in No. 1 and B in No. 2 .

p_{ABC}: Probability that all of the three Events A, B and C will occur in specified Experiments, e.g. A in No. 1, B in No. 2 and C in No. 3 .

The probability p_{ABC} is the one we seek. This is found readily if we note that

$$M_{AB} = p_B M_A \tag{C.8}$$

and

$$M_{ABC} = p_C M_{AB} \ . \tag{C.9}$$

For then we find that

$$p_{ABC} = \frac{M_{ABC}}{M} = p_C \frac{M_{AB}}{M} = p_C p_B \frac{M_A}{M} \tag{C.10}$$

or

$$p_{ABC} = p_A p_B p_C \ . \tag{C.11}$$

Clearly, this result can be generalized, and hence leads to the principle stated in Theorem II .

C.4 DEFINITION OF ARRANGEMENT

An <u>Arrangement</u> is an ordered sequence of objects or symbols. A good example is an English word, composed of letters from the alphabet a, b, c, etc. If all letters in a word are distinct, any interchange generates a new possible "word" (of course, not necessarily an English word!) and hence a new Arrangement. Thus "act" and "cat" are different Arrangements of the objects (a, c, t) . For a word in which some letters are repeated, an interchange of identical letters obviously does not change the word and hence does not lead to a new Arrangement. Thus interchange of the "e's" in the word "seem" does not lead to a new Arrangement of the objects (e, e, m, s) .

C.5 THEOREM III. ARRANGEMENTS OF DISTINCT OBJECTS

There are M! possible Arrangements for M distinct objects.

In proving this theorem we imagine a specific procedure for generating Arrangements: We provide boxes numbered 1, 2, 3, etc., up to M , and fill these boxes in order by assigning objects to them. We then note the following:

Any of the M objects may be chosen for Box No. 1 ,

Any of the remaining (M-1) objects may be chosen for Box No. 2 ,

Any of the remaining (M-2) objects may be chosen for Box No. 3 ,

and so on . For the last two boxes:

Either of two objects may be chosen for Box No. (M-1), and

Only one object remains to fill Box No. M .

Because the objects are distinct any change in assigning them to these boxes generates a new Arrangement. Hence the total number of possible Arrangements is Q_1 , given by

$$Q_1 = M(M - 1)(M - 2) - - - (2)(1) = M! , \tag{C.12}$$

as was to be shown.

As an example consider the 24 "words" which are generated as arrangements of the letters (a, b, c, d) . (See Table C.1). Here M = 4 . There are 4 letters to choose for the first Position. This is seen by the existence of 4 groups designated as I, II, III and IV; here the first letter in group I is "a" , the first in group II is "b", etc. After the first letter is chosen there are 3 possibilities for the second Position. This is seen by the existence of 3 subgroups under each of the headings I, II, etc. Referring to group I: in the upper subgroup the second letter is "b", in the middle one it is "c" and in the lower one "d" . In filling the third Position there are only 2 possibilities, and in the fourth only one; hence each subgroup consists of 2 "words" or Arrangements.

The total number of different Arrangements is 4! or M! as expected.

C.6 THEOREM IV: TWO CLASSES OF OBJECTS

If there are n identical objects of Type A and (N-n) identical objects of Type B the number of possible arrangements of the N objects is Q_2, given by

$$Q_2 = \frac{N!}{n!(N-n)!} \tag{C.13}$$

In proving this theorem we notice that if the Type A objects were modified to make them distinct the number of arrangements of the N objects would be increased from Q_2 to, say, Q'. Specifically Q' would be equal to the product of Q_2 and the number of possible arrangements of the type A objects among themselves (assuming them temporarily distinct). From Theorem III we then see that Q' is just $Q_2 n!$.

Proceeding further with this line of reasoning we let Q'' be the number of arrangements of the N objects which would be possible if the Type B objects were also made distinct. For Q'' we have

$$Q'' = Q'(N - n)! = Q_2 n!(N - n)! \tag{C.14}$$

But Q'' is just the number of possible arrangements for N distinct objects and hence is equal to N! . Setting Q'' equal to N! in Eq. (C.14) and solving for Q_2 we obtain the expected results given in Eq. (C.13) .

C.7 THEOREM V: ANY NUMBER OF CLASSES

Suppose there are N_a identical objects of Type A, N_b identical objects of Type B, etc., making up a total of N objects. Then the number of possible arrangements is Q_3, given by

$$Q_3 = \frac{N!}{N_a!\, N_b! \,---} \tag{C.15}$$

This theorem is a rather obvious generalization of Theorem IV and can be proved by extending the reasoning used there. (Prob. 8.7) .

C.8 PROOF THAT $\sigma^2 = a$ IN POISSON STATISTICS WHEN $a \gg 1$.

By $\underline{a}$ (or $\bar{n}$) we mean the average observed value of n in a counting experiment. (See Eq. (8.5)). By σ^2 we mean the average value of the quantity $(n - a)^2$. We wish to show that $\sigma^2 = \underline{a}$ when the probability W_n^N is given by the Poisson expression, Eq. (8.32).

We first determine the average value of n^2, which we designate as A. From the general expression for an average, Eq. (8.24), we may write

$$A = \sum_{n=0}^{N} n^2 W_n^N . \tag{C.16}$$

Using the Poisson result for W_n^N, Eq. (8.32), this may be written

$$A = e^{-a} \sum_{n=o}^{N} n^2 a^n/n! = e^{-a}[a + \sum_{n=2}^{N} n^2 a^n/n!]. \tag{C.17}$$

A useful trick now is to replace n^2 by

$$n^2 = n(n - 1) + n ; \tag{C.18}$$

when this is done, and the expression for A rearranged we obtain

$$A = e^{-a} [a + a^2 \sum_{n=2}^{N} \frac{a^{(n-2)}}{(n-2)!} + a \sum_{n=2}^{N} \frac{a^{(n-1)}}{(n-1)!}] \tag{C.19}$$

or

$$A = e^{-a}[a^2 \sum_{n=2}^{N} \frac{a^{(n-2)}}{(n-2)!} + a \sum_{n=1}^{N} \frac{a^{(n-1)}}{(n-1)!}] \tag{C.20}$$

or

$$A = e^{-a}[a^2 \sum_{r=o}^{N-2} \frac{a^r}{r!} + a \sum_{s=o}^{N-1} \frac{a^s}{s!}] . \tag{C.21}$$

In proceeding from Eq. (C.20) to Eq. (C.21) we have replaced (n-2) by r in the first sum on the right hand side and have replaced (n-1) by s in the second sum. Since both (N-1) and (N-2) are much greater than a , each of these sums may be approximated by e^a , according to a result from Appendix B. We then obtain

$$A = a^2 + a . \tag{C.22}$$

It is now relatively easy to obtain the desired expressions for σ^2 . We start with an expression in terms of the W_n^N , which we then expand:

$$\sigma^2 = \Sigma(n - a)^2 W_n^N = \Sigma n^2 W_n^N - 2a \Sigma n W_n^N + a^2 \Sigma W_n^N , \tag{C.23}$$

where all sums are from n = 0 to N. From previous definitions and results we then obtain

$$\sigma^2 = A - 2a\bar{n} + a^2 \cdot 1 = (a^2 + a) - 2a^2 + a^2 . \tag{C.24}$$

Hence, as desired, we find

$$\sigma^2 = a . \tag{C.25}$$

TABLE C.1

ARRANGEMENTS OF THE OBJECTS (a, b, c, d)

I	II	III	IV
abcd	bacd	cabd	dabc
abdc	badc	cadb	dacb
acbd	bcad	cbad	dbac
acdb	bcda	cbda	dbca
adbc	bdac	cdab	dcab
adcb	bdca	cdba	dcba

TABLE C.2

POISSON FUNCTION W (n,a) = $a^n e^{-a}/n!$

n \ a	1	2	3	4	5	6	7	8	9	10	11	12
0	.368	.135	.050	.018	.007	.003	.001	.000	.000	.000	.000	.000
1	.368	.271	.149	.073	.034	.015	.006	.003	.001	.001	.000	.000
2	.184	.271	.224	.147	.084	.045	.022	.011	.005	.002	.001	.000
3	.061	.180	.224	.195	.140	.089	.052	.029	.015	.008	.004	.002
4	.015	.090	.168	.195	.176	.134	.091	.057	.034	.019	.010	.005
5	.003	.036	.101	.156	.176	.161	.128	.092	.061	.038	.022	.013
6	.001	.012	.050	.104	.146	.161	.149	.122	.091	.063	.041	.026
7	.000	.003	.022	.060	.104	.138	.149	.140	.117	.090	.065	.044
8		.001	.008	.030	.065	.103	.130	.140	.132	.113	.089	.066
9		.000	.003	.013	.036	.069	.101	.124	.132	.125	.109	.087
10			.001	.005	.018	.041	.071	.099	.119	.125	.119	.105
11			.000	.002	.008	.023	.045	.072	.097	.114	.119	.114
12				.001	.003	.011	.026	.048	.073	.095	.109	.114
13				.000	.001	.005	.014	.030	.050	.073	.093	.106
14					.001	.002	.007	.017	.032	.052	.073	.091
15					.000	.001	.003	.009	.019	.035	.053	.072
16						.000	.001	.005	.011	.022	.037	.054
17							.001	.002	.006	.013	.024	.038
18							.000	.001	.003	.007	.015	.026
19								.000	.001	.004	.008	.016
20									.001	.002	.005	.010
21									.000	.001	.002	.006
22										.000	.001	.003
23											.001	.002
24											.000	.001
25												.000

TABLE C3

THE GAUSSIAN DISTRIBUTION FOR $\sigma = 1$; SEE EQS. (8.45)

$$W(x) = (2\pi)^{-\frac{1}{2}} \exp\left(\frac{-x^2}{2}\right) \qquad \alpha(x) = \int_0^x W(x_0)dx_0$$

x	W	α	x	W	α
0.0	.3989	.0000	2.1	.0440	.4821
0.1	.3970	.0398	2.2	.0355	.4861
0.2	.3910	.0793	2.3	.0283	.4893
0.3	.3814	.1179	2.4	.0224	.4918
0.4	.3683	.1554	2.5	.0175	.4938
0.5	.3521	.1915	2.6	.0136	.4953
0.6	.3332	.2257	2.7	.0104	.4965
0.7	.3123	.2580	2.8	.0079	.4974
0.8	.2897	.2881	2.9	.0060	.4981
0.9	.2661	.3159	3.0	.0044	.4987
1.0	.2420	.3413	3.1	.0033	.4990
1.1	.2179	.3643	3.2	.0024	.4993
1.2	.1942	.3849	3.3	.0017	.4995
1.3	.1714	.4032	3.4	.0012	.4997
1.4	.1497	.4192	3.5	.0009	.4998
1.5	.1295	.4332	3.6	.0006	.4998
1.6	.1109	.4452	3.7	.0004	.4999
1.7	.0940	.4554	3.8	.0003	.4999
1.8	.0790	.4641	3.9	.0002	.5000
1.9	.0656	.4713	4.0	.0001	.5000
2.0	.0540	.4772			

TABLE C.4

THE ERROR FUNCTION. SEE EQ. (9.42)

z	erf(z)	z	erf(z)
0.00	0.000		
0.10	0.112	1.10	0.880
0.20	0.223	1.20	0.910
0.30	0.328	1.30	0.934
0.40	0.428	1.40	0.952
0.50	0.520	1.50	0.966
0.60	0.604	1.60	0.976
0.70	0.678	1.70	0.984
0.80	0.742	1.80	0.989
0.90	0.797	1.90	0.993
1.00	0.843	2.00	0.995

APPENDIX D

THERMODYNAMIC RELATIONS FOR A PERFECT MONATOMIC GAS

D.1 INTRODUCTION

For gases the laws relating temperature, volume and pressure are relatively simple. Because of this, and because important features of the laws can be explained in terms of simple kinetic theory, gases provide convenient test media for illustrating concepts and principles of thermodynamics. In this Appendix we shall treat the simplest situation: a monatomic gas of such low particle density that interactions between atoms can be ignored. For such a medium the "perfect gas law" is applicable, as given for one mole of gas in Eq. (1.2) ; in that equation the molar gas constant R_g is given by kN_a , where k is the Boltzmann constant and N_a is Avogadro's number. The internal energy for a noninteracting gas consists only of translational energy and hence, according to Eq. (8.2) , is given by 3kT/2 per atom. For a mole of gas (N_a atoms) the internal energy U is therefore given by

$$U = \frac{3}{2} R_g T . \tag{D.1}$$

Equation (10.2) provides an appropriate expression of the First Law of Thermodynamics, with dU given by

$$dU = \frac{3}{2} R_g \, dT \quad \text{(per mole)} . \tag{D.2}$$

D.2 ADIABATIC CHANGE IN STATE

Suppose the gaseous system is in an insulated container so that heat cannot pass into or out from the system. Any change in state that occurs under these conditions is said to be adiabatic . In an adiabatic change dQ is zero and the First Law, Eq. (10.2), becomes

$$PdV = -dU = -\frac{3}{2} R_g \, dT. \tag{D.3}$$

Substituting for P from the perfect gas law, Eq. (1.2), yields

$$\frac{dV}{V} = -\frac{3}{2} \frac{dT}{T} , \tag{D.4}$$

showing that the fractional change in volume dV/V is proportional to the fractional change in Kelvin temperature. We now integrate the left hand side of Eq. (D.4) between the limits V_o and V_1 and the right hand side between the corresponsing limits T_o and T_1 ; here T_o is the Kelvin temperature of the system in the "0" state, when the volume is V_o , and T_1 is its temperature in the "1" state when the volume is V_1 . We obtain

$$\ln \frac{V_1}{V_o} = - \frac{3}{2} \ln \frac{T_1}{T_o} . \tag{D.5}$$

We might solve this equation for V_1/V_o , but even in its present form it leads clearly to the following statement: In an adiabatic change between states "0" and "1" the volume ratio V_1/V_o depends only on the temperature ratio T_1/T_o , and vice versa.

The work ΔW_{ad} done by the gaseous system in passing adiabatically from the "0" state to the "1" state is

$$\Delta W_{ad} = \int_o^1 P\, dV = - \int_o^1 dU = U_o - U_1 ; \tag{D.6}$$

here $U_o - U_1$ is the amount by which the internal energy of the system decreases in changing from the "0" to the "1" state. Since U_o depends only on the temperature T_o and U_1 only on the temperature T_1 we see that the work done by this system in an adiabatic change depends only on the initial and final temperatures.

D.3 ISOTHERMAL CHANGE IN STATE

Since the internal energy for our system depends only on the temperature the increment dU is zero in any isothermal change. Hence for such a change the First Law, Eq. (10.2), becomes simply

$$dQ = P\, dV ;$$

the work done by the system under these conditions is just equal to the heat added to it. In a finite change of volume from V' to V" the heat added ΔQ and work done ΔW_{iso} are given by

$$\Delta Q = \Delta W_{iso} = \int_{V'}^{V''} P\, dV . \tag{D.7}$$

Substituting for P from the perfect gas law, Eq. (1.2), yields for temperature T

$$\int_{V'}^{V''} P\,dV = R_g T \int_{V'}^{V''} \frac{dV}{V} = R_g T \,\ell n \frac{V''}{V'} . \tag{D.8}$$

Combining the above two equations we obtain

$$\Delta Q = \Delta W_{iso} = R_g T \,\ell n \frac{V''}{V'} . \tag{D.9}$$

D.4 CARNOT CYCLE

We now use the results in Sections D.2 and D.3 to consider a series of four steps, called the Carnot cycle , in which changes in state are introduced but finally the system is brought back to its starting condition. We shall see that the net amount of heat added to the gas in such a cycle is, in general, not zero but that the change in entropy is zero. We consider a mole of perfect gas whose parameters are initially p_o , V_o and T_o .

D.4.1 Step 1 . The gas undergoes expansion adiabatically ($\Delta Q = 0$) to a new volume V_1 while its temperature reaches the value T_1 . The work done during this step is ΔW_1 , given from Eq. (D.6) by the simple expression

$$\Delta W_1 = U_o - U_1 . \tag{D.10}$$

D.4.2 Step 2 . An isothermal expansion occurs at temperature T_1 to a new volume V_2 . From Eq. (D.9), with suitable translation of symbols, the heat ΔQ_2 added and work ΔW_2 done in this step are given by

$$\Delta Q_2 = \Delta W_2 = R_g T_1 \,\ell n \frac{V_2}{V_1} . \tag{D.11}$$

D.4.3 Step 3. An adiabatic compression takes place reducing the temperature to its starting value T_o , the volume then becoming V_3 . We now recall that the work done by the gaseous system in an adiabatic change depends only on the initial and final temperatures. Hence ΔW_3 is given by

$$\Delta W_3 = U_1 - U_o , \tag{D.12}$$

just the negative of ΔW_1 , Eq. (D.10).

D.4.4 Step 4. At constant temperature T_o the gas is compressed to its starting volume V_o . Comparing with Eq. (D.11) we see that the heat ΔQ_4 added and work ΔW_4 done are given by

$$\Delta Q_4 = \Delta W_4 = R_g T_o \, \ell n \, \frac{V_o}{V_3} \, . \tag{D.13}$$

D.4.5 Totals for the Cycle

We first consider the total heat added to the system during the entire four-step cycle; this is given by the sum of ΔQ_2 and ΔQ_4 since, of course, no heat was added in steps 1 and 3 . Letting the sum be represented by the cyclic integral $\oint dQ$ we obtain

$$\oint dQ = R_g T_1 \, \ell n \, \frac{V_2}{V_1} + R_g T_o \, \ell n \, \frac{V_o}{V_3} \, , \tag{D.14}$$

which it is convenient to rewrite as

$$\oint dQ = R_g (T_1 - T_o) \, \ell n \, \frac{V_2}{V_1} + R_g T_o \, \ell n \, \frac{V_2 V_o}{V_1 V_3} \, ; \tag{D.15}$$

here use has been made of a property of logarithms : the sum of $\ell n\, x$ and $\ell n\, y$ is equal to $\ell n\, xy$. We now recall a fact noticed in connection with Eq. (D.5): In an adiabatic change from T_o to T_1 the ratio V_1/V_o depends only on the ratio T_1/T_o . Since step 3 is an adiabatic change from T_1 to T_o , and since the step is reversible, we see that step 1 leads to the same volume ratio as would the reverse of step 3 , so V_2/V_3 is equal to V_1/V_o . Hence the last term on the right in Eq. (D.15) contains the quantity $\ell n\, 1$, and is zero.

It is now easy to determine the total work $\oint dW$ done during the cycle. We see immediately that the contributions ΔW_1 and ΔW_3 add to zero. Hence $\oint dW$ is the sum of ΔW_2 and ΔW_4 which is the same as the sum $\oint dQ$. We have therefore from Eq. (D.15) (remembering that the second term on the right is zero)

$$\oint dW = \oint dQ = R_g (T_1 - T_o) \, \ell n \, \frac{V_2}{V_1} \, . \tag{D.16}$$

Let us now consider the increase $\oint dS$ in the entropy S of the gaseous system during a Carnot cycle. Since no heat is added during steps 1 and 3 it is clear from its definition in Eq. (10.5) that the entropy is not changed during these steps. Hence $\oint dS$ is given by the sum of $\Delta Q_2/T_1$ and $\Delta Q_4/T_o$. From Eqs. (D.11) and (D.13) we then obtain, using the additive property for logarithms,

$$\oint dS = R_g \ell n \frac{V_2 V_o}{V_1 V_3} = 0 \; ; \qquad \text{(D.17)}$$

here we made use of the fact noted earlier, that V_2/V_3 is equal to V_1/V_o .

The result in Eq. (D.17) is very important. Showing the total change of S in a cycle to be zero gives support to the idea that S is a state function. Admittedly, we have shown the cyclic integral $\oint dS$ to be zero only for a special system and for a special kind of cycle. But it is found from experience, and expressed in the Second Law of Thermodynamics that $\oint dS$ is zero generally and dS is a state function for any system. By contrast the cyclic changes of work W and heat Q are clearly not zero (in general) and these quantities do not qualify as state functions.

APPENDIX E

COMPLEX NUMBERS

We list here definitions, properties and theorems relative to complex numbers, as used in Chaps. 13 and 14.

Definition A complex number, say, $\underline{z}$ is a pair of real numbers, x and y, with certain properties to be described.

Representation We write $\underline{z}$ in either of the following forms: $\underline{z} = x + jy$ or $\underline{z} = (x,y)$.

Definition We call x the real part of $\underline{z}$ and y the imaginary part of $\underline{z}$.

Definition By a pure real number we mean one of the type which may be written either as (x,0) or, simply, as x. By a pure imaginary number we mean one of the type which may be written either as (0,y) or as jy.

Definition Two complex numbers $\underline{z}_1 = (x_1,y_1)$ and $\underline{z}_2 = (x_2,y_2)$ are equal if $x_1 = x_2$, $y_1 = y_2$.

Definition The sum of $\underline{z}_1$ and $\underline{z}_2$ is:

$$\underline{z}_1 + \underline{z}_2 = [(x_1 + x_2) , (y_1 + y_2)] .$$

(The difference is defined similarly)

Definition The product of $\underline{z}_1$ and $\underline{z}_2$ is:

$$\underline{z}_1 \underline{z}_2 = [(x_1x_2 - y_1y_2), (x_1y_2 + x_2y_1)] .$$

From this it follows that the product of a real number A and a complex number $\underline{z} = (x,y)$ is:

$$A \underline{z} = (Ax,Ay) .$$

Definition The reciprocal $\underline{z}^{-1}$ of $\underline{z}$ is that number which, if multiplied by $\underline{z}$ yields the product (1,0) .

Using previously given rules one obtains:

$$\underline{z}^{-1} = (x',y'), \quad \text{where} \quad x' = \frac{x}{x^2 + y^2} ; \ y' = \frac{-y}{x^2 + y^2} .$$

Definition The derivative of $\underline{z}$ with respect to any real parameter t is:

$$\frac{d\underline{z}}{dt} = (\frac{dx}{dt}, \frac{dy}{dt}) .$$

By extension a similar rule applies to a derivative of any order with respect to a real parameter.

APPENDIX E (continued)

Definition The integral of $\underline{z}$ with respect to any real parameter t, is that complex number which when differentiated yields $\underline{z}$. Thus

$$\int \underline{z}\, d t = (\int x dt\ , \int y dt) .$$

An Equivalent Way of Defining Operations on a Complex Variable

It can be shown that all the operations on $\underline{z}$ defined above are just the familiar ones for real variables if $\underline{z}$ is assumed represented by (x + jy) and j is a constant treated like a real number with this unusual property:

$$j^{2n} = (-1)^n\ ,\ j^{2n+1} = (-1)^n\, j ,$$

where n is any integer. (As an aid to remembering this property j is frequently referred to as the "Square root of -1"). When the result of the "real variable" operation on (x + jy), etc., is reduced to the form X + jY the real and imaginary parts are given by (X, Y), respectively, and are the same as if they had been obtained by application of the complex number operations defined above.

Definition The magnitude $|\underline{z}|$ of $\underline{z}$ is a real number given by

$$|\underline{z}| \equiv (x^2 + y^2)^{\frac{1}{2}} .$$

From this and previously given rules it can be shown that the magnitude of a product is equal to the product of the separate magnitudes; thus

$$|\underline{z}_1 \underline{z}_2| = |\underline{z}_1|\cdot|\underline{z}_2| \text{ and } A\,|\underline{z}| = |A\,\underline{z}| .$$

Definition The complex conjugate $\underline{z}^*$ of $\underline{z}$ is given by

$$\underline{z}^* = (x,-y).$$

It follows that

$$\underline{z}\underline{z}^* = |z|^2 .$$

Definition The phase angle θ of $\underline{z}$ is given by

$$\theta \equiv \tan^{-1} (y/x)$$

It can be shown that the phase angle for a product is the sum of the separate phase angles.

APPENDIX E (continued)

<u>Definition</u> By the complex function $e^{j\theta}$ we mean:

$$e^{j\theta} = (\cos\theta, \sin\theta) .$$

It can be shown (using this definition and the rules for operations on complex numbers given above) that:

(1) $e^{j\theta}e^{j\phi} = e^{j(\theta+\phi)}$,

(2) $e^{j.0} = 1;\quad e^{j\pi/2} = j;\quad e^{-j\pi/2} = -j;\quad e^{j\pi} = -1$.

(3) $\frac{d}{dt}(e^{j\omega t}) = j\omega e^{j\omega t}$,

(4) $|e^{j\theta}| = 1$,

(5) Any complex number $\underline{z} \equiv (x,y)$ can be written in the form

$$\underline{z} = Ae^{j\theta},$$

where A is the magnitude of $\underline{z}$ and θ the phase angle.

<u>Theorem</u>

Let $\underline{z}$ be a number which satisfies the equation

$$A\frac{d^2\underline{z}}{dt^2} + B\frac{d\underline{z}}{dt} + C\,\underline{z} = \underline{E} ,$$

where

(i) A, B, C and t are real;

(ii) $\underline{z}$, $\underline{E}$ are complex, given by

$$\underline{z} = (x,y);\quad \underline{E} = (E_1, E_2) .$$

Then it follows from previously given rules that

$$A\frac{d^2x}{dt^2} + B\frac{dx}{dt} + C\,x = E_1$$

$$A\frac{d^2y}{dt^2} + B\frac{dy}{dt} + C\,y = E_2 .$$

APPENDIX E (CONTINUED)

<u>Theorem</u> Let A and B be complex; let Re [u] represent the real part of u ; and let < v > represent the time-average of v. Then it can be shown that

$$\langle \text{Re}\,[A\,e^{j\omega t}]\,\text{Re}\,[Be^{j\omega t}] \rangle = \tfrac{1}{2}\,\text{Re}\,[AB^*]\ .$$

REFERENCES

Study of this text should be supplemented by much outside reading. For readers whose background is primarily physical any of various standard texts on general biology will help in learning terminology and basic facts about living systems. For those whose background is primarily biological, a good general physics text should be available for review of basic principles in physics.

Some guidance is needed in finding suitable material on interdisciplinary topics. Listed below are books, review articles and research papers which bear on various aspects of biological physics. Under General References are several standard textbooks on biophysics, as well as a few other sources which help to suggest the scope of biologically related physics. Following these are references, listed chapter by chapter, which pertain more specifically to material taken up in this book. These latter references are the source of subject matter presented in the text, or they provide further reading on a topic discussed; they are listed here in the order cited in each chapter. An alphabetical listing of authors appears in the Author Index.

General References

1. Molecular Biophysics by R. B. Setlow and E. C. Pollard, Addison-Wesley Publishing Co., Inc., Reading, Mass. 1962.

2. Biophysical Science by Eugene Ackerman, Prentice-Hall, Inc., Englewood Cliffs, New Jersey, 1962.

3. Biophysical Principles of Structure and Function by F. M. Snell, S. Shulman, R. P. Spencer and C. Moos, Addison-Wesley Publishing Co., Inc., Reading, Mass. 1965.

The above three references are textbooks on biophysics for students at the advanced-undergraduate or graduate level. Besides their intended purposes they serve as good sources of basic information on selected biological cells and macromolecules, presented from a point of view appropriate for this course.

4. Physics for Biology and Medicine, I. W. Richardson and E. B. Neergard, Wiley-Interscience, 1972. A text designed for first-year medical students including principles and applications of continuum mechanics, electricity, thermodynamics, acoustics and optics.

5. Biophysical Science - A Study Program, J. Oncley, Ed., John Wiley & Sons, Inc., New York, N. Y. 1959. (Same material as in Review of Modern Physics Volume 31, pages 1-568 (1959).) This is a collection of 61 papers, written by authorities in a variety of areas, designed to help define the wide-ranging subject of biophysics and suggest promising areas for research.

6. What is "Biological Physics"? A Resource Letter, E. K. Hege and M. L. Gupta, Journal of Biological Physics, Volume 1, pp 69-122 (1973). This is an extensive and valuable source, in which are listed and described journals, year-books, reference books, monographs and text-books which relate to Biological Physics. In the authors' words: "The purpose of this resource letter is threefold: To attempt a refinement of the tenuous definition of the term "Biological Physics". To do this *via* a compendium .... of materials which might appropriately be labelled biological physics. To provide a useful introduction to the learning resources in biological physics for college students and their professors"

REFERENCES FOR CHAPTER ONE

1.1 H. Margenau, The Nature of Physical Reality (McGraw-Hill Book Co., Inc., N. Y. 1950). In this book the author defines and discusses many important concepts in the philosophy of science, some of which are alluded to in Section 1.1 of the present chapter. The quotation from Laplace in Section 1.1 comes from Margenau's treatment of "Laplace's demon", pp 397 ff. "Causality", or "determinism", is discussed in his Chapter 19.

1.2 D. R. Wilkie, Muscle, The Institute of Biology's Studies in Biology No. 11 (Edward Arnold, Ltd, 1968). In this small volume of 63 pages basic facts of muscle structure and behaviour are presented in a lucid manner.

1.3 H. E. Huxley, Muscle Cells, Chapter 7 (116 pp) in The Cell, Vol. IV, edited by J. Brachet and A. E. Mirsky (Academic Press, 1960). This is a classic review chapter by a prominent investigator in the subject.

1.4 A. F. Huxley, Muscle Structure and Theories of Contraction, Chapter 6 (pp 255-318) in Progress in Biophysics, Vol. 7 (Pergamon Press, 1957). A specific simple model for the action of a cross bridge during contraction is described and analyzed mathematically in Part IV.

1.5 A. V. Hill, "Work and Heat in a Muscle Twitch", Proc. Roy. Soc. B136 , 220-227 (1949). In this one of a long series of contributions by the author, he measures both the work done and the heat generated during muscle contraction.

1.6 A. M. Gordon, A. F. Huxley and F. J. Julian, "The Variation in Isometric Tension with Sarcomere Length in Vertebrate Muscle Fibers", J. Physiol. 184, 170-192 (1966). This article is the source for Fig. 1.2 .

1.7 D. Fawcett, Cilia and Flagella, Chapter 4 (pp. 217-297) in The Cell, Vol. II, edited by J. Brachet and A. E. Mirsky (Academic Press, New York, 1961). A comprehensive review of the subject.

1.8 M. A. Sleigh, Patterns of Ciliary Beating, in Aspects of Cell Mobility, Symposia of the Society for Experimental Biology, No. 22, pp. 131-150 (Academic Press Inc., New York 1968). Sequences are recorded showing how the shape of a beating cilium changes with time. A sliding-filament hypothesis is advanced.

1.9 B. Afzelius, "Electron Microscopy of the Sperm Tail", J. Biophysic. and Biochem. Cytol. 5, 269-278 (1959). Electron micrographs are presented which show "arms" for the outer filaments.

1.10 I. R. Gibbons, "Studies on the Protein Components of Cilia from Tetrahymena Pyriformis", Proc. Nat. Acad. Sci. USA 50, 1002-1010 (1963). See Reference 1.11 .

1.11 I. R. Gibbons and A. J. Rowe, "Dynein: A Protein with Adenosine Triphosphate Activity from Cilia", Science 149, 424-425 (1965). This paper and that in Reference 1.10 describe experiments in which the "arms" are shown to consist of macromolecules (molecular weight 600,000) to which the name dynein is given.

1.12 D. L. Ringo, "Subunits in Flagellar Fibers", J. Ultrastruct. Res. 17, 266-277 (1967). On the basis of electron micrographs it is proposed that each filament in a cilium or flagellum is composed of near-spherical subunits, each about 45 A in diameter.

1.13 C. J. Brokaw, Mechanisms of Sperm Movement, in Aspects of Cell Mobility, Symposia of the Soc. for Experim. Biol., No. 22, pp. 101-116 (Academic Press, Inc., N. Y. 1968). Energy considerations are discussed and alternative mechanisms for the movement are compared.

1.14 R. Rikmenspoel, "Contractile Processes in Flagella", Biophys. J. 11, 446-463 (1971). A wave equation, developed in earlier papers, is given for flagellar bending waves. By comparing consequences of the equation with observations the author obtains basic information about mechanisms for the motions.

REFERENCES FOR CHAPTER TWO

2.1 R. R. Long, Mechanics of Solids and Fluids (Prentice-Hall, Inc., Englewood Cliffs, N.J., 1961). This is a text of intermediate level in which Cartesian tensors are introduced. Proof that any symmetrical tensor can be put in diagonal form is in Section 1.9. A general justification for the transformation rule of Eq. (2.8) is given in Section 2.3, where it is also shown that the stress tensor is symmetrical.

2.2 P. S. Laplace, Mecanique celeste, Translated from the French, with a commentary, by Nathaniel Bowditch. (Boston, Gray, Little and Wilkins, 1829-39). This translation, itself a set of rare books, is from a very ambitious work by Laplace in which his stated aim was to derive all known phenomena of the system of the world, from strict mathematical principles. His treatment of surface tension, including Eq. (2.30), is in lengthy Supplements to Volume 10.

2.3 N. S. Adam, The Physics and Chemistry of Surfaces (Dover Publications, Inc., New York, 1968).
This classic work, originally published by Oxford University Press (1930, 1938, 1941), contains much valuable information on surface tension, surface films, adsorption and other interfacial phenomena. A derivation of the Laplace law, Eq. (2.30), appears in Chapter 1.

2.4 R. P. Rand and A. C. Burton, "Mechanical properties of the red cell membrane. I. Membrane stiffness and intracellular pressure", Biophys. J. 4, 115-135 (1964). Information about the membrane is obtained by analyzing its response to stress in the form of locally applied suction.

2.5 J. Hoffman, "Physiological Characteristics of Human Red Cell Ghosts", J. Gen. Physiol. 42, 9-28 (1958). Experiments with ghosts of red cells (obtained by emptying the cells of hemoglobin) yield results on the tension required to break the membrane.

2.6 A. Katchalsky, O. Kedem, C. Klibansky and A. de Vries, "Rheological Considerations of the Haemolysing Red Blood Cell", in Flow Properties of Blood, pp. 155-171 (Pergamon Press, 1960). Red cells are hemolyzed (caused to release their hemoglobin) by lowering the external concentration c of NaCℓ to a critical value c_c; it is found that c_c depends on whether c is decreased slowly or rapidly.

2.7 J. M. Mitchison and M. M. Swann, "The Mechanical Properties of the cell surface. I. The Cell elastimeter", J. Exp. Biol. 31, 443-460 (1954). Here the authors describe the "cell elastimeter" which embodies a capillary to which suction is applied.

REFERENCES FOR CHAPTER SIX

6.1 Lamb, Horace, "Hydrodynamics" (Dover Publications, N.Y., 1945). This is a classic advanced treatise. Frictional constants for the sphere and ellipsoid are derived in Paragraphs 337 through 339; references are also given to earlier literature.

6.2 Burgers, J. M. "On the motion of small particles of elongated form, suspended in a viscous liquid", Chap. III (pp. 113-184) in Second Report on Viscosity and Plasticity, prepared by the Committee for the Study of Viscosity of the Academy of Sciences at Amsterdam (Kon. Ned. Acad. Wet., Verhand. (Eerste Sectie), DI XVI, No. 4, pp. 1-287 1938). There is much valuable information here, including theory whose results are quoted in Eqs. (6.18) and (6.19).

6.3 Einstein, A. "A New Determination of Molecular Dimensions" Chapter III in "The Theory of Brownian Movement" by A. Einstein (Dover Publications, Inc. 1956). The mathematical and physical basis is given here in English translation for Einstein's expression giving the viscosity of a suspension containing spheres, Eqs. (6.38) and (6.56). Originally the material was published in the following papers: Ann. Physik 19, 289-306 (1906); Ann. Physik 34, 591-592 (1911).

6.4 Taylor, G. I. "The Viscosity of A Fluid Containing Small Drops of Another Fluid", Proc. Roy Soc. A 138, 41-48 (1932). Here the author extends the Einstein expression (Eq. 6.56) for spheres to include the situation when the spheres are liquid.

6.5 Simha, R. "The Influence of Brownian Movement on the Viscosity of Solutions", J. Phys. Chem. 44, 25-34 (1940). Here the expressions are derived which are plotted in Figs. 6.5 and 6.6; also the limiting expression given in Eq. (6.58) is stated here. See also Reference 6.6.

6.6 Frisch, H. L. and Simha, R. "The Viscosity of Colloidal Suspensions and Macromolecular Solutions", Chapter 14 (pp 525-613) in Rheology, Vol. 1, F. R. Eirich, Ed., (Academic Press, Inc. N.Y. (1956). This is a valuable detailed review of various theories for the viscosity of suspensions containing particles of various shapes and kinds. For example the results for ellipsoids, on which Fig. 6.5 is based, are given in convenient form. Reference is made to the Staudinger work ,

6.7 Morawetz, Herbert "Macromolecules in Solution" (Volume 21 of a series on High Polymers, Interscience Publishers, N.Y., (1965) Chapter VI of this monograph contains much valuable information on "frictional properties of dissolved macromolecules" which is relevant to the present Chapter.

6.8 Debye, P. "The Intrinsic Viscosity of Polymer Solutions", J. Chem Phys. 14, 636-639 (1946). Here the author carries through a relatively simple analysis for the free-draining flexible-chain model of a polymer molecule. His theory forms the basis for the abbreviated treatment presented in Section 6.45 .

REFERENCES FOR CHAPTER SEVEN

GENERAL

7.1 B. S. Guttman, Biological Principles (W. A. Benjamin, Inc., New York, 1971). Here contemporary material on a wide range of topics is treated in a thought-provoking style and in an attractive format.

7.2 T. C. Ruch and H. D. Patton (eds.) Physiology and Biophysics, (W. B. Saunders Co., Phila. 1965). This standard text has 63 chapters on various subjects, each by an author expert on that subject.

7.3 P. L. Altman and D. S. Dittmer (eds.) Blood and other Body Fluids (Federation of American Societies for Experimental Biology, Washington, D.C. 1961). Tables of data are listed on the composition and properties (mostly chemical) of biological fluids.

WATER

7.4 D. Eisenberg and W. Kauzmann, The Structure and Properties of Water (Oxford University Press, 1969). A comprehensive advanced treatise on water properties and their explanations, where possible, in terms of basic principles.

7.5 F. M. Snell, S. Shulman, R. P. Spencer and C. Moos, Biophysical Principles of Structure and Function (General Reference 3) . Chapter 4 of this text is on "Water" and gives an attractive account of the subject in terms of basic principles at an intermediate level.

7.6 Biophysics and Physical Chemistry of Connective Tissue, Proceedings of a conference held at Stowe, Vermont, October 10-16, 1965; J. H. Bland and R. L. Lipson, Chairmen. Published in Federation Proceedings, May-June 1966, pp 939-1145. The articles in these "Proceedings" span a variety of topics relevant to this Chapter, especially on the subject of water, bone and synovial fluid.

7.7 S. Glasstone, K. J. Laidler and H. Eyring, The Theory of Rate Processes (McGraw-Hill Book Co., Inc., N. Y., 1941) An advanced treatise on a method for calculating "rates" for a wide variety of processes; Chapter IX of this volume deals with viscosity and diffusion.

7.8 J. Frenkel, Kinetic Theory of Liquids (Dover Publications, Inc., 1955) An advanced treatise (translated from a text originally written in Russia in 1942) on basic properties of liquids. In this Chapter we have referred mainly to material on viscosity in Frenkel's Chapter IV.

SOLUTIONS

7.9 A. Weissler and V. A. Del Grosso, The velocity of sound in sea water, J. Acoust. Soc. Amer. 23, 219-223 (1951). This paper is a source of data on NaCl solutions for Table 7.2.

7.10 J. Stuehr and E. Yeager, "Propagation of ultrasonic waves in electrolytic solutions", Chap. 6 in Physical Acoustics, Vol II A, W. P. Mason, Ed. (Academic Press, N.Y. 1965). An authoritative review chapter; of special relevance here are data on compressibilities for electrolytes.

7.11 E. L. Carstensen and H. P. Schwan, Acoustic properties of hemoglobin solutions, J. Acoust. Soc. Am. 31, 305-311 (1959). Data on sound velocity are given here from which we calculate compressibilities for Hb solutions.

7.12 L. Ananthanarayanan and A. Veis, The molecular parameters of monomeric and acid-soluble collagens. Low shear gradient viscosity and electric birefringence, Biopolymers 11, 1365-1377 (1972). Viscosity data on collagen are obtained here, and compared with theory for ellipsoids.

7.13 P. Doty, Characterization of macromolecules, and Configurations of macromolecules in solution, in Biophysical Science - A Study Program (General Reference 5), pp 61-68 and pp 107-117, resp. The author has made much use of the parameter [η] in characterizing biological macromolecules. In these articles he summarizes the subject briefly and gives bibliography.

BLOOD

7.14 A. L. Copley and G. Stainsby (eds.) Flow Properties of Blood and Other Biological Systems (Pergamon Press, New York, 1960). There are about forty articles here on mechanical characteristics of biological fluids, based on a conference at Oxford in 1959.

7.15 C. H. Best and N. B. Taylor, The Living Body, (Henry Holt and Company, N. Y., 1958). A lucid and interesting description of the human blood is given here.

7.16 A. C. Burton, Hemodynamics and the physics of the circulation, Chap. 27 in Reference 7.2 . Many interesting features are discussed here of blood flow in the human body. Much reference is made to a Ph.D. thesis by R. H. Haynes (The rheology of blood, University of Western Ontario, 1957) and an article by Haynes and Burton (Amer. J. Physiol. 197, 943-950 (1959)).

7.17 G. W. Scott Blair, Elementary Rheology (Academic Press, New York, 1969). A charming small book introducing rheology ("the study of deformation of materials, including flow") to nonrheologists. There are two chapters on blood and other biological media.

7.18 H. E. Schultze and J. F. Heremans, Molecular Biology of Human Proteins (Elsevier Publishing Co., 1966). This is a source of data on plasma proteins.

7.19 R. J. Urick, A sound velocity method for determining the compressibility of finely divided substances, J. Appl. Phys. 18, 983-987 (1947). Compressibility data for red cells and plasma are obtained here.

7.20 A. S. Ahuja and G. Bugliarello, A note on the compressibility of blood, Biorheology 7, 199-203 (1971). The authors support the method of Urick (Ref. 7.19).

7.21 A. L. Copley, Apparent viscosity and wall adherence of blood systems, pp 97-121 in Ref. 7.14 . It is shown here that the apparent viscosity of plasma in flow through a tube is dependent on the nature of the wall.

7.22 E. L. Carstensen, K. Li and H. P. Schwan, Determination of the acoustic properties of blood and its components, J. Acoust. Soc. Am. 25, 286-289 (1953). Compressibility data for albumin solutions are given.

7.23 E. W. Merrill, E. R. Gilliland, G. Cokelet, H. Shin, A. Britten and R. E. Wells, Jr., Rheology of human blood, near and at zero flow, Biophysic. J. 3, 199-213 (1963). These authors use a Couette arrangement to measure yield-stress and obtain plots of stress *vs* velocity gradient.

7.24 R. H. Haynes, The viscosity of erythrocyte suspensions. A review of theory, Biophysic. J. 2, 95-103 (1962). Concepts of *apparent viscosity*, *differential viscosity*, and *generalized viscosity* are defined and related. An empirical equation is described.

SYNOVIAL FLUIDS

7.25 J. C. Caygill and G. H. West, The rheological behaviour of synovial fluid and its possible relation to joint lubrication, Med. & Biol. Engng 7, 507-516 (1969). Investigations on human synovial fluid are reported.

7.26 B. N. Preston, M. Davies and A. G. Ogston, The composition and physicochemical properties of hyaluronic acids prepared from ox synovial fluid and from a case of mesothelioma, Biochem. J. 96, 449-474 (1965). Results are given of a very extensive investigation of HA.

7.27 H. Morawetz, *Macromolecules in Solution* (Ref. 6.7). An analysis of light scattering by polymers is treated in pp 243-262.

7.28 A. G. Ogston and J. E. Stanier, On the state of hyaluronic acid in synovial fluid, Biochem. J. 46, 364-376 (1950). This is a pioneering study of HA.

7.29 T. C. Laurent, M. Ryan and A. Pietruszkiewicz, Fractionation of hyaluronic acid. The polydispersity of hyaluronic acid from the bovine vitreous body, Biochim. Biophys. Acta 42, 476-485 (1960). After separating HA into fractions the authors study its properties as a function of molecular weight.

7.30 J. H. Bland, Theoretical mechanism of production of the symptom stiffness, Federation Proceedings 28, 1073-1079 (1969). Viscous behaviour of HA in various electrolytes is invoked as a clue to the origin of stiffness.

7.31 C. W. McCutchen, Boundary lubrication by synovial fluid: demonstration and possible osmotic explanation, Federation Proceedings 25, 1061-1068 (1966). Here the author describes ingeneous experiments for testing ideas on biological lubrication.

ELASTICITY OF BIOLOGICAL STRUCTURES

7.32 J. C. Slater, Introduction to Chemical Physics, (McGraw-Hill Book Co., Inc., 1939). Chapter 9 contains material on intramolecular forces which is relevant to the present Section 7.6.

7.33 L. Pauling, The Nature of the Chemical Bond (Cornell University Press, 1940). Basic concepts relating to interatomic forces are defined, developed and applied to a wide range of situations.

7.34 S. Enomoto and S. Krimm, Elastic moduli of helical polypeptide chain structures, Biophysic. J. 2, 317-326 (1962). Calculations are made of Young's modulus for keratin and other macromolecules.

7.35 M. J. Glimcher, Molecular biology of mineralized tissues with particular reference to bone, in Biophysical Science - A Study Program (General Reference 5) pp 359-393. Basic characteristics of bone are reviewed in this readable and informative article.

7.36 F. O. Schmitt, Interaction properties of elongate protein macromolecules with particular reference to collagen, in Biophysical Science - A Study Program (General Reference 5) pp 349-358. Collagen is one of the most studied proteins; its primary characteristics are reviewed here.

7.37 S. B. Lang, Ultrasonic method for measuring elastic coefficients of bone and results on fresh and dried bovine bones, IEEE Trans. Bio-Medical Engineering, BME-17, 101-105 (1970). Data are given on five elastic constants for bone.

7.38 J. L. Katz, Hard tissue as a composite material. 1. Bounds on the elastic behaviour, J. Biomechanics 4, 455-473 (1971). Models are compared for calculating elastic constants of tissues in bone and teeth, knowing those of collagen and apatite.

7.39 J. I. Brash and J. Skorecki, Determination of the modulus of elasticity of bone by a vibration method, Med. & biol. Engng. 8, 389-393 (1970). Specimens of bone from cattle are vibrated in flexural and longitudinal modes.

7.40 C.F. Fox, The structure of cell membranes, Scientific American 226, 30-38 (1972). This is a lucid review for nonspecialists, including recent findings.

7.41 S. J. Singer and G. L. Nicolson, The fluid mosaic model of the structure of cell membranes, Science 175, 720-731 (1972). Evidence is given to support the idea that the interior of a cell membrane has fluid characteristics.

7.42 G. Vanderkool and D. Green, Genesis of a membrane model, New Scientist and Science Journal 50, 22-24 (1971). A "liquid crystal" model is proposed for biomembranes; proteins and lipids are considered to be in constant thermal motion.

7.43 R. P. Rand, Mechanical properties of the red cell membrane. II. Viscoelastic breakdown of the membrane, Biophysic. J. 4, 303-316 (1964). Results are fitted to a Maxwell-Kelvin model.

7.44 R. W. Fields, Mechanical properties of the frog sarcolemma, Biophysic. J. 10, 462-479 (1970). Stress-strain data are obtained.

7.45 A. Mauro and W. R. Adams, The structure of the sarcolemma of the frog skeletal muscle fiber, J. Bioph. Biochem. Cytol. 10, 177-185 (1961). Sarcolemma is shown to consist of four layers.

7.46 U. Dinnar, A note on the theory of deformation in compressed skin, Mathematical Biosciences 8, 71-82 (1970). Results are fitted to a Maxwell-Kelvin model.

REFERENCES FOR CHAPTER EIGHT

8.1 Robert Brown, A brief account of microscopical observations made in the months of June, July and August, 1827, on the particles contained in the pollen of plants; and on the general existence of active molecules in organic and inorganic bodies. The Philosophical Magazine and Annals of Philosophy (New Series) 4, 161-173 (1828). Equipped with a lens of focal length "1/32nd of an inch," the author carried out extensive observations of motion now known as Brownian movement. His "molecules" are small visible particles (diameter about "(1/20,000)th of an inch"); he suggested these as basic elements in the structure of pollen grains and other organic and inorganic bodies. Although his molecules are therefore much too large according to present ideas, his work was indeed fundamental to development of molecular and atomic concepts as now accepted.

8.2 A. Einstein, Theorie der Brownschen Bewegung, Ann. d. Physik 19, 371-381 (1906). Original source for Einstein's derivation of an equation equivalent to Eq. (9.18). A translation appears in Ref. 8.3.

8.3 A. Einstein, Investigations on the Theory of the Brownian Movement (Dover Publications, Inc., 1956). A small paperback volume reprinted from an earlier collection. The material was translated from the German by A. D. Cowper and edited by R. Furth. It contains five articles by Einstein published originally during the period 1905 to 1908, while he was employed in the Patent Office at Berne, Switzerland. Historical notes add further interest to this important volume.

8.4 M. von Smoluchowski, Zur Kinetischen Theorie der Brownschen Molekular-bewegung und der Suspensionen, Ann. d. Physik, 4th Series, 21, 756-780 (1906).

8.5 J. B. Perrin, Atoms (Constable, London, 1920). Basic experiments on Brownian movement and related topics are described here.

8.6 Heisenberg, W. "The Physical Principles of the Quantum Theory" University of Chicago 1930, republished by Dover Publications (184 pages).
Here the author of the "uncertainty principle" discusses the origin of basic fluctuations or uncertainties in experiments.

8.7 Bridgman, P. W. "The Nature of Physical Theory" Princeton University Press 1936, republished by Dover Publication, New York (136 pages).

8.8 Schrodinger, E. "What is Life?" Cambridge University Press 1946
In this famous little book Schrodinger, a physicist, used statistical arguments (referring to a "$\sqrt{n}$ law" such as the expression for σ which results from Eq. (8.10) to make suggestions about basic biological structures involved in heredity.

8.9 Arley, N. and Buch, K. R. "Introduction to Theory of Probability and Statistics," John Wiley and Sons (Science Editions) New York 1966.

8.10 Rosanov, Y. A. "Introductory Probability Theory" translated by R. A. Silverman, Prentice-Hall Inc.; Englewood Cliffs, N.J., 1969.

References 8.9 and 8.10 are examples of texts suitable for reading on basic definitions and theorems of probability theory.

8.11 Baird, D. C. "Experimentation: An Introduction to Measurement Theory and Experiment Design," Prentice-Hall, Inc.; Englewood Cliffs, N.J. 1962.
A treatment of basic procedures in design, evaluation and reporting of experiments. Error theory is given and applications illustrated in detail.

REFERENCES FOR CHAPTER NINE

9.1 A. King, Thermophysics (W. H. Freeman and Co., San Francisco, 1962). A short but useful treatment of Brownian movement is given in Chapter 14.

9.2 P. Langevin, Sur la theorie du mouvement brownien, Comptes Rendus 146 , 530-533 (1908). Original source of theory presented in Section 9.2.1

9.3 H. Lamb, Hydrodynamics (Dover Publications, N.Y. 1945). Viscous flow near a rotating sphere is treated in Section 334.

9.4 H. Morawetz, Reference 6.7 (1965). Information on rotary diffusion, flow birefringence, and disorientation of oriented solutions is contained in Chap. 6 .

9.5 W. Gautschi, "Error function and Fresnel integrals," in Handbook of Mathematical Functions, M. Abramowitz and I. A. Stegun, eds. (Dover Publications, Inc., New York 1965). Properties of the error function and Gaussian distribution are listed and numerical values tabulated.

9.6 D. L. Wise and G. Houghton, Solubilities and diffusivities of oxygen in hemolyzed human blood solutions, Biophysic. J. 9, 36-53 (1969). Section 9.6.1 is based mostly on this paper.

9.7 H. K. Schachman, Ultracentrifugation in Biochemistry (Academic Press, N. Y. 1959). Experimental methods are described and critically analyzed for measuring sedimentation and diffusion coefficients by centrifugation techniques.

9.8 C. H. Chervenka, A manual of Methods for the Analytical Ultracentrifuge (Spinco Division of Beckman Instruments Inc., Palo Alto, California, 1970). Here methods for measuring D and s are described in working detail in a very clear manner.

9.9 Engineering Manual (McGraw-Hill, Inc., N. Y. 1967). Data on diffusion coefficients are cited.

9.10 L. G. Longsworth, Diffusion measurements, at 25°, of aqueous solutions of amino acids, peptides and sugars, J. Am. Chem. Soc. 75, 5705-5709 (1953). See also a related paper by the same author in the same journal, Vol. 74, 4155-4159 (1952); also American Institute of Physics Handbook, 3rd edition, pp. 2-221 to 2-229 (McGraw-Hill Book Co., 1972). In these references data are given on diffusion coefficients for a wide variety of ions, molecules and particles in aqueous media.

9.11 Handbook of Biochemistry, selected data for Molecular Biology, 2nd edition, H. A. Sober, Ed., the Chemical Rubber Co., Cleveland, Ohio 1970. Source of data for Table 9.4 .

9.12 E. J. Cohn and J. T. Edsall, Proteins, Amino Acids and Peptides (Reinhold Publishing Corp., N. Y., 1943). Physical and physical-chemical methods for characterizing macromolecules are emphasized. Ratios f/f_o for ellipsoids are tabulated on p. 406 .

9.13 D. F. Waugh, Blood coagulation-a study in homeostasis, pp. 557-562, in General Reference 5 . Mechanics of the complicated and necessary processes of blood coagulation are reviewed.

9.14 D. F. Waugh, Protein-protein interactions, in Advances in Protein Chemistry Vol. 9 (Academic Press, Inc., 1954). Aggregation characteristics for proteins are described, including those for insulin and fibrinogen.

9.15 C. T. O'Konski and A. J. Haltner, "Characterization of the monomer and dimer of tobacco mosaic virus by transient electric birefringence, Am. Chem. Soc. J. 78, 3604-3610 (1956). Source of data on rotary diffusion coefficients for TMV.

9.16 S. Ananthanarayanan and A. Veis, The molecular parameters of monomeric and acid-soluble collagens. Low shear gradient viscosity and electric birefringence, Biopolymers 11, 1365-1377 (1972).

9.17 I. H. Billick and J. D. Ferry, The conversion of fibrinogen to fibrin. XVII Further studies of electrical birefringence of fibrinogen, Am. Chem. Soc. J. 78, 933-935 (1956). Source of data on D_r for fibrinogen.

9.18 S. Krause and C. T. O'Konski, Electric properties of macromolecules. III. Kerr constants and rotational diffusion of bovine serum albumin in aqueous solutions, Am. Chem. Soc. J. 81, 5082-5088 (1959).

9.19 G. B. Jeffery, The motion of ellipsoidal particles immersed in a viscous liquid. Proc. Roy. Soc. A102, 161-179 (1922-23).

9.20 P. Boeder, Über Strömungsdoppelbrechung, Z. Physik 75, 273-281 (1932). Source of original theory on orientation of ellipsoids in a velocity-gradient field.

9.21 W. Kuhn and H. Kuhn, Die Abhängigkeit der Viscosität von Strömungsgefälle bei hochverdünnten Suspensionen und Lösungen. Helv. Chim. Acta. 28, 97-127 (1945). Original source for Eq. (9.98). The results are quoted by Morawetz, Reference 9.4 .

9.22 H. A. Scheraga, Non-Newtonian viscosity of solutions of ellipsoidal particles, J. Chem. Phys. 23, 1526-1532 (1955).

REFERENCES FOR CHAPTER TEN

10.1 J. C. Slater, Introduction to Chemical Physics (McGraw-Hill Book Co., New York 1939). Concepts of thermodynamics are treated clearly and concisely in Chaps. I and II of this textbook.

10.2 M. W. Zemansky, Heat and Thermodynamics (McGraw-Hill Book Co., New York 1957). In this classic text the ideas of temperature, internal energy, entropy and other thermodynamic quantities are developed carefully, with emphasis on their experimental bases.

10.3 H. J. Morowitz, Entropy for Biologists (Academic Press, New York 1970). As its name suggests,this introduction to thermodynamics has a special orientation toward biological applications. Principles are discussed in a sound manner and interesting problems are provided.

10.4 A. Katchalsky and P. F. Curran, Nonequilibrium Thermodynamics in Biophysics (Harvard University Press, Cambridge, Mass., 1965). Here principles and methods of thermal physics are developed, with emphasis on important biophysical applications of nonequilibrium thermodynamics.

10.5 R. B. Lindsay and H. Margenau, Foundations of Physics (John Wiley and Sons, New York 1936). A statement of the Second Law of thermodynamics in the form quoted in Section 10.3 is given on page 216. The point is made that this statement of the Law is equivalent to forms given by other authors (which at first seem rather different).

10.6 H. E. Blum, Time's Arrow and Evolution (Harper and Brothers, New York 1962). The author considers detailed biochemical processes and speculates on the thermodynamical basis for evolution.

10.7 E. Schrodinger, What is Life? (Macmillan Co., New York 1946). In this famous little volume the author, as physicist, speculates on applications of physical reasoning to biology; specifically, the ideas of statistical physics are applied to the topic of mutations.

10.8 L. Onsager, Reciprocal relations in irreversible processes. I and II. Phys. Rev. 37, 405-426 (1931) and Phys. Rev. 38, 2265-2279 (1931). In these papers the author introduces the phenomenological equations, Eqs. (10.41), and the reciprocal relations, Eqs. (10.43), suggesting that these are valid generally.

11.1 A. K. Solomon, Characterization of biological membranes by equivalent pores, J. Gen. Physiol. 51, 335^{s} - 364^{s} (1968). A review of theory and experimental evidence for pore models. Se also A. K. Solomon, Equivalent pore dimensions in cellular membranes, Proc. 1st Nat. Biophys. Conf. (Yale Univ. Press 1959) pp. 314-322. For a less technical discussion by the same author see "Pores in the cell membrane," Scientific American, December 1960.

11.2 J. L. Anderson and J. A. Quinn, Restricted transport in small pores, Biophysics. J. 14, 130-150 (1974). A detailed analysis is given here of diffusion and flow in small pores.

11.3 O. Kedem and A. Katchalsky, Thermodynamic analysis of the permeability of biological membranes to non-electrolytes, Biochim. et Biophysica Acta 27, 229-246 (1958). Source of theory for Section 11.5. The same approach is described in Katchalsky and Curran, Reference 10.4, 1965, Chapter 10.

11.4 A. J. Staverman, The theory of measurement of osmotic pressure, Rec. Trav. Chim. 70, 344-352 (1951). Origin of the idea of a reflection coefficient and especially its application in Eq. (11.32).

11.5 P. Mueller, D. O. Rudin. H. T. Tien and W. C. Westcott, Reconstitution of excitable cell membrane structure in vivo, Circulation 26, 1167-1177 (1962); see also, by same authors, Methods for the formation of single bimolecular lipid membranes in aqueous solution, J. Phys. Chem. 67, 534-535 (1963).

11.6 A. Cass and A. Finkelstein, Water permeability of thin lipid membranes, J. Gen. Physiol. 50, 1765-1784 (1967).

11.7 T. Hanai and D. A. Haydon, The permeability to water of bimolecular lipid membranes, J. Theoret. Biol. 11, 370-382 (1966).

11.8 G. H. Brown, Liquid crystals and their roles in inanimate and animate systems, American Scientist 60, 64-73 (1972). Principal facts and applications are discussed clearly and a bibliography is provided.

11.9 H. Trauble, The movement of molecules across lipid membranes; a molecular theory, J. Membrane Biol. 4, 193-208 (1971).

11.10 K. H. Stenzel, A. L. Rubin, W. Yamayoshi, T. Miyata, T. Suzuki, T. Sohde and M. Nishizawa, Optimization of collagen dialysis membranes, Trans. Amer. Soc. Artif. Int. Organs XVII, 293-298 (1971).

11.11 E. M. Renkin, Filtration, diffusion and molecular sieving through porous cellulose membranes, J. Gen, Physiol. 38, 225-243 (1954). Source of data and results on cellulose membranes, Section 11.6.2 and gels, Section 11.6.6.

11.12 J. C. Bugher, Characteristics of collodion membranes for ultra-filtration, J. Gen. Physiol. 36, 431-448 (1953). Electron micrographs in this paper show the pore structure of collodion membranes.

11.13 C. V. Paganelli and A. K. Solomon, The rate of exchange of tritiat water across the red cell membrane, J. Gen. Physiol. 41, 259-277 (1957). Source of data on P_d for red cell membrane in Section 11.6.3.

11.14 V. W. Sidel and A. K. Solomon, Entrance of water into human red ce under an osmotic pressure gradient, J. Gen. Physiol. 41, 243-257 (1957). Source of data on P_f for red cell membrane in Section 11.6.3.

11.15 A. K. Solomon and C. M. Gary-Bobo, Aqueous pores in lipid bilayers and red cell membranes, Biochim. Biophys. Acta 255, 1019-1021 (1972).

11.16 S. J. Singer and G. L. Nicolson, The fluid mosaic model of the structure of cell membranes, Science 175, 720-731 (1972).

11.17 M. A. Lauffer, Theory of diffusion in gels, Biophysic. J. 1, 205-213 (1961).

11.18 R. C. Lehman and E. Pollard, Diffusion rates in disrupted bacterial cells, Biophys. J. 5, 109-119 (1965).

11.19 I. W. Richardson and E. B. Neergaard, Physics for Biology and Medicine (Wiley-Interscience, New York, 1972). This general textbook has a useful and readable chapter on molecular transport.

REFERENCES FOR CHAPTER TWELVE

12.1 Tissue Elasticity, J. W. Remington, ed. (American Physiological Society , Washington, D. C. 1957). This small volume is based on a conference held at Dartmouth College in 1955.

12.2 M. Landowne and R. W. Stacy, Glossary of terms, in Tissue Elasticity, Reference 12.1 .

12.3 A. L. King, Law of elasticity for an ideal elastomer, Am. J. Physics 14, 28-30 (1946). Source of theory which underlies the simplified treatment of flexible chains in Section 12.3 . Bibliography to earlier work is also given.

12.4 A L. King and R. W. Lawton, The elasticity of soft body tissues, The Scientific Monthly 71, 258-260 (1950). Examples of S-shaped stress-strain curves are shown for various tissues.

12.5 Kinetic Theory of Liquids , J. Frenkel (Dover Publications, Inc. , N.Y. 1955 ; originally written in 1943) . The folding-ruler model of a macromolecule is employed on pp. 457-459 .

12.6 The Physics of Rubber Elasticity, L. R. G. Treloar (Clarendon Press, Oxford, 1958). Much of the discussion of rubber in Section 12.4.1 is based on this book. Equation (12.36) is derived in Treloar's Chapter IV.

12.7 P. Hallock and I. C. Benson , Studies on the elastic properties of human isolated aorta, J. Clinical Invest. 16, 595-602 (1937). Source of data for Section 12.4.2 .

12.8 A. L. King, Pressure-volume relation for cylindrical tubes with elastomeric walls; the human aorta, J. Applied Physics 17, 501-505 (1946) . Source for material discussed in Section 12.4.2 .

12.9 R. Ross and P. Bornstein, Elastic fibers in the body, Scientific American, June, 1971 pp. 44-52 . Source of information on elastin.

12.10 R. W. Fields, Mechanical properties of the frog sarcolemma, Biophysical J. 10, 462-480 (1970). Source of data for Section 12.4.3 .

12.11 R. W. Fields and J. J. Faber , Biophysical analysis of the mechanical properties of the sarcolemma, Canadian Journal of Physiology and Pharmacology. 48, 394-404 (1970) .

12.12 J. B. Speakman, J. Textile Inst. Trans. 17, 431 (1926). Source of tension-length data for wool in Fig. 12.11. Cited by David, et. al. (1967).

12.13 C. W. David, H. B. Haukaas, J. G. Kalnins and R. Schor, Statistical-mechanical studies of the α-β transformation in keratins. II. The tension-length isotherms, Biophysic. J. 7, 505-510 (1967).

12.14 H. B. Bull, Elasticity of keratin fibers, J. Am. Chem. Soc. 66, 1253-1258 (1944); Elasticity of keratin fibers, II, J. Am. Chem. Soc. 67, 533-536 (1945); Thermal and elastic properties of α-keratin, J. Phys. Chem. 58, 101-103 (1954).

12.15 M. M. Breuer, The binding of small molecules to hair - I: The hydration of hair and the effect of water on the mechanical properties of hair, J. Soc. Cosmet. Chem. 23, 447-470 (1972). This publication includes a bibliography with earlier papers on related subjects, by the author and others.

12.16 R. Schor, H. B. Haukaas and C. W. David, Statistical-mechanical studies of the α-β transformation in keratins. III. A Monte Carlo simulation, J. Chem. Phys. 49, 4726-4729 (1968).

12.17 M. M. Breuer, Stress-strain isotherm of keratin fibers, Biopolymers 6, 1503-1506 (1968).

REFERENCES FOR CHAPTER THIRTEEN

13.1 K. F. Herzfeld and T. A. Litovitz, Absorption and Dispersion of Ultrasonic Waves (Academic Press, New York, 1959). An advanced and comprehensive treatment of theoretical and experimental findings on propagation constants for sound waves in various media.

13.2 Leonard Hall, "The origin of ultrasonic absorption in water", Phys. Rev. 73, 775-781 (1948). Source of theory discussed in Section 13.3.

13.3 F. Dunn, P. D. Edmonds and W. J. Fry, "Absorption and dispersion of ultrasound in biological media" in Biological Engineering, H. P. Schwan, Ed. (McGraw-Hill Book Co., N. Y. 1969).

13.4 J. Stuehr and E. Yeager, "The propagation of ultrasonic waves in electrolytic solutions", in W. P. Mason (ed.) Physical Acoustics, Vol. 2, Part A., Chap. 6, pp 351-462 (Academic Press, Inc., N. Y., 1965).

13.5 R. E. Barrett (1966), cited by Dunn, Edmonds and Fry in Reference 13.3 as source of data on ultrasonic absorption in solutions of amino acids in water.

13.6 M. Eigen and G. G. Hammes, Elementary steps in enzyme reactions, in Advances in Enzymology, Vol. 25, pp 1-38 (1963) (Interscience Publishers, N. Y.). Source of data suggesting a simple relaxation process in glycine solutions, Fig. 13.4 .

13.7 E. L. Carstensen and H. P. Schwan, Acoustic properties of hemoglobin solutions, J. Acoust. Soc. Am. 31, 305-311 (1959).

13.8 F. W. Kremkau, Macromolecular interaction in the absorption of ultrasound in biological material, Ph.D. Thesis, Dept. of Electrical Engineering, Univ. of Rochester, 1972. Source for results quoted in Section 13.5. A summary of this work appears in Reference 13.9.

13.9 F. W. Kremkau and E. L. Carstensen, Macromolecular interaction in sound absorption, pp 37-42 in Reference 13.10, 1972. Here the authors provide a concise summary of the material in Reference 13.8 .

13.10 Interaction of Ultrasound and Biological Tissues, edited by J. M. Reid and M. R. Sikov, Proceedings of a Workshop held in Seattle, Washingto Nov. 8-11, 1971. This report is available from the U. S. Government as DHEW Publication (RDA) 73-8008 (September, 1972) and contains numerous sho articles on various aspects of biophysical ultrasound.

13.11 S. A. Hawley and F. Dunn, Ultrasonic absorption in aqueous solution of dextran, J. Chem. Phys. 50, 3523-3526 (1969). See also W. D. O'Brien, and F. Dunn, Ultrasonic absorption by biomacromolecules, pp 13-19 in Reference 13.10, 1972.

13.12 T. J. Bauld and H. P. Schwan, Attenuation and reflection of ultraso in canine lung tissue, J. Acoust. Soc. Am. 56, 1630-1637 (1974); F. Dunn, Attenuation and speed of ultrasound in lung, J.Acoust. Soc. Am. 56, 1638-1639 (1974).

13.13 A. J. Barlow, A. Erginsav and J. Lamb, Viscoelastic relaxation of supercooled liquids: Part II, Proc. Roy. Soc. Lond. A298, 481-494 (1967). Source of the BEL equation for the shear impedance in viscoelastic media, Eq. (13.76). See also Reference 13.14 .

13.14 M. C. Phillips, A. J. Barlow and J. Lamb, Relaxation in liquids: a defect-diffusion model of viscoelasticity, Proc. Roy. Soc. Lond. A329, 193-218 (1972). A physical justification is given here for the BEL equation, Eq. (13.76).

13.15 J. Lamb, Physical properties of fluid lubricants; rheological and viscoelastic behaviour, Proc. Inst. Mech. Engrs. 182, 293-310 (1967-8). Here the author summarizes many researches by himself and associates on shear impedance data for many viscoelastic liquids; he discusses the BEL equation, Eq. (13.76).

13.16 P. E. Rouse, Jr., A theory of the linear viscoelastic properties of dilute solutions of coiling polymers, J. Chem. Phys. 21, 1272-1280 (1953). Source of theory discussed in Section 13.7 .

13.17 J. D. Ferry, Viscoelastic Properties of Polymers (John Wiley and Sons, N. Y., 1961). General reference for Section 13.7.

13.18 G. B. Thurston, Viscoelasticity of human blood, Biophysic. J. 12, 1205-1217 (1972). Source of data for Section 13.8 .

13.19 A. L. Dounce, Nuclear gels and chromosomal structure, American Scientist 59, 74-83 (1971). Source of information for Section 13.9.1.

13.20 E. R. Fitzgerald and A. E. Freeland, Viscoelastic response of intervertebral discs at audiofrequencies, Med. & biol. Engng. 9, 459-478 (1971).

13.21 B. S. Mather, Comparison of two formulae for in vivo prediction of strength of the femur, Aerospace Medicine 38, 1270-1272 (1967).

13.22 J. M. Jurist, In vivo determination of the elastic response of bone, I. and II, Phys. Med. Biol. 15, 417-426 and 427-434, resp. (1970).

13.23 J. N. Campbell and J. M. Jurist, Mechanical impedance of the femur: a preliminary report, J. Biomechanics 4, 319-322 (1971).

13.24 J. W. Pugh, R. M. Rose, I. L. Paul, E. L. Radin, Mechanical resonance spectra in human cancellous bone, Science 181, 271-272 (1973). See also J. Black, Comment on "Mechanical resonance spectra in human cancellous bone", Science 181, 273 (1973).

13.25 E. R. Fitzgerald, Particle Waves and Deformation in Crystalline Solids, (Interscience Publishers, New York 1966).

REFERENCES FOR CHAPTER FOURTEEN

14.1 F. H. Johnson, H. Eyring and M. J. Polissar, The Kinetic Basis of Molecular Biology (John Wiley and Sons, Inc., N. Y., 1954). Applications of quantum and statistical mechanics to biology. Chapters 8 and 9 are valuable sources on temperature and pressure as environmental factors.

14.2 S. Licht, Therapeutic Heat and Cold, (Waverly Press, Inc., Baltimore, Md. 1965). In this book there are 21 chapters by various authors on topics relating to medical use of heating and cooling.

14.3 J. F. Lehmann, Ultrasound therapy, Chapter 13 in Therapeutic Heat and Cold, Reference 14.2, 1965. An extensive literature on physiological effects of ultrasound is reviewed.

14.4 C. J. Thompson, Mathematical Statistical Mechanics (The Macmillan Co., N. Y. 1972). Applications of the Ising model to biology are taken up in Chapter 7, including a review of treatments of helix-coil transitions in DNA.

14.5 E. Blair, Clinical Hypothermia (McGraw-Hill Book Co., N. Y. 1964). Here are described medical applications for procedures in which the general body temperature is lowered.

14.6 H. v. Leden and W. G. Cahan, Cryogenics in Surgery (Medical Examination Publishing Co., Inc., Flushing, N. Y. 1971). Principles and techniques for locallized use of extreme cold are taken up.

14.7 Cryobiology, H. T. Meryman, Ed. (Academic Press, N. Y. 1966). This volume contains 15 chapters written by authorities in the field, all dealing with freezing of biological systems.

14.8 H. T. Meryman, Review of biological freezing, Chapter 1 in Cryobiology, Reference 14.7 , 1966.

14.9 H. T. Meryman, Freezing injury and its prevention in living cells, in Annual Review of Biophysics and Bioengineering 3, 341-363 (1974).

14.10 D. A. Marsland, Protoplasmic streaming in relation to gel structure in the cytoplasm, in The Structure of Cytoplasm, W. Seifriz, Ed. (Monograph of the Society of Plant Physiologists, Iowa State College Press, Ames, 1942). Source of data on "viscosity" of cytoplasm in pressurized cells. See also numerous references cited by JEP (1954, Reference 14.1).

14.11 D. E. S. Brown, The pressure coefficient of "viscosity" in the eggs of Arbacia punctulata, J. Cell Comp. Physiol. 5, 335-346 (1934-35). Source of Fig. 14.3.

14.12 D. C. Pease and D. A. Marsland, The cleavage of Ascaris eggs under exceptionally high pressure, J. Cell. Comp. Physiol 14, 407-408 (1939).

14.13 J. F. Fulton, Decompression Sickness (W. B. Saunders Co., Philadelphia, 1951). This is a collection of articles compiled under the auspices of the National Research Council. Both scientific and practical aspects of the subject are treated by the various authors.

14.14 E. N. Harvey, D. K. Barnes, W. D. McElroy, A. H. Whiteley, D. C. Pease and K. W. Cooper, Bubble formation in animals, J. Cell. & Comp. Physiol. 24, 1-22 (1944). See also five other articles on the same topic in the same volume, all authored by Harvey and associates.

14.15 E. N. Harvey, Physical factors in bubble formation, in Decompression Sickness (Reference 14.13, 1951), Chap. IV, pp 90-114. A summary of relevant physics, with emphasis on cavitation nuclei.

14.16 E. N. Harvey, L. A. Blinks, V. C. Twitty and D. M. Whitaker, Animal experiments on bubble formation, in Decompression Sickness (Reference 14.13, 1951), Chap. V, pp. 115-164.

14.17 F. E. Fox and K. F. Herzfeld, Gas bubbles with organic skin as cavitation nuclei, J. Acoust. Soc. Am. 26, 984-989 (1954).

14.18 R. Gramiak and P. M. Shah, Detection of intracardiac blood flow by pulsed echo-ranging ultrasound, Radiology 100, 415-418 (1971). Evidence is shown that microbubbles are present normally in the human cardiac chambers.

14.19 R. P. Rand, Mechanical properties of the red cell membrane. II. Viscoelastic breakdown of the membrane, Biophysic. J. 4, 303-316 (1964).

14.20 J. Hoffman, M. Eden, J. S. Barr, Jr. and R. S. Bedell, The hemolytic volume of human erythrocytes, J. Cell. Comp. Physiol. 51, 405-414 (1958).

14.21 A. Katchalsky, O. Kedem, C. Klibansky and A. deVries, Rheological considerations of the hemolyzing red blood cell, in Flow Properties of Blood and Other Biological Systems, A. L. Copley and G. Stainsbys, eds., (Pergamon Press, New York 1960).

14.22 W. Bonfield, Mechanisms of deformation and fracture in bone, Composites, pp 173-175, Sept., 1971. Source of data on tensile strength of bone.

14.23 A. Casale, R. S. Porter and J. F. Johnson, The mechanochemistry of high polymers, Rubber Chemistry and Technology 44, 534-577 (1971). An extensive review of methods and results for degradation of (mostly nonbiological) polymers by application of shearing stresses.

14.24 C. Levinthal and P. Davison, Degradation of deoxyribonucleic acid under hydrodynamic shearing forces, J. Mol. Biol. 3, 674-683 (1961). Section 14.7.2 is based on this paper.

14.25 S. E. Charm and B. L. Wong, Shear degradation of fibrinogen in the circulation, Science 170, 466-468 (1970). Section 14.7.3 is based on this paper.

14.26 G. I. Taylor, The viscosity of a fluid containing small drops of another fluid, Proc. Roy. Soc. (London) A138, 41-48 (1932), and: The formation of emulsions in definable fields of flow, Proc. Roy. Soc. (London) A146, 501-523 (1934). Source of theory for deformation of droplets in a field of hydrodynamic shear, referred to in Section 14.7.4.

14.27 F. D. Rumscheidt and S. G. Mason, Particle motions in sheared suspensions. XII.Deformation and burst of fluid drops in shear and hyperbolic flow , J. Colloid Sci. 16 , 238-261 (1961). This is one of many articles by Mason and co-workers on the behaviour of particles in well-defined flow fields.

14.28 Lord Rayleigh, On the instability of jets, Proc. London Math. Soc. 10, 4-13 (1879), and: On the capillary phenomena of jets, Roy. Soc. 29, 71-97 (1879). These papers are reprinted in Scientific Papers by Lord Rayleigh (Dover Publications,Inc. , N. Y. 1964), Articles 58 and 60.

14.29 J. A. Rooney, Hydrodynamic shearing of biological cells, J. Biological Physics 2, 26-40 (1973) .This paper was a source for Section 14.7.6 .

14.30 H. Schmid-Schönbein and R. Wells, Fluid drop-like transition of erythrocytes under shear, Science 165 , 288-291 (1969).

14.31 H. L. Goldsmith and J. Marlow, Flow behaviour of erythrocytes. I. Rotation and deformation in dilute suspensions, Proc. Roy. Soc. (London) B182 , 351-384 (1972).

14.32 C. G. Nevaril, E. C. Lynch, C. P. Alfrey, Jr. and J. D. Hellums, Erythrocyte damage and destruction induced by shearing stress, J. Lab. Clin. Med. 71, 784-790 (1968).

14.33 J. V. Champion, P. F. North, W. T. Coakley and A. R. Williams, Shear fragility of human erythrocytes, Biorheology 8, 23-29 (1971).

14.34 L. B. Leverett, J. D. Hellums, C. P. Alfrey and E. C. Lynch, Red blood cell damage by shear stress, Biophysic. J. 12, 257-273 (1972).

14.35 A. R. Williams, Viscoelasticity of the human erythrocyte membrane, Biorheology 10 , 303- (1973).

14.36 J. Krizan and A. R. Williams, Biological membrane rupture and a phase transition model, Nature New Biology 246 , 121-123 (1973).

14.37 A. C. Burton, The Stretching of "pores" in a membrane, in Permeabili and Function of Biological Membranes, Bolis, et al., Eds. (North-Holland Publishing Co., Amsterdam, 1970), pp 1-19.

14.38 B. N. Nanjappa, H. K. Chang , and C. A. Glomski, Trauma of the erythrocyte membrane associated with low shear stress, Biophysic. J. 13, 1212-1222 (1973).

14.39 J. C. R. Licklider, Basic correlates of the auditory stimulus, Chap. 25 in Handbook of Experimental Psychology , S. S. Stevens, Ed. (John Wiley and Sons, New York, 1951). Source of Fig. 14.8 .

14.40 K. D. Kryter, The Effects of Noise on Man (Academic Press, N.Y. 1970). Critical analysis of an extensive literature on environmental noise and its consequences to man. Source of information for Section 14.8 .

14.41 A. Cohen, J. R. Anticaglia and H. H. Jones, Noise induced hearing loss - exposures to steady-state noise, Presented at the American Medical Association Sixth Congress on Environmental Health, Chicago, Illinois, 1969 . Source of data in Fig. 14.9 , cited by Kryter, Reference 14.40 .

14.42 National Research Council of Canada, Snowmobile noise: its sources, hazards and control, Report APS-477 (Acoustics Section, Division of Physics, Ottawa 1970).

14.43 G. v. Békésy, Experiments in Hearing (McGraw-Hill Book Co., N.Y. 1960). This book includes results of basic experiments by the author over a period of 34 years.

14.44 E. Ackerman, Biophysical Science (Prentice-Hall, Inc., Englewood Cliffs, N.J, 1962) . Chapters 1, 3 and 6 deal with the ear and hearing.

14.45 J. Tonndorf, In memoriam Georg von Békésy 1899-1972, J. Acoust. Soc. Amer. 55, 576-577 (1974). This is an introduction to a series of eight articles dealing with experiments and theories on hearing. The articles are significant in themselves, and are also an excellent source of references to earlier work.

14.46 H. E. v. Gierke, Biodynamic models and their applications, J. Acoust. Soc. Amer. 50, 1397-1413 (1971). A tutorial on the mechanical response of man to environmental vibrations and impacts.

14.47 I. E. El'piner, Ultrasound: Physical, Chemical and Biological Effec (Consultants Bureau, N.Y. 1964). This book, translated from the Russian, is an extensive review of investigations covering a wide range of topics involving ultrasound as an agent.

14.48 P. P. Lele, Production of deep focal lesions by focused ultrasound - current status, Ultrasonics 5, 105-112 (1967). The author describes advantages and problems in surgical use of focussed ultrasound.

14.49 T. C. Robinson and P. P. Lele, An analysis of lesion development in the brain and in plastics by high-intensity focused ultrasound at low-megahertz frequencies, J. Acoust. Soc. Amer. 51, 1333-1351 (1972). Evidence is given that ultrasonically produced lesions correspond to regions where the induced temperature exceeds a minimum value of about 55°C

14.50 H. G. Flynn, Physics of acoustic cavitation in liquids, Chap. 9 in Physical Acoustics Vol. IB, W. P. Mason, Ed. (Academic Press, N.Y., 1964). A review of the extensive literature on theory and experimental results on cavitation, including applications to chemistry and biology.

14.51 D-Y. Hsieh and M. S. Plesset, Theory of rectified diffusion of mass into gas bubbles, J. Acoust. Soc. Am. 33, 206-215 (1961).

14.52 L. A. Crum and A. I. Eller, Motion of bubbles in a stationary sound field, J. Acoust. Soc. Amer. 48, 181-189 (1969). Experimental tests are reported of theory for the acoustic radiation force on a gas bubble.

14.53 S. A. Elder, Cavitation microstreaming, J. Acoust. Soc. Amer. 31, 54-64 (1959). A study of small-scale acoustic streaming hear an oscillating gas bubble.

14.54 W. L. Nyborg, Radiation pressure on a small rigid sphere, J. Acoust. Soc. Amer. 42, 947-952 (1967).

14.55 D. L. Storm, Interfacial distortions of a pulsating gas bubble. In Bjørnø, L. (Ed.), Proc. 1973 Symp. on Finite-Amplitude Wave Effects in Fluids, Copenhagen. (IPC Science and Technology Press Ltd., 1974) pp 234-239.

14.56 E. A. Neppiras and E. E. Fill, A cyclic cavitation process, J. Acoust. Soc. Amer. 46, 1264-1271 (1969).

14.57 B. E. Noltingk and E. A. Neppiras, Cavitation produced by ultrasonics, Proc. Phys. Soc. (London) B63, 674-685 (1950); E. A. Neppiras and B. E. Noltingk, Cavitation produced by ultrasonics: theoretical conditions for the onset of cavitation, Proc. Phys. Soc. (London) B64, 1032-1038 (1951).

14.58 D. E. Goldman and W. W. Lepeschkin, Injury to living cells in standing sound waves, J. Cell. Comp. Physiol. 40, 255-268 (1952).

14.59 F. G. Blake, Jr., Bjerknes forces in stationary sound fields, J. Acoust. Soc. Amer. 21, 551 L (1949).

14.60 J. A. Rooney, Hemolysis near an ultrasonically pulsating gas bubble, Science 169, 869-871 (1970).

14.61 E. N. Harvey, E. B. Harvey and A. L. Loomis, Further observations on the effect of high frequency sound waves on living matter, Biol. Bull. 55, 459-469 (1928). An observation is included here, on "whirling" motions near gas-filled spaces in plant tissues.

14.62 W. L. Nyborg , D. L. Miller and A. Gershoy, Physical consequences of ultrasound in plant tissues and other bio-systems, to be published in Proceedings of the Seventh Rochester Conference on Environmental Toxicity: Fundamental and Applied Aspects of Non-Ionizing Radiation Rochester, N.Y., 1974 (Plenum Publishing Corporation , N.Y.) Here evidence is given for the significance of gas-filled spaces in plant tissues under sonation.

14.63 W. T. Coakley, D. Hampton and F. Dunn, Quantitative relationships between ultrasonic cavitation and effects upon amoebae at 1 MHz, J. Acoust. Soc. Amer. 50 , 1546-1553 (1971).

14.64 W. L. Nyborg and H. J. Dyer, Ultrasonically induced motions in single plant cells. In Medical Electronics: Proceedings of the Second International Conference on Medical Electronics , Paris, France, 24-27 June 1959. (Iliffe and Sons, Ltd., London, 1960) pp 391-396.

14.65 A Gershoy and W. L. Nyborg, Perturbation of plant-cell contents by ultrasonic micro-irradiation, J. Acoust. Soc. Amer. 54 , 1356-1367 (1973).

14.66 F. O. Schmitt, Ultrasonic micromanipulation, Protoplasma 7, 332-340 (1929).

14.67 W. L. Wilson, F. J. Wiercinski, W. L. Nyborg, R. M. Schnitzler and F. J. Sichel, Deformation and motion produced in isolated living cells by localized ultrasonic vibration, J. Acoust. Soc. Amer. 40, 1363-1370 (1966).

14.68 W. L. Wilson and R. M. Schnitzler, The effect on cleavage of Arbacia eggs of ultrasound applied to a small area of the cell surface, Biol. Bull. 125, 397(A) (1963).

14.69 H. J. Dyer, Changes in behaviour of moss treated with ultrasound, J. Acoust. Soc. Amer. 37, 1195A (1965).

14.70 M. J. Ravitz and R. M. Schnitzler, Morphological changes induced in the frog semitendinosus muscle fiber by localized ultrasound, Experimental Cell Research 60 , 78-85 (1970).

14.71 H. M. Frost, Action of ultrasound on a viscoelastic solid, Ph.D. Thesis, University of Vermont, 1974 . See also H. M. Frost and W. L. Nyborg, Action of ultrasound on a viscoelastic solid, in Ultrasonics International 1973 Conference Proceedings (IPC Science and Technology Press, Ltd., Guildford , Surrey, England) pp 81-88 .

14.72 A. R. Williams, D. E. Hughes and W. L. Nyborg, Hemolysis near a transversely oscillating wire, Science 169 , 871-873 (1970).

14.73 J. A. Rooney, Shear as a mechanism for sonically induced biological effects, J. Acoust. Soc. Amer. 52, 1718-1724 (1972)

14.74 A. R. Williams, Release of serotonin from human platelets by acoustic microstreaming. J. Acoust.Soc.Am.56, 1640-1643 (1974)

14.75 J. Crowell, B. Kusserow and W. L. Nyborg, Functional changes in white blood cells after sonation. In Press (1975).

14.76 W. J. Fry, V. J. Wulff, D. Tucker and F. J. Fry, Physical factors involved in ultrasonically induced changes in living systems: I. Identification of non-temperature effects, J. Acoust. Soc. Amer. 22, 867-876 (1950). And, by the same authors: II. Amplitude duration relations and the effect of hydrostatic pressure for nerve tissue, J. Acoust. Soc. Amer. 23, 364-368 (1951).

14.77 W. J. Fry and F. Dunn, Ultrasonic irradiation of the central nervous system at high sound levels, J. Acoust. Soc. Amer. 28 , 129-131 (1956).

14.78 C. C. Connolly, Ph.D. Dissertation, University of London, 1968. See also J. Acoust. Soc. Amer. 50 , 100A (1971).

14.79 C. C. Connolly and J. Pond, The possibility of harmful effects in using ultrasound for medical diagnosis, Bio-Medical Engineering, March 1967, pp 112-115 .

14.80 J. B. Pond, B. Woodward and M. A. Dyson, A microscope viewing ultrasonic irradiation chamber, Phys. Med. Biol. 16 , 521-524 (1971).

14.81 F. Dunn and F. J. Fry, Ultrasonic threshold dosages for the mammalian central nervous system, IEEE Trans. on Bio-Med. Engin. BME-18 , 253-256 (1971).

14.82 Interaction of Ultrasound and Biological Tissues, J. M. Reid and M. R. Sikov (Eds.), DHEW Publication (FDA) 73-8008, (Bureau of Radiological Health, Rockville, Md., 1972). Proceedings of a workshop held in Seattle, Washington in November 1971. Brief summaries and lists of references are included for a wide variety of bioeffects of ultrasound.

14.83 Ultrasonic Energy, E. Kelly (Ed) (University of Illinois Press , Urbana, 1965). Articles are contained here by various authors on ultrasonic bioeffects and other subjects, based on talks at a conference held in 1961.

AUTHOR INDEX

In the following list, the name of each individual is followed by the numbers of General References (written Gen. Ref. 1, etc.) and References (written 2.13, etc.) which identify articles or books of which he or she is author, co-author or editor. All of these have been referred to in the text.

SUBJECT INDEX